Modeling, Computation and Optimization

Statistical Science and Interdisciplinary Research

Series Editor: Sankar K. Pal *(Indian Statistical Institute)*

Description:
In conjunction with the Platinum Jubilee celebrations of the Indian Statistical Institute, a series of books will be produced to cover various topics, such as Statistics and Mathematics, Computer Science, Machine Intelligence, Econometrics, other Physical Sciences, and Social and Natural Sciences. This series of edited volumes in the mentioned disciplines culminate mostly out of significant events — conferences, workshops and lectures — held at the ten branches and centers of ISI to commemorate the long history of the institute.

Vol. 1 Mathematical Programming and Game Theory for Decision Making
edited by S. K. Neogy, R. B. Bapat, A. K. Das & T. Parthasarathy
(Indian Statistical Institute, India)

Vol. 2 Advances in Intelligent Information Processing:
Tools and Applications
edited by B. Chandra & C. A. Murthy
(Indian Statistical Institute, India)

Vol. 3 Algorithms, Architectures and Information Systems Security
edited by Bhargab B. Bhattacharya, Susmita Sur-Kolay,
Subhas C. Nandy & Aditya Bagchi
(Indian Statistical Institute, India)

Vol. 4 Advances in Multivariate Statistical Methods
edited by A. SenGupta (Indian Statistical Institute, India)

Vol. 5 New and Enduring Themes in Development Economics
edited by B. Dutta, T. Ray & E. Somanathan
(Indian Statistical Institute, India)

Vol. 6 Modeling, Computation and Optimization
edited by S. K. Neogy, A. K. Das and R. B. Bapat
(Indian Statistical Institute, India)

Platinum Jubilee Series

Statistical Science and
Interdisciplinary Research — Vol. 6

Modeling, Computation and Optimization

Editors

S. K. Neogy
A. K. Das
R. B. Bapat

Indian Statistical Institute, India

Series Editor: **Sankar K. Pal**

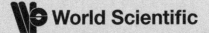

NEW JERSEY · LONDON · SINGAPORE · BEIJING · SHANGHAI · HONG KONG · TAIPEI · CHENNAI

Published by

World Scientific Publishing Co. Pte. Ltd.
5 Toh Tuck Link, Singapore 596224
USA office: 27 Warren Street, Suite 401-402, Hackensack, NJ 07601
UK office: 57 Shelton Street, Covent Garden, London WC2H 9HE

Library of Congress Cataloging-in-Publication Data
Modeling, computation, and optimization / edited by S.K. Neogy, A.K. Das & R.B. Bapat.
   p. cm. -- (Statistical science and interdisciplinary research ; v. 6)
   Includes bibliographical references.
   ISBN-13: 978-9814273503 (hardcover : alk. paper)
   ISBN-10: 9814273503 (hardcover : alk. paper)
   1. Mathematical models. 2. Numerical analysis. 3. Mathematical optimization.
   4. Decision making--Mathematical models. I. Neogy, S. K. II. Das, A. K.
   III. Bapat, R. B. IV. Title. V. Series.

QA401.M5365 2009
511'.8--dc22

                                                                                           2008054107

British Library Cataloguing-in-Publication Data
A catalogue record for this book is available from the British Library.

Copyright © 2009 by World Scientific Publishing Co. Pte. Ltd.

All rights reserved. This book, or parts thereof, may not be reproduced in any form or by any means, electronic or mechanical, including photocopying, recording or any information storage and retrieval system now known or to be invented, without written permission from the Publisher.

For photocopying of material in this volume, please pay a copying fee through the Copyright Clearance Center, Inc., 222 Rosewood Drive, Danvers, MA 01923, USA. In this case permission to photocopy is not required from the publisher.

Printed in Singapore.

Foreword

The Indian Statistical Institute (ISI) was established on December 17, 1931 by a great visionary Prof. Prasanta Chandra Mahalanobis to promote research in the theory and applications of statistics as a new scientific discipline in India. In 1959, Pandit Jawaharlal Nehru, the then Prime Minister of India introduced the ISI Act in the parliament and designated it as an Institution of National Importance because of its remarkable achievements in statistical work as well as its contribution to economic planning.

Today, the Indian Statistical Institute occupies a prestigious position in the academic firmament. It has been a haven for bright and talented academics working in a number of disciplines. Its research faculty has done India proud in the arenas of Statistics, Mathematics, Economics, Computer Science, among others. Over seventy five years, it has grown into a massive banyan tree, like the institute emblem. The Institute now serves the nation as a unified and monolithic organization from different places, namely Kolkata, the Head Quarters, Delhi, Bangalore and Chennai, three centers, a network of five SQC-OR Units located at Mumbai, Pune, Baroda, Hyderabad and Coimbatore, and a branch (field station) at Giridih.

The platinum jubilee celebrations of ISI have been launched by Honorable Prime Minister Prof. Manmohan Singh on December 24, 2006, and the Govt. of India has declared June 29 as the "Statistics Day" to commemorate the birthday of Prof. Mahalanobis nationally.

Prof. Mahalanobis, was a great believer in interdisciplinary research, because he thought that this will promote the development of not only Statistics, but also the other natural and social sciences. To promote interdisciplinary research, major strides were made in the areas of computer science, statistical quality control, economics, biological and social sciences, physical and earth sciences.

The Institute's motto of 'unity in diversity' has been the guiding principle of all its activities since its inception. It highlights the unifying role of statistics in relation to various scientific activities.

In tune with this hallowed tradition, a comprehensive academic programme, involving Nobel Laureates, Fellows of the Royal Society, Abel prize winner and other dignitaries, has been implemented throughout the Platinum Jubilee year, highlighting the emerging areas of ongoing front line research in its various scientific divisions, centers, and outlying units. It includes international and national-level seminars, symposia, conferences and workshops, as well as series of special lectures. As an outcome of these events, the Institute is bringing out a series of comprehensive volumes in different subjects under the title Statistical Science and Interdisciplinary Research, published by the World Scientific Publishing Co., Singapore.

The present volume titled "Modeling, Computation and Optimization" is the sixth one in the series. The volume consists of twenty chapters, written by some pioneers of the field as well as younger researchers from different parts of the world, and it is unique in bringing topics like modeling, computation and optimization together. The chapters demonstrate different emerging research trends in theory, computation and applications of mathematical modeling to problems in statistics, economics, optimization and game theory. Articles focusing on thrust areas like model for wireless networks, model of Nash networks, reliability models in economics, support vector machines, complementarity modeling and games, and addressing the challenging issues related to interface of modeling and computation are included. I believe the state-of-the art studies presented in this book will be very useful to both researchers and practitioners.

Thanks to the contributors for their excellent research contributions, and to the volume editors, Dr. S. K. Neogy, Dr. A. K. Das and Prof. R. B. Bapat, for their sincere effort in bringing out the volume nicely in time. The editorial supports rendered by Prof. Dilip Saha and Dr. Barun Mukhopadhyay are appreciated. Initial design of the cover as well as technical assistance by Mr. Indranil Dutta is acknowledged. Thanks are also due to World Scientific for their initiative in publishing the series and being a part of the Platinum Jubilee endeavor of the Institute.

October 2008
Kolkata

Sankar K. Pal
Series Editor and
Director

Preface

This volume is devoted to the presentation and discussion of state of the art studies in Mathematical Modeling, Computation and Optimization presented in the form of twenty chapters. It is a peer reviewed volume under the Platinum Jubilee Volume Series of Indian Statistical Institute. The topics of this volume display the emerging research trends in theory, computation and application of Mathematical Modeling to problems in statistics, economics, optimization and game theory. The collection of research articles in this volume exhibit the rich versatility of theories and a lively interplay between theory and significant applications in different areas. This volume contains some research articles which focuses on some exciting areas like model for wireless networks, model of Nash networks, dynamic model of advertising, application of reliability models in economics, support vector machines, optimization, complementarity modeling and games. This volume addresses issues associated with the interface of modeling and computation and it is hoped that the research articles will significantly aid in the dissemination of research efforts in these areas. Some pioneers of the field as well as some prominent younger researchers have contributed articles in this volume which are briefly mentioned below.

In Chapter 1, Andrey Garnaev considers modeling a Jamming Game for wireless networks in the framework of zero-sum games with linearized Shannon capacity utility function. The base station has to distribute the power fairly among the users in the presence of a jammer. The jammer in turn tries to distribute its power among the channels to produce as much harm as possible. This game can also be viewed as a minimax problem against nature. The author shows that the game has a unique equilibrium and also develops an efficient algorithm for this game which can find the optimal strategies in finite number of steps.

Chapter 2 by J. Derks, J. Kuipers, M. Tennekes and F. Thuijsman is about one-way flow connections model of unilateral network formation. The existence of Nash networks is proved for games where the corresponding payoff functions allow for heterogeneity among the profits that agents gain by the network. However when link costs are heterogeneous, it is shown by a counterexample that Nash networks do not always exist.

The aim of Chapter 3 by Reinoud Joosten is to model and to analyze strategic interaction over time in a duopolistic market in which advertising causes several types of externalities. Each period the firms independently and simultaneously choose whether to advertise or not. A new dynamic model of advertising in very general terms is formulated. A broad variety of long and short term externalities can be modeled by altering the (restrictions on the) parameters chosen. The author determines feasible rewards and (subgame perfect) equilibria for the limiting average reward criterion using methods inspired by the repeated games literature. Uniqueness of equilibrium is by no means guaranteed but Pareto efficiency may serve very well as a refinement criterion for wide ranges of the advertisement costs.

Unaware of the developments in each others areas, researchers in Reliability theory and Economics have been working independently. While some of the concepts from one area have direct interpretation in the other, there are many other notions which need further investigation. In Chapter 4, Subhash Kochar and Maochao Xu point out some interesting relationships that exist between some of the notions in Reliability Theory and Economics.

In Chapter 5, Theo S.H. Driessen introduces the notions of semi-null players and semi-dummy players in order to provide a new axiomatization for the Shapley value. A semi-null player is powerful as a singleton, but powerless by joining another nonempty coalition. According to the axiomatic approach to solutions, semi-null players receive the egalitarian payoff. It is shown that the Shapley value is the unique solution verifying semi-null player property, symmetry, efficiency, and linearity.

The purpose of Chapter 6 is to characterize operators on the set of real valued functions on a finite set which is coextrema additive. $\mathcal{E}$-coextrema additivity model by Atsushi Kajii, Hiroyuki Kojima and Takashi Ui provides a rich framework for analyzing effects of optimism and pessimism in economic problems.

Chapter 7 by T. S. Arthanari identifies a very important gap in economic/marketing modeling. In this chapter, the author critically looks at our capitalist market system, or the models used to predict market behaviour and make market decisions, and wondering what happened to the main player, the consumer, in the system. Consumption is a key driver of the economy and assumptions are made about consumers in the models and the market place. This leads one to consider thought experiments, social dialogues, and models that stipulate the rightful place for the consumers as main players.

Drawing up an optimal advertising plan over time for a new product is an important field of study in marketing. The problem becomes more complex for new products that are part of the technological generations. Chapter 8 by A. K. Bardhan and Udayan Chanda deals with the determination of optimal advertising expenditure for two generation consumer durables. The model considers intergenerational diffusion effect and also introduces a framework for modeling innovation diffusion for two competing generations.

In Chapter 9, Anjana Gupta, Aparna Mehra and Davinder Bhatia employ a nonconvex separation theorem to scalarize the vector minimization problem subject to the constraint given in the form of set inclusion. A new Lagrange function is formulated for the scalarized problem. Saddle point criteria are developed which ensures the existence of the Lagrange multipliers. Lagrange duality results are also discussed in this chapter.

Semi infinite Programming refers to the class of problems involving infinite number of constraints while the number of variables remain finite. This model naturally arises in an abundant number of applications in different fields of mathematics, economics and engineering. In Chapter 10, Deepali Gupta and Aparna Mehra define the concept of approximate optimal solutions for semi infinite programming problems. The KKT type necessary optimality conditions, characterizing approximate optimal solutions are derived using the exact penalty function approach. Finally, the authors provide the bound on the penalty parameter in terms of dual variables so as to obtain an almost approximate solution for the primal semi infinite programming problem.

In Chapter 11, Oğuz Solyalı and Haldun Süral consider a one supplier - multiple retailers system over a finite planning horizon. Retailers have external demands for a single product and their inventories are controlled by the supplier based on order-up-to level inventory policy. The problem is to determine the time and the quantity of product to order for the supplier, the retailers to be visited in any period, the quantity of product to be delivered in these visits and the vehicle routes for deliveries so as to minimize system-wide inventory and routing costs. The authors present a Lagrangian relaxation based solution procedure and implement the procedure on test instances.

In Chapter 12, K. K. Thampi and M. J. Jacob consider the probability and severity of ruin for a renewal class of risk process in which the claim inter occurrence times is generalized exponential. A closed form expression for the distribution of the deficit at ruin is obtained and the application of the results are illustrated with several examples.

Free riding occurs when players benefit from the actions and efforts of others without contributing to the costs incurred in generating the benefits. Typical situations where free riding may occur in the international arena are global warming abatement and management of international fish resources. Chapter 13 by Kim Hang Pham Do and Henk Folmer presents the feasible proportional allocation rule to discourage free riding for a special class of free riding problems. Some theoretical and practical properties of the rule are discussed. Applications to the management of the Baltic sea cod fishery and the Norwegian spring-spawning herring fishery are also presented.

The problem of resource extraction activity is of great interest and the competition among extractors can be modeled as a game. In Chapter 14, Luca Grilli studies a differential game for the extraction activity of a renewable good, in which players are overlapping generations. The framework of overlapping generations allows to consider intragenerational (players in the same generation) and intergenerational (players in different generations) game equilibria. The author introduces a Stackelberg differential game with asynchronous time horizons and non-fixed role structure. The overlapping generations framework results in the presence of two different behaviours, the myopic and the non-myopic behaviour. A possible solution for the myopic case is presented in this chapter.

Chapter 15 by Yogesh K. Agarwal and Prabha Sharma is devoted to the study of Optimal Communication Spanning Tree Problem (OCSTP). OCSTP is formulated as a mixed integer programming problem and Benders partitioning approach is applied for solving it. It is shown that after fixing the values of integer variables for defining a given tree, the dual of the resulting problem is very easy to solve. This dual solution is used to generate a cut for the Benders master problem. Rather than solving the master problem directly as an integer program, the authors use the standard local search algorithm to obtain an approximate solution. The algorithm proposed in this chapter evaluates the Benders objective function at each neighboring tree, and moves to the neighbor that minimizes this objective function. A cut for the master problem is generated from the new solution and added to the master problem. It is shown that average solution quality produced by Benders search is significantly better than that produced by standard local search.

Sperners lemma is a well-known result in combinatorial topology and its most significant application is in proving Brouwer's fixed point theorem. An analogue of Sperners lemma with multiple, unrestricted labels is proved in Chapter 16 by R. B. Bapat. This extends a result due to Hochberg, McDiarmid and Saks obtained in the context of determining the bandwidth of the triangulated triangle.

In Chapter 17, Reshma Khemchandani, Jayadeva and Suresh Chandra propose a new incremental technique called Incremental Twin Support Vector Machines for training in batch mode. This technique is based on a newly developed classifier called Twin Support Vector Machines (TWSVM) classifier. Numerical implementation on several benchmark datasets has shown that the Incremental Twin SVM is not only fast, but also has good generalization.

Portfolio diversification (i.e., possessing shares of many companies at the same time for reducing risks) is considered to be an important task in the investors community to reduce the risk of a portfolio without not necessarily reducing the returns. Chapter 18 by Sanjeet Singh presents a classification study of different categories of companies on the basis of their various financial attributes using a quadratic optimization based classifier namely Support Vector Machine (SVM). This model is also used for sector wise classification of the company. To validate the performance, the results are compared with the ratings for companies provided by ICICI direct, a well-known trading website in Indian stock market. The comparison shows that the model generated by SVM is efficient and the results obtained using this technique are quite impressive.

In Chapter 19, Yasunori Kimura deals with a generalized proximal point algorithm for a sequence of m-accretive operators in Banach spaces. The author investigates the condition of coefficients more deeply, and obtains weak convergence of an iterative scheme with a weaker coefficient condition.

Complementarity model provides a unifying framework for several optimization problems. Modeling using a complementarity framework arises naturally for games, economics, engineering and management decision making problems. In Chapter 20, S. K. Neogy and A. K. Das present a survey on complementarity models in non-cooperative games and certain classes of structured stochastic game problems.

The 20 refereed articles contained in this volume are selected among the state of the art papers presented in the International Conference on Modeling, Computation and Optimization held during January 9–10, 2008 at Indian Statistical Institute, Delhi Centre. The conference was inaugurated by Professor S. K. Pal, Director, Indian Statistical Institute who also delivered the welcome address. There have been important new developments in the computational techniques of optimization and game problems. This conference was organized during the period of Platinum Jubilee Celebrations of the Indian Statistical Institute and it aimed at discussing new developments in the methods of decision making and to build an interaction between the academic model developers and practitioners by bringing them together to address the important issues in modeling, computation and optimization. It is the hope of the editors that this edited volume will have a seminal role to play in the journey of the development of theory and application of modeling computation and optimization.

S. K. Neogy, A. K. Das and R. B. Bapat
(Editors)

Acknowledgments

The editors are thankful to the following referees who have helped in reviewing the articles for this volume.

- E. Hernández, Universidad Nacional de Educación a Distancia, Spain.
- Christiane Tammer, Martin-Luther-University Halle-Wittenberg, Institute of Mathematics, Germany.
- M. Durea, Al. I. Cuza University, Iasi, Romania.
- Adedeji Badiru, The University of Tennessee, USA.
- Satoshi Ito, Center for Development of Statistical Computing, Institute of Statistical Mathematics, Japan.
- Francisco Guerra Vazquez, Universidad de las Americas, Mexico.
- Ton Storcken, University of Maastricht, Netherlands.
- Anirban Kar, University of Warwick, UK.
- Hans Hermann Haller, Virginia Polytechnic Institute and State University, USA.
- Jurjen Kamphorst, Utrecht University,The Netherlands.
- Marko Lindroos, University of Helsinki, Finland.
- Gregory Gutin, University of London, UK.
- Yeol Je Cho, Gyeongsang National University, Korea.
- S. Plubtieng, Naresuan University, Thailand.
- Meijuan Shang, Shijiazhuang University, PR China.
- Yongfu Su, Tianjin Polytechnic University, PR China.
- Hayato Waki, Tokyo Institute of Technology, Japan.
- Sunyoung Kim, Ewha Women's University, Korea.
- Steffen Jørgensen, University of Southern Denmark, Denmark.
- Florian O. O. Wagener, Universiteit van Amsterdam, The Netherlands.

- R.van den Brink, Free University, De Boelelaan 1105, The Netherlands.
- Félix Belzunce, Universidad de Murcia, Spain.
- Franco Pellerey, Politecnico di Torino, Italy.
- Gert Cauwenberghs, University of California San Diego, USA.
- L. Lambertini, Università di Bologna, Italy.
- R. Cellini, Università di Catania, Italy.
- Prabal Roy Chowdhury, Indian Statistical Institute, Delhi Centre.
- Jean-Philippe Lefort, University of Heidelberg, Germany.
- Martin Jacobsen, University of Copenhagen, Denmark.
- Florin Avram, Heriot-Watt University, UK.
- Ismail Serdar Bakal, Middle East Technical University, Turkey.
- John E.Tyworth, The Pennsylvania State University, USA.
- Suresh Chandra, Indian Institute of Technology Delhi, India
- Dinko Dimitrov, University of Bayreuth, Germany.
- J. Arin, University of the Basque Country, Spain.
- Chuan Yi Tang, National Tsing Hua University, Taiwan.
- Chih-Jen Lin, National Taiwan University, Taiwan.
- Gábor Ivanyos, Informatics Laboratory, Computer and Automation Research Institute, Hungarian Academy of Sciences, Hungary
- T. S. Arthanari, University of Auckland, New Zealand.
- Jordi Massó, Universitat Autònoma de Barcelona, Spain.
- Reinoud Joosten, University of Twente, The Netherlands.
- Klaus Kultti, University of Helsinki, Finland.
- Antonio Frangioni, Università di Pisa, Italy.
- Ivo Nowak, Humboldt-University, Institute for Mathematics, Berlin.
- John E. Mitchell, Rensselaer Polytechnic Institute, USA.
- Stefano Vannucci, University of Siena, Italy.
- Hans Keiding, University of Copenhagen, Denmark.
- Xi Yin Zheng, Yunnan University, PR China.
- A. Shapiro, Georgia Institute of Technology, USA.

We are grateful to our colleagues for preparation of this Platinum Jubilee volume and Mrs. Simmi Marwah, Indian Statistical Institute, Delhi Centre for converting some papers in LATEX format. Finally, we thank World Scientific for their cooperation at all stages in publishing this volume.

S. K. Neogy, A. K. Das and R. B. Bapat
(Editors)

Contents

Foreword		v
Preface		vii
Acknowledgments		xiii
1.	Modeling a Jamming Game for Wireless Networks *Andrey Garnaev*	1
2.	Existence of Nash Networks in the One-way Flow Model of Network Formation *J. Derks, J. Kuipers, M. Tennekes and F. Thuijsman*	9
3.	Strategic Advertisement with Externalities: A New Dynamic Approach *Reinoud Joosten*	21
4.	Connections between Some Concepts in Reliability and Economics *Subhash Kochar and Maochao Xu*	45
5.	A New Axiomatization of the Shapley Value for TU-games in terms of Semi-null Players Applied to 1-concave Games *Theo S.H. Driessen*	57

6. Coextrema Additive Operators 73
    Atsushi Kajii, Hiroyuki Kojima and Takashi Ui

7. Models without Main Players 97
    T. S. Arthanari

8. Dynamic Optimal Advertising Expenditure Strategies for
    Two Successive Generations of High Technology Products 115
    Udayan Chanda and A. K. Bardhan

9. Nonconvex Vector Minimization with Set Inclusion Constraint 137
    Anjana Gupta, Aparna Mehra and Davinder Bhatia

10. Approximate Optimality in Semi Infinite Programming 155
    Deepali Gupta and Aparna Mehra

11. A Relaxation Based Solution Approach for the
    Inventory Control and Vehicle Routing Problem in
    Vendor Managed Systems 171
    Oğuz Solyalı and Haldun Süral

12. The Distribution of Deficit at Ruin on a Renewal Risk Model 191
    K. K. Thampi and M. J. Jacob

13. Fair Allocations to Discourage Free Riding Behavior 205
    Kim Hang Pham Do and Henk Folmer

14. A Stackelberg Differential Game with Overlapping
    Generations for the Management of a Renewable Resource 221
    Luca Grilli

15.	Benders' Partitioning Approach for Solving the Optimal Communication Spanning Tree Problem	237
	Yogesh K. Agarwal and Prabha Sharma	
16.	Sperner's Lemma with Multiple Labels	257
	R. B. Bapat	
17.	Incremental Twin Support Vector Machines	263
	Reshma Khemchandani, Jayadeva and Suresh Chandra	
18.	Portfolio Risk Management Using Support Vector Machine	273
	Sanjeet Singh	
19.	Weak Convergence of an Iterative Scheme with a Weaker Coefficient Condition	287
	Yasunori Kimura	
20.	Complementarity Modeling and Game Theory: A Survey	299
	S. K. Neogy and A. K. Das	

Chapter 1

Modeling a Jamming Game for Wireless Networks

Andrey Garnaev[1]
Department of Computer Modelling and Multiprocessor Systems,
Faculty of Applied Mathematics and Control Processes,
St Petersburg State University
Universitetskii prospekt 35,
Peterhof, St Petersburg, Russia 198504,
e-mail: agarnaev@rambler.ru

Abstract

We consider jamming in wireless networks in the framework of zero-sum games with linearized Shannon capacity utility function. The base station has to distribute the power fairly among the users in the presence of a jammer. The jammer in turn tries to distribute its power among the channels to produce as much harm as possible. This game can also be viewed as a minimax problem against the nature. We show that the game has the unique equilibrium and investigate its properties and also we developed an efficient algorithm which allows to find the optimal strategies in finite number of steps.

Key Words: Zero-sum game, equilibrium, allocation resources, jamming

1.1 Introduction

Power control in wireless networks became an important research area. Since the technology in the current state cannot provide batteries which have small weight and large energy capacity, the design of algorithms for efficient power control is crucial. For a comprehensive survey of recent results on power control in wireless networks an interested reader can consult Tse and Viswanath (2005). It turns out that game theory provides a

---
[1] This work is party supported by joint RFBR and NNSF Grant no.06-01-39005

convenient framework for approaching the power control problem see for instance Lai and El Gamal (2006) and references therein. Most of the work on application of game theory to power control considers mobile terminals as players of the same type. Here we consider the jamming problem with two types of players. The first type of players is a regular one (base station) which want to use the available wireless channels in the most efficient way. The second type of players is jammer who want to prevent or to jam the communication of the regular users.

Jorswieck and Boche (2004) and Suarez-Real (2006) have analyzed the worst case wireless channel capacity when the noise variances are fixed (possibly unknown at the transmitter) and the carrier gains are allowed to vary while verifying a certain constraint. In that case, transmission at the worst rate guarantees error free communication under any possible conditions of the channel, although it might give a pessimistic result. This formulation leads to a minimax problem. Other problem formulations involving jamming in which one wireless terminal wishes to maximize the mutual information and the other tries to minimize it, can be found at Kashyap and Basar (2004). Altman, Avrachenkov and Garnaev (2006) considered a jamming problem where transmission cost is involved.

In this chapter we consider the following jamming problem. There is a base station which needs to allocate the power resource $\bar{T}$ to n users. We assume that for each user there is a channel and there is an interference among the channels. The pure strategy of the base station is $T = (T_1, \ldots, T_n)$ where $T_i \geq 0$ for $i \in [1, n]$ and $\sum_{i=1}^n T_i = \bar{T}$ where $\bar{T} > 0$ for $i \in [1, n]$. The component T_i can be interpreted as the power level dedicated to user i. The pure strategy of the jammer is $J = (J_1, \ldots, J_n)$ where $J_i \geq 0$ for $i \in [1, n]$ and $\sum_{i=1}^n J_i = \bar{J}$ where $\bar{J} > 0$. We consider the linearized Shannon capacity utility as payoff to base station given as follows

$$v(T, J) = \sum_{i=1}^n \frac{g_i T_i}{N_i^0 + h_i J_i},$$

where N_i^0 is the power level of the uncontrolled noise of the environment, and $g_i > 0$ and $h_i > 0$ are fading channel gains for user i.

We consider zero-sum game, so the payoff to jammer is $-v(T, J)$. We will look for the optimal solution, that is, we want to find $(T^*, J^*) \in A \times B$ such that

$$v(T, J^*) \leq v(T^*, J^*) \leq v(T^*, J) \text{ for any } (T, J) \in A \times B,$$

where A and B are the sets of all the strategies of the base station and jammer, respectively.

1.2 The Main Results

In this section we will find solution of the game in closed form.

Note that $v(T, J)$ is linear on T so the optimal base station strategy will employ only the best quality channels. Also,

$$\frac{\partial^2 v}{\partial J^2} = \frac{2 g_i T_i h_i^2}{(N_i^0 + h_i J_i)^3}.$$

Thus, $v(T, J)$ is concave on J and the problem of finding the optimal jammer strategy we can reduce to the problem of finding a Lagrangian multiplier. This conclusions are summed up in the following theorem.

Theorem 1.1. (T, J) *is the equilibrium if and only if there are ω and ν such that*

$$T_i \begin{cases} \geq 0, & \text{if } \frac{g_i}{N_i^0 + h_i J_i} = \omega, \\ = 0, & \text{if } \frac{g_i}{N_i^0 + h_i J_i} < \omega \end{cases} \quad (1.1)$$

and

$$\frac{g_i h_i T_i}{(N_i^0 + h_i J_i)^2} \begin{cases} = \nu, & \text{if } J_i > 0, \\ \leq \nu, & \text{if } J_i = 0. \end{cases} \quad (1.2)$$

Parameter ω is specified by linear nature of $v(T, J)$ on T, meanwhile parameter ν is the Lagrangian multiplier. It is clear ω and ν have to be positive.

Analyzing the results of the previous theorem we can produce more precise way of describing of the the optimal solution.

Theorem 1.2. *Let (T, J) be an equilibrium.*
(a) If $T_i = 0$ then

$$J_i = 0 \text{ and } \frac{g_i}{N_i^0} \leq \omega,$$

(b) if $T_i > 0$ and $J_i = 0$ then

$$\frac{g_i}{N_i^0} = \omega \text{ and } T_i \leq \nu \frac{(N_i^0)^2}{g_i h_i},$$

(c) if $T_i > 0$ and $J_i > 0$ then

$$J_i = \frac{g_i}{h_i}\left(\frac{1}{\omega} - \frac{N_i^0}{g_i}\right), \quad (1.3)$$

$$T_i = \frac{\nu}{\omega^2} \frac{g_i}{h_i} \qquad (1.4)$$

and

$$\frac{g_i}{N_i^0} > \omega. \qquad (1.5)$$

Proof. Since there are i and k such that $T_i > 0$ and $J_k > 0$, then $\omega > 0$ and $\nu > 0$ by (1.2) and (1.1).

(a) Since $T_i = 0$, by (1.2), $J_i = 0$. So, by (1.1), $g_i/N_i^0 \leq \omega$.

(b) Since $T_i > 0$ and $J_i = 0$, by (1.1), $g_i/N_i^0 = \omega$ and, by (1.2), $g_i h_i T_i/(N_i^0)^2 \leq \nu$.

(c) Since $T_i > 0$ by (1.1),

$$\frac{g_i}{N_i^0 + h_i J_i} = \omega.$$

So, J_i is given by (1.3) and also the inequality (1.5) has to hold. Since $J_i > 0$, by (1.3)

$$\frac{g_i h_i T_i}{(N_i^0 + h_i J_i)^2} = \nu.$$

Substituting in this formula J_i given by (1.3) implies that T_i is given by (1.4). □

Based on Theorem 1.2 we can find the optimal solution in explicit form. Namely, the next first theorem tells that always there is the unique equilibrium where the optimal strategies are positive for the same users. The second theorem tells that in some very rare cases the game has infinitive number of equilibrium

Theorem 1.3. *There is unique equilibrium (T, J) such that T and J are positive for the same users. Namely, the equilibrium has the form $(T(\omega_*, \nu_*), J(\omega_*))$ where*

$$J_i(\omega) = \frac{g_i}{h_i} \left[\frac{1}{\omega} - \frac{N_i^0}{g_i} \right]_+ \quad \text{for } i \in [1, n]$$

and

$$T_i(\omega, \nu) = \begin{cases} \frac{\nu}{\omega^2} \frac{g_i}{h_i} & \text{for } i \in I(\omega), \\ 0 & \text{otherwise,} \end{cases}$$

where $I(\omega) = \{i \in [1,n] : J_i(\omega) > 0\}$ and ω_* is the unique root of the equation:

$$\sum_{i=1}^{n} \frac{g_i}{h_i} \left[\frac{1}{\omega} - \frac{N_i^0}{g_i}\right]_+ = \bar{J}$$

and

$$\nu_* = \frac{\bar{T}}{\sum_{i \in I(\omega_*)} g_i/h_i} \omega_*^2.$$

Proof. Let T and J are positive for the same users. Then, by Theorem 1.2, T and J are given by (1.4) and (1.3). So, the problem of finding of the optimal strategies is reduced to the problem of finding two positive parameters ω and ν. Let

$$H(\omega) = \sum_{i=1}^{n} \frac{g_i}{h_i} \left[\frac{1}{\omega} - \frac{N_i^0}{g_i}\right]_+.$$

Then, since $\sum_{i=1}^{n} J_i = \bar{J}$, the optimal ω is the root of the equation $H(\omega) = \bar{J}$. It is clear that $H(\omega) = 0$ for $\omega \geq \max_i g_i/N_i^0$, $H(\omega)$ is strictly positive and decreasing in $(0, \max_i g_i/N_i^0)$ and $H(+0) = \infty$. So, such root exists and it is unique. Then, ν also can be defined from the condition $\sum_{i=1}^{n} T_i = \bar{T}$ and (1.4). $\square$

Theorem 1.4. Let there exist an $i_* \in [1,n]$ such that

$$\sum_{i=1}^{n} \frac{g_i}{h_i} \left[\frac{N_{i_*}^0}{g_{i_*}} - \frac{N_i^0}{g_i}\right]_+ = \bar{J}.$$

Then (T, J) is the equilibrium where

$$J_i = \frac{g_i}{h_i} \left[\frac{N_{i_*}^0}{g_{i_*}} - \frac{N_i^0}{g_i}\right]_+ \quad \text{for } i \in [1,n]$$

and

$$T_i \begin{cases} = \nu \left(\frac{N_{i_*}^0}{g_{i_*}}\right)^2 \frac{g_i}{h_i} & \text{for } i \in I \setminus \{i_*\}, \\ \leq \nu \left(\frac{N_{i_*}^0}{g_{i_*}}\right)^2 \frac{g_{i_*}}{h_{i_*}} & \text{for } i = i_*, \\ = 0 & \text{otherwise}, \end{cases}$$

for any

$$\nu \in \left[\frac{1}{\sum_{i \in I \setminus \{i_*\}} g_i/h_i}, \frac{1}{\sum_{i \in I} g_i/h_i}\right] \times \bar{T} g_{i_*}^2/(N_{i_*}^0)^2.$$

Without loss of generality we can assume that the users are arranged such that
$$N_1^0/g_1 \leq N_2^0/g_2 \leq \ldots \leq N_n^0/g_n.$$
Then, following the approach developed by Altman, Avrachenkov and Garnaev (2007) for water-filling optimization problem we can present solution in closed form as given in the following theorem.

Theorem 1.5. *The solution* (T^*, J^*) *of the jamming game with linear utility function is given by*

$$J_i^* = \begin{cases} \dfrac{g_i}{h_i} \dfrac{\bar{T} + \sum_{t=1}^{k}(g_t/h_t)(N_t^0/g_t - N_i^0/g_i)}{\sum_{t=1}^{k}(g_t/h_t)}, & \text{if } i \leq k, \\ 0, & \text{if } i > k, \end{cases}$$

$$T_i^* = \begin{cases} \dfrac{\bar{T} g_i/h_i}{\sum_{t=1}^{k} g_t/h_t}, & \text{if } i \leq k, \\ 0, & \text{if } i > k, \end{cases}$$

where k can be found from the following conditions:
$$\varphi_k < \bar{T} \leq \varphi_{k+1},$$
where
$$\varphi_t = \sum_{i=1}^{t}(g_i/h_i)(N_t^0/g_t - N_i^0/g_i) \text{ for } t \in [1, n]$$
and $\varphi_{n+1} = \infty$.

Proof. It is clear that $H(\omega) = 0$ for $\omega \geq g_1/N_1^0$, $H(\omega)$ is strictly positive and decreasing in $(0, g_1/N_1^0)$.
Let $k \in [1, n]$ be such that
$$\frac{g_k}{N_k^0} > \omega_* \geq \frac{g_{k+1}}{N_{k+1}^0},$$
where $N_{n+1}^0/g_{n+1} = \infty$ and ω_* is given by Theorem 1.3.

Then, $[1/\omega_* - N_i^0/g_i]_+ = 1/\omega_* - N_i^0/g_i$ for $i \in [1,k]$ and $[1/\omega_* - N_i^0/g_i]_+ = 0$ $i \in [k+1, n]$. So,

$$H(\omega_*) = \sum_{i=1}^{k}(g_i/h_i)(1/\omega_* - N_i^0/g_i).$$

Since $H(\omega_*) = \bar{T}$ we have that

$$\omega^* = \frac{\sum_{i=1}^{k}(g_i/h_i)}{\bar{T} + \sum_{i=1}^{k}(N_i^0/h_i)}. \qquad (1.6)$$

Because of strictly decreasing of H in $(0, g_1/N_1^0)$ we can find k from the following conditions:

$$H(g_k/N_k^0) < \bar{T} \le H(g_{k+1}/N_{k+1}^0).$$

Since

$$\sum_{i=1}^{k}(g_i/h_i)(N_{k+1}^0/g_{k+1} - N_i^0/g_i) = \sum_{i=1}^{k+1}(g_i/h_i)(N_{k+1}^0/g_{k+1} - N_i^0/g_i),$$

the switching point k can be found from the following equivalent conditions:

$$\varphi_k < \bar{T} \le \varphi_{k+1}, \qquad (1.7)$$

where

$$\varphi_t = \sum_{i=1}^{t}(g_i/h_i)(N_t^0/g_t - N_i^0/g_i) \text{ for } t \in [1, n].$$

Then, by Theorem 1.3,

$$\frac{\nu_*}{\omega_*^2} = \frac{\bar{T}}{\sum_{t=1}^{k} g_t/h_t}.$$

The last relation, Theorem 1.3, (1.6) and (1.7) imply Theorem 1.5. □

1.3 Conclusion

In this chapter we considered a jamming game for wireless networks which is a development of the game suggested by Altman, Avrachenkov and Garnaev (2006) for the case of linearized Shannon capacity utility. We showed that this game, which can be described as a game of base station versus nature, is a very natural one since it turns out that in the optimal behaviour the base station as well as the nature plays on the same channels. Also, for this game we developed an efficient algorithm which can find the optimal strategies in finite number of steps. Let us to demonstrate this algorithm on an example. Let $n = 5$, $g_i = h_i = 1$, $i \in [1,5]$ and the noises are distributed by the Rayleigh law $N_i^0 = \kappa^{i-1}$, $i \in [1,5]$, with $\kappa = 1.7$, $\bar{J} = 5$ and $\bar{T} = 10$. Then, $\varphi_t = (0, 0.7, 3.08, 9.149, 22.9054)$. So, $k = 3$ and the optimal strategy of jammer is $(2.53, 1.83, 0.64, 0, 0)$ and the optimal strategy of the base station is $(10/3, 10/3, 10/3, 0, 0)$

Bibliography

Altman, E., Avrachenkov, K. and Garnaev, A. (2007). Jamming game in wireless networks with transmission cost, *Lecture Notes in Computer Science* **4465**, pp. 1–12,

Altman, E., Avrachenkov, K. and Garnaev, A. (2007). *Closed form solutions for water-filling problems in optimization and game frameworks*, in *Proc. Workshop on Game Theory in Communication Networks (GameComm2007)*, (Nantes, France).

Kashyap, A., Basar, T. and Srikant R. (2004). Correlated jamming on MIMO Gaussian fading channels, *IEEE Transactions on Information Theory* **50**, pp. 2119–2123.

Jorswieck, E. A. and Boche, H. (2004). Performance analysis of capacity of MIMO systems under multiuser interference based on worst case noise behavior, *EURASIP Journal on Wireless Communications and Networking* **2**, pp. 273–285.

Lai, L. and Gamal, H. EI (2006). *Fading Multiple Access Channels: A Game Theoretic Perspective*, in *IEEE International Symposium on Information Theory (ISIT)*, pp. 1334–1338, (Seattle, WA).

Suarez-Real, A. (2006). *Robust Waterfilling strategies for the fading channe*, Master SICOM Thesis, INRIA.

Tse, D. and Viswanath, P. (2005). *Fundamentals of Wireless Communication*, (Cambridge University Press).

Chapter 2

Existence of Nash Networks in the One-way Flow Model of Network Formation

J. Derks, J. Kuipers, M. Tennekes and F. Thuijsman[*1]
Maastricht University, Department of Mathematics,
P.O. Box 616, 6200 MD Maastricht, The Netherlands
e-mail[*]: frank@micc.unimaas.nl

Abstract

We study a one-way flow connections model of unilateral network formation. We prove the existence of Nash networks for games where the corresponding payoff functions allow for heterogeneity among the profits that agents gain by the network. Furthermore, we show by a counterexample that, when link costs are heterogeneous, Nash networks do not always exist.

Key Words: Non-cooperative games, network formation, Nash networks

2.1 Introduction

Consider a group of agents who share certain profits by a network. In this network, the agents are represented as nodes. We consider one-way flow networks, where the links between the agents are directed and therefore depicted as arcs. The direction of the arcs corresponds to the flow of profits, i.e., a link between agents i and j which points at i means that i receives profits from being connected to j.

We study the formation of these one-way flow networks. We define a non-cooperative game in which agents have the opportunity to form costly links. Each agent can only form links pointing at him. All formed links together define the outcome network. We define a payoff function which assigns a payoff for each agent given the outcome network in the following

---
[1]We would like to thank two anonymous referees for constructive comments.

way: each agent pays certain costs for each link that he formed and each agent gains certain profits from each other agent from whom a directed path to him exists in the outcome network. A network is called a Nash network if no agent can gain a stricty higher payoff by deviating from his set of formed links.

Our model is based on the one-way flow connections model proposed by [Bala and Goyal (2000a)]. They characterize and prove the existence of Nash networks for games where profits and link costs are homogeneous, i.e., all links are equally expensive and all agents have equal profits. [Galeotti (2006)] studies heterogeneity among profits and link costs and he characterizes the architecture of (strict) Nash networks for various settings while assuming such Nash networks exist.

In this chapter we prove the existence of Nash networks for games with heterogeneous profits and owner-homogeneous link costs, i.e. all links have equal costs with respect to the agent who forms them. Furthermore, we provide a counterexample of a game with heterogeneous link costs for which Nash networks do not exist. The link costs of this game can be chosen arbitrarily close to the situation of owner-homogeneity.

Independently of us, and using a different approach, [Billand et al. (2007)] also proved the existence of Nash networks for games with heterogeneous profits and owner-homogeneous link costs. [Derks and Tennekes (2008b)] provide yet another alternative proof based directly on the ideas of [Billand et al. (2007)], but reducing the analysis to a short and elementary proof.

Several other models of network formation has been studied extensively in literature. Two-way flow models, i.e. models where profits can flow in both directions of a link, have been studied by [Bala and Goyal (2000a)], [Bala and Goyal (2000b)], [Galeotti et al. (2006)], and [Haller and Sarangi (2005)]. [Haller et al. (2007)] show the existence of Nash networks for two-way flow games with heterogeneous profits and homogeneous link costs and provide a counterexample with heterogeneous link costs where a Nash network does not exist. However, the results of [Haller et al. (2007)] do not imply ours, since they study two-way flow games while we focus on one-way flow games.

A model that is close to these one-way and two way flow models of network formation is the connections model introduced by [Jackson and Wolinsky (1996)]. Here, agents form links bilaterally instead of unilaterally. In other words, a link is only formed if both agents choose that link.

For an overview of literature on models of network formation we refer to [Jackson (2005)] and [Van den Nouweland (2005)].

2.2 Model and Notations

Let N denote a finite set of agents. We define a one-way flow network g on the agent set N as a set of links $g \subseteq N \times N$, where loops are not allowed, i.e. $(i,i) \notin g$ for all $i \in N$. A *path* from i to j in g is a sequence of distinct agents $i_1, i_2, \ldots, i_k$ with, $k \geq 1$, such that $i = i_1, j = i_k$ and $(i_s, i_{s+1}) \in g$ for each $s = 1, 2, \ldots, k-1$. Notice that for $k = 1$ we have that $i = i_1$ is a trivial path without links from i to himself.

Let $N_i(g) = \{j \in N \mid$ a path from j to i exists in $g\}$ and let $N_i^d(g) = \{j \mid (j,i) \in g\}$. Note that $i \in N_i(g)$, and $i \notin N_i^d(g)$.

For each agent i, let $\pi_i : \mathcal{G}_N \to \mathbb{R}$ be a payoff function, where $\mathcal{G}_N$ is the set of all possible one-way flow networks on N. We will use the following payoff function, which has been proposed by [Galeotti (2006)].

$$\pi_i(g) = \sum_{j \in N_i(g)} v_{ij} - \sum_{j \in N_i^d(g)} c_{ij} \tag{2.1}$$

Here v_{ij} is the profit that agent i receives from being connected to j and c_{ij} is the cost of link (j,i) for agent i. The profits and costs are assumed to be non-negative throughout this chapter.

We follow other literature on one-way flow models in the sense that the direction of the links indicates information flow. Consequently link $j \to i$, which is denoted by (j,i), is owned by agent i.

For convenience we will use the symbol '+' for the union of two networks as well as for the union of a network with a single link, e.g. $g \cup g' \cup \{(j,i)\}$ equals $g + g' + (j,i)$.

We say that link costs are *homogeneous* if there is a constant c with $c_{ij} = c$ for all $i, j \in N$. We say that link costs are *owner-homogeneous* if for each agent i there is a constant c_i with $c_{ij} = c_i$ for all $j \in N$. Otherwise, the link costs are *heterogeneous*.

In this chapter we study a non-cooperative game. This game is played by the agents in N. Simultaneously and independently, each agent i chooses a, possibly empty, set S of agents he wants to connect to by creating the links (j,i), for each $j \in S$. Together, the links of all agents form a network $g \in \mathcal{G}_N$. Then, each agent i receives a payoff $\pi_i(g)$. Since each agent wants to maximize his payoff in response to what the other agents are doing, the focus of this chapter is on Nash networks, i.e., networks in which no agent can profit from a unilateral deviation.

It is standard in literature to consider the set of agents, the costs, and the profits as fixed. However, our approach requires the comparison of different game situations. To facilitate this aproach we define a (non-cooperative) *network formation game* to be a triple (N, v, c) on agent set N with payoff functions π_i, $i \in N$, based on the profits $v = (v_{ij})_{i,j \in N}$ and costs $c = (c_{ij})_{i,j \in N}$, as described in Equation 2.1.

We define an *action* of agent i to a network g in network formation game (N, v, c) by a set of agents $S \subseteq N \setminus \{i\}$. The network that results after i chooses to link up with the agents in S, is described by

$$g_{-i} + \{(j, i) : j \in S\},$$

with g_{-i} denoting the network obtained from g after removing the links $(j, i) \in g$ owned by i. An action S^* of agent i is called a *best response* if

$$\pi_i\big(g_{-i} + \{(j,i) : j \in S^*\}\big) \geq \pi_i\big(g_{-i} + \{(j,i) : j \in S\}\big)$$

for all actions $S \subseteq N \setminus \{i\}$.

A network g is a Nash network in the game (N, v, c) if $N_i^d(g)$ is a best response for all $i \in N$, i.e., if for each agent i

$$\pi_i(g) \geq \pi_i\big(g_{-i} + \{(j,i) : j \in S\}\big)$$

for all actions $S \subseteq N \setminus \{i\}$.

Observe that for an agent i with costs sufficiently high, more specifically $c_{ik} > \sum_{j \in N, j \neq i} v_{ij}$ for all agents $k \neq i$, the only best response for agent i is $S = \emptyset$. Then, his payoff is v_{ii}, and any other action yields a smaller payoff (here we essentially need the fact that the profits are non-negative).

An agent k with no own links in a network g is only of interest for those agents i with $c_{ik} \leq v_{ik}$ since

$$\begin{aligned}
\pi_i(g + (k,i)) &= \sum_{j \in N_i(g+(k,i))} v_{ij} - \sum_{j \in N_i^d(g+(k,i))} c_{ij} \\
&= \sum_{j \in N_i(g)} v_{ij} + v_{ik} - \sum_{j \in N_i^d(g)} c_{ij} - c_{ik} \\
&= \pi_i(g) + v_{ik} - c_{ik}.
\end{aligned} \qquad (2.2)$$

2.3 Owner-homogeneous Costs

In this section we will prove existence of Nash networks in network formation games where the costs are owner-homogeneous.

For costs sufficiently small, the so-called *cycle* networks are Nash networks. Cycle networks consist of one cycle joining all agents (see Figure 2.1).

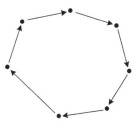

Fig. 2.1 Cycle network

Lemma 2.1. *For an owner-homogeneous cost network formation game (N, v, c), with $c_i \leq \sum_{j \in N, j \neq i} v_{ij}$ for all $i \in N$, all cycle networks are Nash networks.*

Proof. Without loss of generality, consider $N = \{1, 2, \ldots, n\}$ and the cycle network $g = \{(i, i+1) : i = 1, \ldots, n-1\} + (n, 1)$.

Any agent i obtains $\pi_i(g) = \sum_{j \in N} v_{ij} - c_i$, and there is no other network with a larger payoff, implying that g is a Nash network. $\square$

In the owner-homogeneous costs case, we also observe the following: if link (j, k) is present in g, then linking up with agent k is at least as good for an agent $i \neq j, k$, as linking up with j:

$$\pi_i(g + (k, i)) \geq \pi_i(g + (j, i)) \quad \text{whenever } (j, k) \in g. \tag{2.3}$$

In the next theorem, we prove the existence of Nash networks for games with owner-homogeneous link costs. This proof is constructive in nature. Either any cycle network constitutes a Nash network or there is an agent that is not interested in being 'involved'. In the latter case, there might be an agent i who is interested in linking up with this uninvolved agent. In that case the profit values are adapted as described in the proof, and a Nash network is searched in the situation without the uninvolved agent; next, this network is extended by connecting the uninvolved agent with agent i.

The case where no agent is interested in linking up with this uninvolved agent is somewhat simpler, as any Nash network on the set of agents excluding the uninvolved agent is also a Nash network on the full set of agents.

Theorem 2.1. *Nash networks exist for any network formation game with owner-homogeneous costs.*

Proof. We will prove the theorem by induction. 1-agent network formation games trivially have a Nash network. Suppose that (N, v, c) with $N = \{1, 2, \ldots, n\}$ is an owner-homogeneous cost network formation game that does not have a Nash network, while all network formation games with less than n agents do have Nash networks. According to Lemma 2.1, this implies that there is an agent i with $c_i > \sum_{j \in N, j \neq i} v_{ij}$.

Without loss of generality assume $i = n$. Observe that the best response of agent n in any network is the empty set. Consider the owner-homogeneous cost network formation game (N', v', c'), with $N' = N \setminus \{n\}$, and v' and c' equal to v and c restricted to the agents in N'. Let π'_i denote the payoff function for agent i in (N', v', c'). It is clear that $\pi'_i(g) = \pi_i(g)$ for each network g on N', and $i \neq n$.

Since N' has $n - 1$ agents, (N', v', c') has a Nash network, say g'. Consider g' as a network on N, and recall the assumption that (N, v, c) does not have a Nash network. Therefore, there is an agent i in (N, v, c) who does not play his best response in g'. Of course $i \neq n$, as $N_n^d(g') = \emptyset$. Let $T \subseteq N \setminus \{i\}$ be a best response of i in g', and suppose $n \notin T$. Then $(g')_{-i} + \{(j, i) : j \in T\}$ is a network in N' so that

$$\pi'_i(g') = \pi_i(g')$$
$$< \pi_i\big((g')_{-i} + \{(j, i) : j \in T\}\big)$$
$$= \pi'_i\big((g')_{-i} + \{(j, i) : j \in T\}\big),$$

which is a contradiction with g' being a Nash network for (N', v', c').

Now suppose that $n \in T$. Without loss of generality assume $i = 1$. From $n \in T$ and $N_n^d(g') = \emptyset$ we conclude that $c_1 \leq v_{1n}$ must hold (see 2.2). Consider the following adapted profits $v^* = (v^*_{ij})_{i,j \in N'}$:

$$v^*_{ij} = \begin{cases} v_{ij} & \text{if } j \neq 1, \\ v_{i1} + v_{in} & \text{if } i \neq 1, \, j = 1, \\ v_{11} + v_{1n} - c_1 & \text{if } i, j = 1. \end{cases}$$

Observe that these values are non-negative. Let π^*_i denote the payoff functions in (N', v^*, c'). The profits v^*_{ij} are chosen such that $\pi^*_i(g) = \pi_i(g + (n, 1))$ holds for all networks g on N', and for all $i \in N'$.

By assumption, the network formation game (N', v^*, c') has a Nash network, say g^*; since (N, v, c) does not have a Nash network, there is an agent i in N who can improve in the network $g^* + (n, 1)$, in the context of the game (N, v, c), say by choosing the links with the agents in $S \subseteq N\backslash\{i\}$. This agent is not n because $N_n^d(g^* + (n,1)) = \emptyset$.

Suppose $i \neq 1$. If $n \in S$, then according to (2.3), the action $S\backslash\{n\} \cup \{1\}$ is at least as good as S; therefore we may assume $n \notin S$. The resulting network $(g^* + (n,1))_{-i} + \{(j,i) : j \in S\}$, after i performs the improvement, yields a higher payoff for agent i. Then

$$\pi_i^*(g^*) = \pi_i(g^* + (n,1))$$
$$< \pi_i\big((g^* + (n,1))_{-i} + \{(j,i) : j \in S\}\big)$$
$$= \pi_i\big((g^*_{-i} + \{(j,i) : j \in S\}) + (n,1)\big)$$
$$= \pi_i^*(g^*_{-i} + \{(j,i) : j \in S\})$$
$$\leq \pi_i^*(g^*),$$

where the latter inequality holds because of g^* being a Nash network for (N', v^*, c'). Thus, we arrived at a contradiction, so that we must have $i = 1$.

Due to $c_1 \leq v_{1n}$, and agent n having no own links, we may assume $n \in S$ (see the observation concerning (2.2)). Then

$$\pi_1^*(g^*) = \pi_1(g^* + (n,1))$$
$$< \pi_1(g^*_{-1} + \{(j,1) : j \in S\})$$
$$= \pi_1\big(g^*_{-1} + \{(j,1) : j \in S\backslash\{n\}\} + (n,1)\big)$$
$$= \pi_1^*\big(g^*_{-1} + \{(j,1) : j \in S\backslash\{n\}\}\big)$$
$$\leq \pi_1^*(g^*);$$

a contradiction. We conclude that (N, v, c) must have a Nash network. $\square$

Observe that the Nash networks we obtain, have at most one cycle, in case the Nash networks excluding the uninvolved agent also have at most one cycle. The same applies when considering networks where agents have an outdegree of at most one. The following corollary is now easily established.

Corollary 2.1. *Nash networks exist, with at most one cycle and maximum outdegree of at most 1, for any network formation game with owner-homogeneous costs.*

There may also exist Nash networks with multiple cycles and with outdegrees higher than 1. Consider the following example.

Example 2.1. Let $n = 7$, and let $c_{ij} = 1$ and $v_{ij} = 1$ for all $i, j \in N$. Consider the network with two cycles that is depicted in Figure 2.2. Notice that agent i has two outgoing links. It can be verified that this network is a Nash network. We revisit this network in section 2.5.

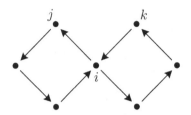

Fig. 2.2 A Nash network with two cycles

2.4 Heterogeneous Costs

For network formation games with heterogeneous costs, Nash networks do not always exist as we will see by the next example. The link costs in this example can be chosen arbitrarily close to the situation of owner-homogeneity.

Example 2.2. Consider a network formation game (N, v, c) where $N = \{1, 2, 3, 4\}$, where the profits are owner-homogeneous and normalized to 1 (i.e., $v_{ij} = 1$ for all i, j), and where the costs are heterogeneous. The numbers next to the links in Figure 2.3 indicate the costs of these links.

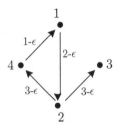

Fig. 2.3 The link costs

Here, ϵ is a strictly positive number which can be chosen arbitrarily close to 0. The costs of the links that are not depicted in this figure are the following:

- links directed to agent 1 have costs $1 + \epsilon$,
- links directed to agent 2 have costs $2 + \epsilon$,
- links directed to agents 3 and 4 have costs $3 + \epsilon$,

The best response of agent 4 to any network is either $\{2\}$ or $\emptyset$, since those are the only actions for which agent 4 has a non-negative payoff. First, suppose that agent 4 plays $\{2\}$ as a best response in a Nash network.

Consequently, the unique best response of agent 1 is {4}. Agent 2 has one unique best response to this situation: {1}. Finally, agent 3 has one unique best response, which is {2}. The obtained network is the same as depicted in Figure 2.3. It follows that {2} is not a best response of agent 4, since ∅ gives a higher payoff. This contradicts our assumption. Hence, there is no Nash network in which agent 4 plays {2}.

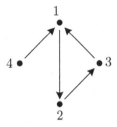

Fig. 2.4   Network obtained in Example 2.2

Now suppose that agent 4 plays ∅ as a best response in a Nash network. Agent 1 will include 4 in every best response to this situation. Then, the unique best response for agent 2 is {1}. To this situation, the unique best response of agent 3 is {2}. Hence, the unique best response of agent 1 is {3, 4} (see Figure 2.4). Now agent 4 has a unique best response to this new situation, which is {2}. This contradicts our assumption of agent 4 playing ∅ in a Nash network. Hence no Nash networks exist.

As is clear from this example, Nash networks do not always exist for network formation games with heterogeneous costs. However, some of the arguments in the proof of our main result (Theorem 2.1) can be generalized to hold also for a specific class of payoff functions with heterogeneous costs. For example, expression (2.3) is also fulfilled when the following conditions are met:

$$c_{ij} \geq c_{ik} - v_{ik} \quad \text{for all different agents } i, j, k.$$

Unfortunately, not all arguments have their counterpart in the heterogeneous case; especially the relation between the Nash networks and cycles is not apparent. Therefore, a generalization of our main result needs a different approach, and in our subsequent article [Derks et al. (2008)] we show the existence of Nash networks for a more general class of network formation games. The adapted proof is also constructive in nature, and starts with a framework of properties that the payoff functions have to obey. These properties provide a generalization of our main result, which is further discussed in [Derks and Tennekes (2008a)].

2.5 Strict Nash Networks

A network g is a *strict-Nash network* if $N_i^d(g)$ is a unique best response for each agent i. [Galeotti (2006)] studies the architecture of strict-Nash networks in detail. He shows that the maximum outdegree of strict-Nash networks is at most 1 in network formation games with (owner-)homogeneous link costs. This is confirmed by Example 2.1. The network depicted in Figure 2.2 is not strict-Nash, because agent j can deviate by forming (k,j) instead of (i,j) which gives him the same payoff due to link costs owner-homogeneity. Notice that in the newly obtained network, agent i can deviate by removing link (k,i) which gives him a higher payoff. The cycle network that we now obtain is both Nash and strict-Nash.

The following example shows that strict-Nash networks do not always exist for games with owner-homogeneous link costs.

Example 2.3. Let again $N = \{1, \ldots, n\}$ be the set of agents. Let $c_1 = n-1$, and $c_i = 1$ for all $i \neq 1$. Let $v_{ij} = 1$ for all $i, j \in N$.

It is easily seen that in each strict-Nash network, all agents in $N \setminus \{1\}$ are contained in one cycle.

Either agent 1 is also contained in this cycle or not. Suppose he is. Then by his one link he receives $n-1$ profits, and the link itself costs $n-1$. Hence, he is indifferent about maintaining this link. Thus, a cycle network cannot be strict-Nash.

Now suppose that agent 1 is not contained in the cycle on $N \setminus \{1\}$. Then, by forming a link with one of the other agents, agent 1 receives $n-1$ profits, and pays $n-1$. Therefore, he is indifferent about forming such a link. Hence, again the network cannot be strict-Nash. Therefore, we conclude that strict-Nash networks do not exist for this game.

Bibliography

Bala, V. and Goyal, S. (2000a). A non-cooperative model of network formation, *Econometrica* **68**, pp. 1181–1229.

Bala, V. and Goyal, S. (2000b). A strategic analysis of network reliability, *Review of Economic Design* **5**, pp. 205–228.

Billand, P., Bravard, C. and Sarangi, S. (2007). Existence of Nash networks in one-way flow models, forthcoming in *Economic Theory*.

Derks, J., Kuipers, J., Tennekes, M. and Thuijsman, F. (2008). *Local dynamics in network formation*, Tech. Rept., Maastricht Univerisity, Department of Mathematics.

Derks, J. and Tennekes, M. (2008a). *A characterization of proper payoff functions in network formation games*, Tech. Rept., Maastricht Univerisity, Department of Mathematics.

Derks, J. and Tennekes, M. (2008b). A note on the existence of Nash networks in one-way flow models, forthcoming in *Economic Theory*.

Galeotti, A. (2006). One-way flow networks: the role of heterogeneity, *Economic Theory* **29**, pp. 163–179.

Galeotti, A., Goyal, S. and Kamphorst, J. (2006). Network formation with heterogeneous players, *Games and Economic Behavior* **54**, pp. 353–372.

Haller, H., Kamphorst, J. and Sarangi, S. (2007). (Non-)existence and scope of Nash networks, *Economic Theory* **31**, pp. 597–604.

Haller, H. and Sarangi, S. (2005). Nash networks with heterogeneous links, *Mathematical Social Sciences* **50**, pp. 181–201.

Jackson, M. O. (2005). *A survey of the formation of networks: stability and efficiency*, in *Group Formation in Economics; Networks, Clubs and Coalitions*, eds: G. Demange and M. Wooders, Ch. 1., (Cambridge University Press).

Jackson, M. O and Wolinsky, A. (2005). A strategic model of social and economic networks, *Journal of Economic Theory* **71**, pp. 44–74.

Nouweland, A. van den (2005). *Models of network formation in cooperative games*, in *Group Formation in Economics; Networks, Clubs and Coalitions*, eds: G. Demange and M. Wooders, Ch. 2., (Cambridge University Press).

Chapter 3

Strategic Advertisement with Externalities: A New Dynamic Approach

Reinoud Joosten[1]
FELab & Department of Finance & Accounting
University of Twente
POB 217, 7500 AE Enschede, The Netherlands.
e.mail: r.a.m.g.joosten@utwente.nl

Abstract

We model and analyze strategic interaction over time in a duopolistic market. Each period the firms independently and simultaneously choose whether to advertise or not. Advertising increases the own immediate sales, but may also cause an externality, e.g., increase or decrease the immediate sales of the other firm ceteris paribus.

There exists also an effect of past advertisement efforts on current sales. The 'market potential' of each firm is determined by its own but also by its opponent's past efforts. A higher effort of either firm leads to an increase of the market potential, however the impact of the own past efforts is always stronger than the impact of the opponent's past efforts. How much of the market potential materializes as immediate sales, then depends on the current advertisement decisions.

We determine feasible rewards and (subgame perfect) equilibria for the limiting average reward criterion using methods inspired by the repeated-games literature. Uniqueness of equilibrium is by no means guaranteed, but Pareto efficiency may serve very well as a refinement criterion for wide ranges of the advertisement costs.

Key Words: Advertising, externalities, average rewards, equilibria

---

[1]This contribution improved from comments by Luca Lambertini, Bas Donkers, Maarten Janssen, audiences in the EARIE 2006 conference, at a Game Theory Day at CentER\Tilburg University and a Tinbergen Institute seminar, and anonymous referees.

3.1 Introduction

The aim of this chapter is to model and to analyze strategic interaction over time in a duopolistic market in which advertising causes several types of 'externalities'. For this purpose, we design a so-called game with joint frequency dependent stage payoffs, JFD-game for short, which allows us to model rather complex relationships, and analyze it with modifications of techniques traditionally used for infinitely repeated games. JFD-games generalize games with frequency dependent stage payoffs, or FD-games, introduced by Brenner & Witt [2003], and classified and analyzed in Joosten et al. [2003]. In an FD-game, current stage payoffs depend on the relative frequencies with which all *actions* were played in the past. In a JFD-game, current stage payoffs depend on the relative frequencies with which all *action pairs* were played in the past. Hence, for the former type of games only the 'marginal' frequencies matter, in the newer type of games the joint frequencies matter. So, this chapter offers two innovations, it brings a novel way of modelling to the advertisement literature, and for this purpose it brings a newly designed game-theoretical approach.[2]

Each period both firms independently and simultaneously first choose whether to advertise or not. The firms know the advertisement decisions and then produce the pertaining Cournot-Nash quantities. If a firm decides to advertise, it pays a fixed amount at the beginning of the period. Hence, for the Cournot competition phase the advertisement costs can be regarded as sunk. It is not our aim to model the Cournot competition explicitly, we are merely interested in its outcome in sales.

Advertising has two effects separated in the time dimension[3], it affects immediate sales directly and future sales in a cumulative manner (cf., e.g., Friedman [1983]). With respect to the short run effects, advertising increases the own immediate sales given the action of the other firm. Advertising may also cause immediate externalities. Friedman [1983] distinguishes predatory and cooperative advertising. An increase in advertising efforts of one firm leads to a sales decrease of the other in the former type, to an increase in the second type of advertising.

There is also an effect of *current* advertisement on *future* sales. We

---

[2] An initial investment for researchers to acquint themselves with the current framework, will, in our view, be compensated easily by considerable gains in modelling flexibility and the convenience with which results can be obtained.

[3] In our view, Lambertini and co-authors define a dichotomy precisely here, cf. e.g., Cellini & Lambertini [2003], Lambertini [2005], Cellini et al. [2008].

use the notion of market potential, both with respect to the total market and the individual firms, in order to capture these effects. The *current* market potential of each firm is determined by its own but also by its opponent's *past* efforts. A higher effort of either firm leads to an increase of the market potentials, but the impact of the own past efforts on the own market potential is always stronger than the impact of the opponent's past efforts. Advertising is therefore cooperative in its cumulative effects on the market potentials. How much of the market potential materializes as immediate sales, depends on the current advertisement decisions.

Dorfman & Steiner [1954] examine the effects of advertising in a static monopoly and derive necessary conditions for the optimal level of advertising. In a dynamic monopolistic model, Nerlove & Arrow [1962] treat advertisement expenditures similar to investments in a durable good. This durable good is called goodwill which is assumed to influence current sales. Historical investments in advertisement increase the stock of goodwill, but simultaneously goodwill depreciates over time. Nerlove and Arrow derive necessary conditions for optimal advertising, thus generalizing the Dorfman and Steiner result. Friedman [1983] in turn generalizes the Nerlove-Arrow model to allow oligopolistic competition in advertising and derives necessary conditions for the existence of a noncooperative equilibrium (Nash [1951]).

Our notion of market potential is quite close to goodwill in e.g., Nerlove & Arrow [1962] and Friedman [1983]. The modeling of the changes in time in the former model follows the work of Vidale & Wolfe [1957], though the authors quote Waugh [1959] as a main source of inspiration. Vidale & Wolfe [1957] present an interesting field study giving empirical evidence of the positive effects of past advertising on current sales. Furthermore, once advertising expenditures are stopped, current sales do not collapse, but slowly deteriorate over time. Though Friedman quotes the work of Nerlove and Arrow as a source for the notion of goodwill, his technical treatment of the changes of the level of goodwill in time is inspired by Prescott [1973].

Economics has produced a large body of work on advertising featuring a broad variety of approaches. One source of variety is the modeling of time-related aspects. For instance, is the model static (e.g., Dorfman & Steiner [1954]), or is it dynamic in the sense that the strategic environment may change (e.g., Nerlove & Arrow [1962])? Another source of variety is the market under consideration, e.g., monopoly (Nerlove & Arrow [1962]), oligopoly (Friedman [1983]), leader-follower oligopoly (Kydland [1977]). A third one is possible combinations of advertising with other marketing instruments, e.g., Schmalensee [1978] combines advertising and quality. A fourth one is the entity to be influenced by advertising, for instance sales (e.g., Nerlove & Arrow [1962]) or market shares[4] (e.g., Fershtman [1984]).

Another dimension is based on the distinction by Nelson [1970] between search and experience goods. The characteristics of former kind are known by-and-large before purchase, whereas the characteristics of the latter can be determined only after purchase. Advertising differs for the two types of goods because the information conveyed to the consumers differs. *Informative* advertising provides information on e.g., the price, availability or characteristics of a product; *persuasive* advertising tries to generate consumer interest for a product, often by association or through rather indirect 'channels'. An example of persuasive advertising would be a famous athlete shown drinking a certain beverage, or eating some kind of cereal. Credence goods (Darby & Karni [1973]) can be regarded as an extreme type of experience good, as it is hard to determine their characteristics even after purchase. The quality of a certain brand of toothpaste can only be determined in the very long run after a visit to a dentist. Informative advertising is directed at search goods, persuasive advertising aims at experience or credence goods.

For dynamic optimal control models of advertisement Sethi [1977] performed a Herculean task by coming up with a classification distinguishing four types. The task may prove to be Sisyphean in this burgeoning field, as a more recent survey by Feichtinger *et al.* [1994] already features six classes. Three new categories were introduced, categories present in the earlier classification were renamed and expanded, and merely one category survived in its original form. The reader interested in differential games on advertising is referred to Dockner *et al.* [2000] and Jørgensen & Zaccour [2004].

FD-games are stochastic games with finite action spaces and infinite state spaces. The basic idea for FD-games stems from the work of Herrnstein on experimental 'games against nature' (cf., Herrnstein [1997]), Brenner & Witt [2003] used frequency-dependent payoffs in a multi-person game. Joosten *et al.* [2003] showed that the analysis of infinitely repeated games (cf., e.g., Van Damme [1991]) can be generalized to this type of games.

Relating this chapter to the literatures mentioned: in our dynamic deterministic duopolistic leaderless model we restrict ourselves to the effects of cooperative persuasive advertising with long and short run externalities on

---

[4]Relevant if market size is 'fixed', e.g., Telser [1962], Schmalensee [1978].

sales. We assume that the firms wish to maximize the average profits over an infinite time-horizon. We determine equilibria for all 'realistic' ranges of advertisement costs employing modifications of techniques traditionally used to analyze infinitely repeated games. We find that a continuum of rewards may exist which can be supported by an equilibrium involving 'threats'.

In the next section, we introduce the advertisement model. Section 3.3 deals with strategies and rewards in our model, whereas section 3.4 deals with the notion of threats and with equilibria. Section 3.5 deals with Pareto efficiency as a refinement criterion, section 3.6 concludes. The Appendix contains issues not fully treated in the main text.

3.2 The Rules of the Game

The advertisement game is played by two firms (players) A and B at discrete moments in time called stages. Each player has two actions and each stage the players independently and simultaneously choose an action. Action 1 for either player denotes 'advertise', action 2 denotes 'not advertise'. We denote the action set of player A (B) by $J^A = \{1, 2\}$ $(= J^B)$ and $J \equiv J^A \times J^B$.

The payoffs at stage $t' \in \mathbb{N}$ of the play depend on the choices of the players at that stage, and on the relative frequencies with which all actions where actually chosen until then. In our model we have two types of externality effects from advertising, an immediate one and one which develops gradually in time. We start by describing the immediate externalities, then we formalize the externalities in time, and finally we connect these effects.

3.2.1 *Immediate Effects, Stage Games*

The effects of advertising on the immediate sales by the two firms can be expressed by the following **sales matrix**

$$\begin{bmatrix} a^A, a^B & b^A, c^B \\ c^A, b^B & d^A, d^B \end{bmatrix}. \tag{1}$$

The top row (left column) of player A (B) corresponds to action 1, i.e., 'to advertise', the bottom row (right column) corresponds to action 2, i.e., 'not to advertise'. If, e.g., player A advertises and player B does not, then A has sales b^A and B has sales c^B. The following restrictions are assumed to

hold for player k, $k = A, B$:
$$a^k, b^k > c^k > d^k > 0,$$
$$a^k + a^{\neg k} > b^k + c^{\neg k} > d^k + d^{\neg k}, \qquad (2)$$
$$\left(b^k + c^{\neg k}\right) - \left(d^k + d^{\neg k}\right) > \left(a^k + a^{\neg k}\right) - \left(b^k + c^{\neg k}\right),$$

where $\neg k$ denotes player k's opponent. Hence, advertising increases the immediate sales of an agent given the action of the opponent. Observe that if firm A advertises and B does not, then B also gets an increase in current sales relative to the situation in which neither advertises. This increase can be regarded as a positive externality from A's effort on B's immediate sales.

We assume that there exist increasing returns to advertising on total immediate sales, but decreasing marginal returns. Total immediate sales are ranked $a^k + a^{\neg k} > b^k + c^{\neg k} > d^k + d^{\neg k}$, so they are highest if both firms advertise, second highest if only one firm advertises, and lowest if no firm advertises. The final inequality in (2) implies that total current sales increase more while moving from the situation in which no firm advertises to the situation in which one firm advertises, than while moving from the latter situation to the one in which both firms advertise. If $a^k > b^k$, then advertising is completely cooperative, otherwise it is indeterminate.

Advertising is not for free in general. We assume the costs of advertising to be fixed in time, independent of the size of the sales and identical for both firms. So, the introduction of these costs $\kappa \geq 0$ defines the game in terms of profits, which yields the **payoff matrix**

$$\begin{bmatrix} a^A - \kappa, a^B - \kappa & b^A - \kappa, c^B \\ c^A, b^B - \kappa & d^A, d^B \end{bmatrix}. \qquad (3)$$

Since 'not advertising' does not induce any cost, player A's (B's) entries in the bottom row (right column) do not change relative to (1).

Example 3.1. Let the sales matrix be determined by $a^k = 100$, $b^k = 104$, $c^k = 88$, and $d^k = 84$. Different values of the advertisement costs may alter the character of the stage game. For instance, taking $\kappa = 6$ induces the payoff matrix

$$\begin{bmatrix} 94, 94 & 98, 88 \\ 88, 98 & 84, 84 \end{bmatrix}.$$

Here, advertising is the 'obvious thing to do'. The payoffs associated with the unique pure Nash equilibrium in which both firms advertise are $(94, 94)$. Furthermore, for $\kappa = 14$, the stage game of the numerical example is given by

$$\begin{bmatrix} 86,86 & 90,88 \\ 88,90 & 84,84 \end{bmatrix}.$$

Here, two pure Nash equilibria exist in which one player advertises and the other one does not. Furthermore, a mixed Nash equilibrium exists with symmetric payoffs. For $\kappa = 25$, the payoff matrix is given by

$$\begin{bmatrix} 75,75 & 79,88 \\ 88,79 & 84,84 \end{bmatrix}.$$

Here, not to advertise is a strictly dominant action for both players. Costs are so high that the increase in sales does not make up.

3.2.2 Long Term Effects, Market Potentials

Another type of externalities accumulates gradually over time. We assume that advertisement at any point in time has two effects in the future. First, the advertisement efforts have a cumulative effect on the way the total market increases and second, the firm showing more cumulative advertisement efforts gets a larger share of this (potentially) expanded market. In order to introduce these externalities, we need several notations.

Let $h_{t'}^A = \left(j_1^A, ..., j_{t'-1}^A\right)$ be the sequence of actions[5] chosen by player A until stage $t' \geq 2$ and let $h_{t'}^B$ be defined similarly for the other player. Let $\Delta^{m \times n}$ denote the set of real-numbered non-negative $m \times n$-matrices such that all components add up to unity, i.e.,

$$\Delta^{m \times n} = \left\{ z \in \mathbb{R}^{m \times n} | z_{ij} \geq 0 \text{ for all } i, j, \text{ and } \sum_{ij} z_{ij} = 1 \right\}.$$

Let matrix $U(i', j') \in \Delta^{2 \times 2}$ be defined by:

$$U_{ij}(i', j') = \begin{cases} 1 & \text{if } (i,j) = (i', j'), \\ 0 & \text{otherwise.} \end{cases}$$

Then, let $q \geq 0$, and define matrix $\rho_t \in \Delta^{2 \times 2}$ recursively for $t \leq t'$ by

$$\rho_1 = \widetilde{\rho} \in \Delta^{2 \times 2}, \text{ and}$$
$$\rho_t = \frac{q+t-1}{q+t}\rho_{t-1} + \frac{1}{q+t}U\left(j_{t-1}^A, j_{t-1}^B\right).$$

---

[5] We assume that each player produces the Cournot quantity in the second phase of the stage in the game that arises from the advertisement decisions taken in the first phase. Hence, we omit the reference to the second phase actions.

Taking $q \gg 0$ moderates 'early' effects on the stage payoffs. Recall that j_{t-1}^A denotes the action chosen by A at stage $t-1$. The interpretation of this matrix is that entry ij of ρ_t 'approximates' the relative frequency with which action pair (i,j) was used before stage $t \geq 2$, as it can be shown that

$$\rho_t = \frac{q+1}{q+t}\widetilde{\rho} + \frac{t-1}{q+t} U\left(h_t^A, h_t^B\right).$$

Here, $U\left(h_t^A, h_t^B\right) = \frac{1}{t-1}\sum_{k=2}^t U\left(j_{k-1}^A, j_{k-1}^B\right)$ for all $t \geq 2$. Clearly, the influence of $\widetilde{\rho}$ and q disappears in the long run.

At stage $t \in \mathbb{N}$, the players have chosen action sequences h_t^A and h_t^B inducing the matrix ρ_t. The latter determines the state in which the play is at stage t.[6] Observe that there exist four possible successor states to any state depending on the action pair chosen at stage t.

Given ρ_t, the **market potential of player** k, $k = A, B$ at stage t is

$$MP_t^k = \mu^k\left(\rho_t\right), \qquad (4)$$

where $\mu^k(\cdot)$, $k = A, B$, is a continuous function from $\Delta^{2\times 2}$ to $\mathbb{R}$. The market potential of a firm at a certain stage is influenced by the firm's own advertisement efforts before, but also by the other player's past advertisement efforts. Own past efforts are always positive ceteris paribus, i.e., the market potential is always higher if the own advertisement efforts have been higher in the past. Also, the own past efforts have a stronger impact on the firm's market potential than the other firm's have. Mathematically, this means that for $\rho_t = \rho$

$$\partial \mu^A / \partial \rho_{11} + \partial \mu^A / \partial \rho_{12} \geq 0, \ \partial \mu^B / \partial \rho_{11} + \partial \mu^B / \partial \rho_{21} \geq 0,$$
$$\partial \mu^A / \partial \rho_{11} + \partial \mu^A / \partial \rho_{12} \geq \partial \mu^A / \partial \rho_{11} + \partial \mu^A / \partial \rho_{21},$$
$$\partial \mu^B / \partial \rho_{11} + \partial \mu^B / \partial \rho_{21} \geq \partial \mu^B / \partial \rho_{11} + \partial \mu^B / \partial \rho_{12}.$$

It can be seen easily that this further implies $\partial \mu^1 / \partial \rho_{12} \geq \partial \mu^1 / \partial \rho_{21}$ as well as $\partial \mu^2 / \partial \rho_{21} \geq \partial \mu^2 / \partial \rho_{12}$. Let the **market potential** at stage t be given by $MP_t \equiv MP_t^A + MP_t^B$. Hence, the market potential increases with the rate of advertising by either firm. So, in this sense, advertising has a public good character as in Fershtman [1984].

Example 3.2. Given ρ_t, the market potentials of the firms are given by

$$MP_t^A = 50 + 110\left([\rho_t]_{11} + [\rho_t]_{12}\right) + 40\left([\rho_t]_{11} + [\rho_t]_{21}\right),$$
$$MP_t^B = 50 + 110\left([\rho_t]_{11} + [\rho_t]_{21}\right) + 40\left([\rho_t]_{11} + [\rho_t]_{12}\right).$$

---

[6]Slightly more formal, we will denote the state at stage $t \in \mathbb{N}$ by ρ_t from now on.

The interested reader may confirm the above inequalities easily. The impact of the own advertisement efforts on the own market potential is weighted by a factor which is nearly three times the factor connected to the other firm's advertisement efforts. We give a graphical illustration below. For $\rho_t = \begin{bmatrix} 0 & 0 \\ 0 & 1 \end{bmatrix}$ the market potentials for both firms are equal to 50; for $\rho_t = \begin{bmatrix} 1 & 0 \\ 0 & 0 \end{bmatrix}$ the market potentials are equal to 200. Hence, always advertising by both firms generates an expansion of the market potentials for both firms by 300%. Another interesting extreme case is $\rho_t = \begin{bmatrix} 0 & 1 \\ 0 & 0 \end{bmatrix}$, i.e., player A always advertises and player B never advertises. Then, A has a market potential equal to 160, whereas B has a market potential of 90. Here, the total market potential increases by 150%, but the increase for the player advertising is 220%, and 80% for the other.

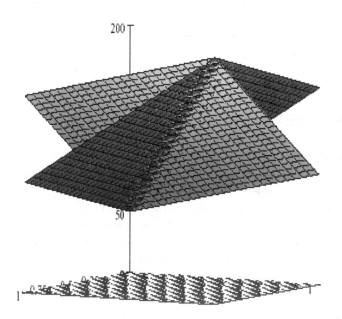

3.2.3 Combining Immediate and Long Term Effects

Now, we intend to be more precise about a^k, b^k, c^k, d^k, $k = A, B$ in (1). For a given state ρ_t the numbers are fixed indeed, but the stage games vary over time as the play proceeds going from one state to another, and the 'numbers' change along. In general, we have numbers $a_t^k, b_t^k, c_t^k, d_t^k$ connected to the set of states $\rho_t \in [0,1]^2$. The following links the latter numbers to the market potentials of the respective firms:

$$a_t^k = (1 - \beta) MP_t^k + a_0^k \beta MP_t,$$
$$b_t^k = (1 - \beta) MP_t^k + b_0^k \beta MP_t,$$
$$c_t^k = (1 - \beta) MP_t^k + c_0^k \beta MP_t,$$
$$d_t^k = (1 - \beta) MP_t^k + d_0^k \beta MP_t.$$

We assume that the inequalities (2) hold for the parameters $a_0^k, b_0^k, c_0^k, d_0^k$ instead of a^k, b^k, c^k, d^k, as well as $a_0^1 + a_0^2 = 1$, $\beta \in [0,1]$. Note that for $\beta = 0$ no strategic interaction on the short run exists. Moreover, the height of the own market potential is exclusively decisive for the current sales. For increasing β the influence of the total market potential increases and then the size of the opponent's market potential increases in importance for the own immediate sales.

The numbers can be motivated as follows. Suppose that a fraction of $1 - \beta$ of firm k's market potential will be realized as actual sales independent from the advertisement decisions taken by the firms at that stage. Part of the public namely is already committed to purchasing the product from one firm. However, another part of the potential buyers are still not committed to buy from either firm, or are interested in buying but they may be thinking about postponing their purchase. Hence, the total market potential which is at stake, i.e., to be influenced by the advertisement behavior of both firms, is βMP_t. If both firms advertise, then each gets half of the market potential at stake. If only firm k advertises, then a fraction b_0^k (c_0^k) of the market potential at stake ends up with the (non-)advertising firm as sales, the remaining fraction of $\left(1 - b_0^k - c_0^{-k}\right)$ will not materialize as sales. If neither firm advertises, then a fraction $\left(1 - d_0^k - d_0^{-k}\right)$ of the market potential at stake will not materialize as sales, the remainder is split equally.

Example 3.3. (Continued) The general expression for the sales game at stage t given ρ_t is

$$\begin{bmatrix} a_t^1, a_t^2 & b_t^1, c_t^2 \\ c_t^1, b_t^2 & d_t^1, d_t^2 \end{bmatrix}$$

where for $\beta = \frac{1}{5}$, $a_0^k = \frac{1}{2}$, $b_0^k = \frac{3}{5}$, $c_0^k = \frac{1}{5}$, and $d_0^k = \frac{1}{10}$, we have

$$a_t^k = \frac{4}{5}MP_t^k + \frac{1}{10}MP_t, \ b_t^k = \frac{4}{5}MP_t^k + \frac{3}{25}MP_t,$$

$$c_t^k = \frac{4}{5}MP_t^k + \frac{1}{25}MP_t, \ d_t^k = \frac{4}{5}MP_t^k + \frac{1}{50}MP_t,$$

$$MP_t^A = 50 + 110\left([\rho_t]_{11} + [\rho_t]_{12}\right) + 40\left([\rho_t]_{11} + [\rho_t]_{21}\right),$$

$$MP_t^B = 50 + 110\left([\rho_t]_{11} + [\rho_t]_{21}\right) + 40\left([\rho_t]_{11} + [\rho_t]_{12}\right),$$

$$MP_t = 100 + 150\left(1 + [\rho_t]_{11} - [\rho_t]_{22}\right),$$

$$\rho_1 = \widetilde{\rho} \in \Delta^{2\times 2}, \text{ and}$$

$$\rho_t = \frac{q+t-1}{q+t}\rho_{t-1} + \frac{1}{q+t}U\left(j_{t-1}^A, j_{t-1}^B\right).$$

To obtain the stage game with respect to the profits, advertisement costs must be subtracted from entries in the first row (column) of player A (B). The effects of advertising on the stage payoffs depend on the costs of advertising, hence, as in Subsection 3.2.1, the game changes with every value of κ.

3.3 What Can They Do, What Can They Get?

At stage t, both players know the current state and the history of play, i.e., the state visited and actions chosen at stage $u < t$ denoted by $\left(\rho_u, j_u^A, j_u^B\right)$. A **strategy** prescribes at all stages, for any state and history, a mixed action to be used by a player. The sets of all strategies for A respectively B will be denoted by $\mathcal{X}^A$ respectively $\mathcal{X}^B$, and $\mathcal{X} \equiv \mathcal{X}^A \times \mathcal{X}^B$. The payoff to player k, $k = A, B$, at stage t, is stochastic and depends on the strategy-pair $(\pi, \sigma) \in \mathcal{X}$; the **expected stage payoff** is denoted by $R_t^k(\pi, \sigma)$.

The players receive an infinite stream of stage payoffs during the play, and they are assumed to wish to maximize their average rewards. For a given pair of strategies (π, σ), player k's **average reward**, $k = A, B$, is given by $\gamma^k(\pi, \sigma) = \liminf_{T\to\infty} \frac{1}{T}\sum_{t=1}^T R_t^k(\pi, \sigma)$; $\gamma(\pi, \sigma) \equiv \left(\gamma^A(\pi, \sigma), \gamma^B(\pi, \sigma)\right)$.

It may be quite hard to determine the **set of feasible (average) rewards** F, directly. It is not uncommon in the analysis of repeated or stochastic games to limit the scope of strategies on the one hand, and to focus on rewards on the other. Here, we will do both, we focus on rewards from strategies which are pure and jointly convergent. Then, we extend our analysis to obtain more feasible rewards.

A strategy is **pure**, if at *each* stage a **pure action** is chosen, i.e., the action is chosen with probability 1. The set of pure strategies for player k is $\mathcal{P}^k$, and $\mathcal{P} \equiv \mathcal{P}^A \times \mathcal{P}^B$. The strategy pair $(\pi, \sigma) \in \mathcal{X}$ is **jointly convergent** if and only if $z^{\pi,\sigma} \in \Delta^{m \times n}$ exists such that for all $\varepsilon > 0$ and all $(i,j) \in J$:

$$\limsup_{t \to \infty} \Pr_{\pi,\sigma} \left[\left| \frac{\#\{j_u^A = i \text{ and } j_u^B = j | \, 1 \leq u \leq t\}}{t} - z_{ij}^{\pi,\sigma} \right| \geq \varepsilon \right] = 0$$

where $\Pr_{\pi,\sigma}$ denotes the probability under strategy-pair (π, σ). $\mathcal{JC}$ denotes the set of jointly-convergent strategy pairs. Under a pair of jointly-convergent strategies, the relative frequency of each action pair $(i,j) \in J$ converges with probability 1 to $z_{ij}^{\pi,\sigma}$ in the terminology of Billingsley [1986, p.274], i.e., this implies $\lim_{t \to \infty} E_{\pi,\sigma}\{U\left(h_t^A, h_t^B\right)\} = z^{\pi,\sigma}$. However, this implies also $\lim_{t \to \infty} E_{\pi,\sigma}\{\rho_t\} = z^{\pi,\sigma}$.

The **set of jointly-convergent pure-strategy rewards** is given by

$$P^{\mathcal{JC}} \equiv cl\left\{\left(x^1, x^2\right) \in \mathbb{R}^2 | \, \exists_{(\pi,\sigma) \in \mathcal{P} \cap \mathcal{JC}} : \left(\gamma^k(\pi, \sigma), \gamma^k(\pi, \sigma)\right) = \left(x^1, x^2\right)\right\},$$

where $cl\, S$ is the closure of the set S. The interpretation is that for any pair of rewards in this set, we can find a pair of jointly-convergent pure strategies that yield rewards arbitrarily close to the original pair of rewards.

The set $P^{\mathcal{JC}}$ can be determined rather conveniently, as we will show now. For a jointly-convergent pair of strategies (π, σ), we have that the expected market potentials for the firms and the total market potential converge to respectively $\mu^A(z^{\pi,\sigma})$, $\mu^B(z^{\pi,\sigma})$ and $\mu^A(z^{\pi,\sigma}) + \mu^B(z^{\pi,\sigma})$ because the functions involved are continuous (cf., e.g., Billingsley [1986]). The 'expected long run stage game' for this pair of jointly convergent strategies (π, σ) is

$$\begin{bmatrix} a_\infty^A(\pi,\sigma) - \kappa, a_\infty^B(\pi,\sigma) - \kappa & b_\infty^A(\pi,\sigma) - \kappa, c_\infty^B(\pi,\sigma) \\ c_\infty^A(\pi,\sigma), b_\infty^B(\pi,\sigma) - \kappa & d_\infty^A(\pi,\sigma), d_\infty^B(\pi,\sigma) \end{bmatrix}$$

where, $a_\infty^A(\pi,\sigma) = (1-\beta)\mu_t^A(z^{\pi,\sigma}) + a_0^k\beta(\mu_t^A(z^{\pi,\sigma}) + \mu_t^B(z^{\pi,\sigma}))$ and the other parameters are defined analogously. Furthermore, under (π, σ) the relative frequency of action pair $(1,1)$ being chosen is $z_{11}^{\pi,\sigma}$ and each time the players receive an expected payoff $\left(a_\infty^A(\pi,\sigma) - \kappa, a_\infty^B(\pi,\sigma) - \kappa\right)$ in the long run. The following notion is then to be interpreted as the sum of all long-run expected payoffs connected to the action pairs chosen weighted by the relative frequencies of those action pairs occurring in the long run. Let

$$\varphi(z^{\pi,\sigma}) \equiv z_{11}^{\pi,\sigma}\left(a_\infty^A(\pi,\sigma) - \kappa, a_\infty^B(\pi,\sigma) - \kappa\right) + z_{12}^{\pi,\sigma}\left(b_\infty^A(\pi,\sigma) - \kappa, c_\infty^B(\pi,\sigma)\right)$$
$$+ z_{21}^{\pi,\sigma}\left(c_\infty^A(\pi,\sigma), b_\infty^B(\pi,\sigma) - \kappa\right) + z_{22}^{\pi,\sigma}\left(d_\infty^A(\pi,\sigma), d_\infty^B(\pi,\sigma)\right).$$

Alternatively, the players receive a long run average payoff equal to $\varphi(z^{\pi,\sigma})$. This finalizes the argument that for this jointly-convergent pair of strategies (π, σ), we have $\gamma(\pi, \sigma) = \varphi(z^{\pi,\sigma})$.

The following result, illustrated in Figures 2 and 3, can be found in Joosten et al. [2003] for FD-games. Related ideas were designed for the analysis of repeated games with vanishing actions (cf., Joosten [1996, 2005], Schoenmakers et al. [2002]). Let $CP^{\mathcal{JC}}$ denote the convex hull of $P^{\mathcal{JC}}$.

Theorem 3.1. *For any FD-game, we have $P^{\mathcal{JC}} = \bigcup_{z \in \Delta^{m \times n}} \varphi(z)$. Moreover, each pair of rewards in $CP^{\mathcal{JC}}$ is feasible.*

From the formulation of Theorem 3.1 it may not be immediately apparent, but an implication is that the set of feasible rewards can be visualized rather elegantly. For this purpose, several algorithms have been designed, involving the computation of a pair of feasible rewards for a significant number of *'frequency-matrices'* $z \in \Delta^{m \times n}$. To obtain Figures 1 and 2, we have used such a program.

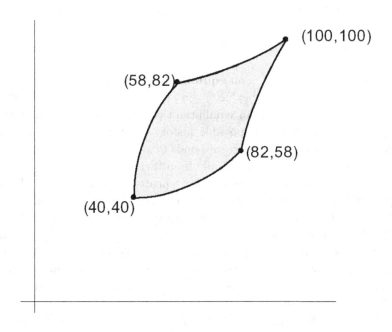

Figure 1: The jointly-convergent pure-strategy rewards for $\kappa = 100$. By Theorem 3.1, all rewards in the convex hull of this set are feasible, as well.

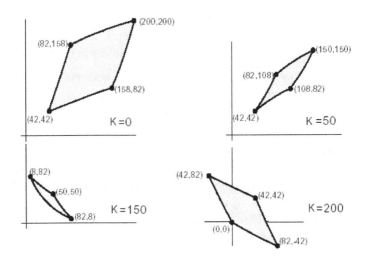

Figure 2: Jointly-convergent pure-strategy rewards for different κ.

3.4 What is 'Rational' to Do?

The strategy pair (π^*, σ^*) is an **equilibrium**, if no player can improve by unilateral deviation, i.e., $\gamma^A(\pi^*, \sigma^*) \geq \gamma^A(\pi, \sigma^*)$, $\gamma^B(\pi^*, \sigma^*) \geq \gamma^B(\pi^*, \sigma)$ for all $\pi \in \mathcal{X}^A, \sigma \in \mathcal{X}^B$. An equilibrium is called **subgame perfect** if for each possible state and possible history (even unreached states and histories) the subsequent play corresponds to an equilibrium, i.e., no player can improve by deviating unilaterally from then on.

In the construction of equilibria for repeated games, 'threats' play an important role. A threat specifies the conditions under which one player will punish the other, as well as the subsequent measures. We call $v = (v^A, v^B)$ the **threat point**, where $v^A = \min_{\sigma \in \mathcal{X}^B} \max_{\pi \in \mathcal{X}^A} \gamma^A(\pi, \sigma)$, and $v^B = \min_{\pi \in \mathcal{X}^A} \max_{\sigma \in \mathcal{X}^B} \gamma^B(\pi, \sigma)$. So, v^A is the highest amount A can get if B tries to minimize A's average payoffs. Under a pair of **individually rational** rewards each player receives at least his threat-point reward.

To present the general idea of the result of Joosten et al. [2003] to come, we adopt terms from Hart [1985], Forges [1986] and Aumann & Maschler [1995]. First, there is a 'master plan' which is followed by each player as long as the other does too; then there are 'punishments' which come into effect if a deviation from the master plan occurs. The master plan is a sequence of 'intra-play communications' between the players, the purpose of which is to decide by which equilibrium the play is to continue. The outcome of the communication period is determined by a 'jointly controlled lottery', i.e., at each stage of the communication period the players randomize with equal probability on both actions; at the end of the communication period one sequence of pairs of action choices materializes.

Detection of deviation from the master plan *after* the communication period is easy as both players use pure actions on the equilibrium path from then on. Deviation *during* the communication period by using an *alternative randomization* on the actions is impossible to detect. However, it can be shown that no alternative unilateral randomization yields a higher reward. So, the outcome of the procedure is an equilibrium. For more details, we refer to Joosten *et al.* [2003]. We restate here the major result which applies to general games with frequency-dependent stage payoffs.

Theorem 3.2. *Each pair of rewards in the convex hull of all individually-rational pure-strategy rewards can be supported by an equilibrium. Moreover, each pair of rewards in the convex hull of all pure-strategy rewards giving each player strictly more than the threat-point reward, can be supported by a subgame-perfect equilibrium.*

The following is visualized in Figure 3 and illustrated in Example 3.4.

Corollary 3.1. *Let $E' = \{(x,y) \in P^{\mathcal{JC}}|\, (x,y) \geq v\}$ be the set of all individually rational jointly-convergent pure-strategy rewards. Then, each pair of rewards in the convex hull of E' can be supported by an equilibrium. Moreover, all rewards in E' giving A strictly more than v^A and B strictly more than v^B can be supported by a subgame-perfect equilibrium.*

Example 3.4. (Continued) For the threat points we have the following:
$$v = \begin{cases} (158 - \kappa, 158 - \kappa) & \text{for } 0 \leq \kappa \leq 116, \\ (42, 42) & \text{for } \kappa \geq 116. \end{cases}$$
So, for $\kappa = 100$, $E' = \{(x,y) \in P^{\mathcal{JC}}|\, (x,y) \geq (58, 58)\}$.

To obtain any reward in E' an equilibrium in pure strategies with threats suffices. Any deviation from the equilibrium path will be punished, hence no player has a incentive to deviate unilaterally from the equilibrium path.

Observe that E' is not convex. We will now show how a pair of rewards in conv E' can be obtained which does not belong to E'. Observe that $\frac{1}{4}(100, 100) + \frac{3}{4}(58, 82) \notin E'$. Let (π, σ) be given by
$$\begin{aligned} \pi_t = \sigma_t &= \left(\tfrac{1}{2}, \tfrac{1}{2}\right) \quad &&\text{for } t = 1, 2 \\ \pi_t = \sigma_t &= 1 \quad &&\text{for } t \geq 3 \quad \text{if } j_1^A = j_1^B \text{ and } j_2^A = j_2^B \\ \pi_t = 1,\ \sigma_t &= 2 \quad &&\text{for } t \geq 3 \quad \text{otherwise.} \end{aligned}$$

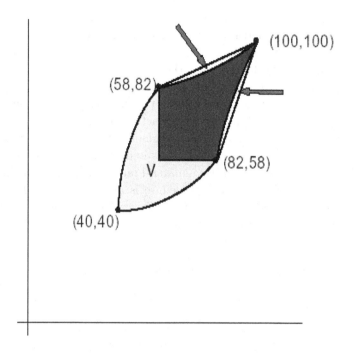

Figure 3: The set of jointly-convergent pure-strategy equilibrium rewards E', is shown in red. The arrows point at $conv\ E' \backslash E'$ ($\kappa = 100$).

At the first two stages, each player randomizes with equal probability on both actions. Then, the play continues with probability $\frac{1}{4}$ with $((1,1),(1,1),(1,1),...)$ and with the complementary probability with $((1,2),(1,2),(1,2),...)$, because the event $j_1^A = j_1^B$ and $j_2^A = j_2^B$ has probability $\frac{1}{4}$. Furthermore, the first sequence gives long run average payoffs of $(100,100)$ and the second one $(58,82)$. Hence, $\gamma(\pi,\sigma)$ yields rewards $\frac{1}{4}(100,100) + \frac{3}{4}(58,82)$. Here, the communication phase has length 2, the jointly controlled lottery consists of the randomization in the first two periods as presented. The play proceeds according to the outcome determined by the jointly controlled lottery. Similarly all convex combinations of $(58,82)$ and $(100,100)$ being multiples of $\frac{1}{4}$ can be obtained. Moreover, for a communication period consisting of three periods we can generate all multiples of $\frac{1}{8}$; in general, for T periods one gets all multiples of $\frac{1}{2}^T$.
To show that $\gamma(\pi,\sigma)$ is an equilibrium reward, we construct (π^e, σ^e) such that

on the equilibrium path the play is exactly identical to the play induced by (π, σ):

$$\pi_t^e = \begin{cases} 2 & \text{if } j_{t'}^B \neq \sigma_{t'} \text{ for any } t' = 3, ..., t-1 \\ \pi_t & \text{otherwise.} \end{cases}$$

$$\sigma_t^e = \begin{cases} 2 & \text{if } j_{t'}^A \neq \pi_{t'} \text{ for any } t' = 3, ..., t-1 \\ \sigma_t & \text{otherwise.} \end{cases}$$

Then, $\gamma(\pi^e, \sigma^e) = \gamma(\pi, \sigma) = \frac{1}{4}(100, 100) + \frac{3}{4}(58, 82)$. Firm B 'threatens' to play action 2 forever if firm A deviates after $t = 2$ from π, inducing average payoffs of $v^A < \gamma^A(\pi^e, \sigma^e)$ to firm A. A similar statement holds for a unilateral deviation by the other firm. No unilateral deviation during the communication period can be detected, but any alternative unilateral randomization yields the same probabilities of continued play for $t \geq 3$. Hence (π^e, σ^e) is an equilibrium.

3.5 Pareto Efficiency

Appreciation for Folk Theorems as presented in the previous section, varies widely among the profession. Osborne & Rubinstein [1994] see as a great advantage that equilibria of the infinitely repeated game exist which are Pareto-superior to any equilibrium of the one-shot game. However, Gintis [2000] comments (p.129): 'By explaining practically anything, the model in fact explains nothing'. Gintis then expresses a preference for Pareto efficiency as a refinement criterion, as well as for more realistic punishments in case of unilateral deviation from an equilibrium path, which should be more forgiving, e.g., allow for repair.

Joosten [2004] showed that rewards from certain subgame perfect equilibria which are 'forgiving', almost entirely coincide with the set of rewards from jointly-convergent pure-strategy equilibria. An implication is that the equilibrium rewards which fulfill the properties desired by Gintis discussed above, is the set of Pareto-efficient equilibrium rewards, which may then be obtained by a pair of 'forgiving' strategies instead of 'grim trigger' strategies.

Example 3.5. (Continued)
Let $PE = \{(x, y) \in conv\ E' |\ \nexists_{(x', y') \in conv E'}(x', y') > (x, y)\}$ denote the set of Pareto-efficient rewards in the convex hull of E'. Hence, for any pair of rewards in PE it holds that there exists no alternative pair of rewards in $conv\ E'$ such that both firms are better off.

We introduce a couple of notations, let for given $\kappa \in [0, 158]$

$$\vartheta^\kappa \equiv \min\left\{82, \left(\frac{4520 - 21\kappa}{2}\right) - \left(\frac{244 - \kappa}{6}\right)\sqrt{3417 - 12\kappa}\right)\right\},$$
$$M^\kappa \equiv \max\{200 - \kappa, 42\},$$
$$P^+ \equiv bd\left(conv\, E'\right) \setminus \left(conv\{v, \left(v^A, \vartheta^\kappa\right)\} \cup conv\{v, \left(\vartheta^\kappa, v^B\right)\}\right).$$

Here, $bd(S)$ denotes the boundary of the set S. Then, we have

$$PE = P^+ \qquad \text{for } \kappa \in [118\tfrac{1}{2}, 158],$$
$$PE = \{(M^\kappa, M^\kappa)\} \text{ otherwise.}$$

So, the criterion of Pareto efficiency reduces the number of equilibria significantly.[7] If $\kappa \in [118\tfrac{1}{2}, 158]$, certain boundary elements of $conv\, E'$ are Pareto efficient. Otherwise, there is a unique Pareto efficient element in $conv\, E'$. Figure 4 visualizes the first case. Note that for $\kappa \in [118\tfrac{1}{2}, 158]$ individually rational rewards exist which Pareto dominate Pareto-efficient equilibrium rewards in E'.

The stimulating idea of the focal point introduced by Schelling [1960] might be useful to reduce the abundance of equilibria. The literature is quite inconclusive about which direction to take even in 'simpler' games, see e.g., Janssen [1998, 2001], Sugden [1995] and Bacharach [1993]. The unique symmetric Pareto-efficient equilibrium in the example above, has one of the necessary properties of a focal point namely salience, also refered to as prominence or conspicuousness.

3.6 Conclusion

We have formulated a new dynamic model of advertising in very general terms. A broad variety of long and short term externalities can be modeled by altering the (restrictions on the) parameters chosen. We have analyzed one family of models rather completely, a full classification and analysis of the *general* setting must be reserved for later. Distinction in this family is made by the height of the advertisement costs which are assumed fixed and equal for both agents. The analysis shows that three subcases can be discerned. For high advertisement costs, only one equilibrium exists namely never to advertise at all. For low advertisement costs, the unique Pareto efficient equilibrium is to advertise always, and a continuum of equilibria exists which are not Pareto efficient. For intermediate costs, a continuum of equilibria exists and a continuum of Pareto efficient equilibria exists. Our

---

[7] See also the Appendix.

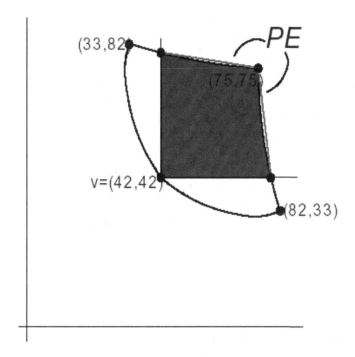

Figure 4: E' is indicated in red, $PE \subset bd(conv\ E')$ is shown as a green kinked line ($\kappa = 125$).

analysis precisely determines the boundary values for the different cases, as well as the (Pareto-efficient) equilibrium rewards.

We have made concessions to reality to obtain first results on advertisement games with frequency-dependent stage payoffs. We modeled advertising as an either-or decision, not taking into account that various budgets might be attributed to it. Our model is easily generalized as any one-period budget in a finite range with a smallest monetary unit can be modeled as a separate action. Vidale & Wolfe [1957] described the interaction of advertising and sales using a simple differential equation in terms of three parameters, the sales decay constant, the saturation level, and the response constant. Some of the phenomena these parameters are meant to capture, are present in our model, albeit implicitly. Further research must reveal whether such empirical findings can be approximated to a higher degree. The building blocks of our model are easily adapted to accomodate input from empirics.

The class of FD-games is rather new, hence the tools for analysis are far from mature yet. Some results beyond the framework of this chapter have been established for so-called Small Fish Wars (cf., e.g., Joosten [2007a,b]). Much is still to be done. Extending the model to allow an n-firm advertisement game is high on the agenda.[8] Large parts of the approach, most importantly Theorem 3.2 and its corollary, seem generalizable, but a comprehensive formal generalization is still pending. Obviously, visualizing the relevant sets of rewards will be difficult for $n \geq 3$.

A crucial step in our approach is finding the set of jointly-convergent pure-strategy rewards, another one is determining the threat point. For the first step continuity of the functions determining the average payoffs on the relevant domains of the stochastic variables involved, suffices. Unfortunately, there exists no general theory on (finding) threat points in FD-games, yet.

3.7 Appendix

Proof of the claim made in Example 3.4. We intend to show that $v^A = \max\{158 - \kappa, 42\}$, divided into two steps. We show that the amount which Firm A can guarantee itself, is identical to the amount to which Firm B can keep Firm A. Now, Firm A can always guarantee itself $\max\{158 - \kappa, 42\}$ by using

$$\pi = \begin{cases} (1,1,1,...) \text{ if } \kappa \leq 116, \\ (2,2,2,...) \text{ if } \kappa \geq 116. \end{cases}$$

If namely $\kappa \leq 116$, then π induces long term stage payoffs determined by

$$[153 - \kappa + 47\rho_t^B, 158 - \kappa + 50\rho_t^B]$$

depending on firm B's actions. Let σ be firm B's strategy, $\underline{\rho} = \liminf_{t \to \infty} \rho_t^B$, and $\overline{\rho} = \limsup_{t \to \infty} \rho_t^B$ where ρ_t^B corresponds to σ. Then in the long run

$$[153 - \kappa + 47\underline{\rho}, 158 - \kappa + 50\underline{\rho}] \leq \left[153 - \kappa + 47\rho_t^B, 158 - \kappa + 50\rho_t^B\right]$$
$$\leq [153 - \kappa + 47\overline{\rho}, 158 - \kappa + 50\overline{\rho}]$$

---

[8]Recently, the framework for 2-person Fish Wars of Joosten [2008] was extended to 3-person ones.

Observe that $x < y$ implies that for $t \to \infty$, $x\underline{\rho} + y(1-\underline{\rho}) \geq x\rho_t^B + y(1-\rho_t^B) \geq x\overline{\rho} + y(1-\overline{\rho})$. So, for large t :

$$\left(153 - \kappa + 47\underline{\rho}\right)\underline{\rho} + \left(158 - \kappa + 50\underline{\rho}\right)(1-\underline{\rho})$$
$$\geq \left(153 - \kappa + 47\rho_t^B\right)\rho_t^B + \left(158 - \kappa + 50\rho_t^B\right)\left(1 - \rho_t^B\right)$$
$$\geq \left(153 - \kappa + 47\overline{\rho}\right)\overline{\rho} + \left(158 - \kappa + 50\overline{\rho}\right)(1-\overline{\rho}).$$

Minimizing the upper bound occurs for $\underline{\rho} = 0$, minimizing the lower bound occurs for $\overline{\rho} = 0$. Hence, firm A can guarantee itself at least $158 - \kappa$, and it can be kept at that amount by firm B if σ satisfies $\underline{\rho} = \overline{\rho} = 0$. (The case $\kappa \geq 116$ is similar.) Firm B can keep Firm A's rewards at $\max\{158 - \kappa, 42\}$ by using $\sigma = (2, 2, 2, ...)$. Let π be firm A's strategy, $\underline{\rho} = \liminf_{t \to \infty} \rho_t^A$ and $\overline{\rho} = \limsup_{t \to \infty} \rho_t^A$ where ρ_t^A corresponds to σ. Then, Firm A faces long term stage payoffs of

$$\begin{bmatrix} 52 + 106\rho_t^A - \kappa \\ 42 + 91\rho_t^A \end{bmatrix} \leq \begin{bmatrix} 52 + 106\overline{\rho} - \kappa \\ 42 + 91\overline{\rho} \end{bmatrix}.$$

Observe that $x > y$ implies that for $t \to \infty$, $x\rho_t^B + y(1-\rho_t^B) \leq x\overline{\rho} + y(1-\overline{\rho})$. Hence, A's long term average stage payoffs are smaller than or equal to $(52 + 106\overline{\rho} - \kappa)\overline{\rho} + (42 + 91\overline{\rho})(1-\overline{\rho}) = 15\overline{\rho}^2 + (101 - \kappa)\overline{\rho} + 42$. The latter takes its minimum at $\overline{\rho}^* = \frac{\kappa - 101}{30}$. It follows easily that for $\overline{\rho}^* \leq \frac{1}{2}$ the maximizing element in the range $[0, 1]$ is $\overline{\rho} = 1$ and the associated rewards are $158 - \kappa$, and for $\overline{\rho}^* \geq \frac{1}{2}$ the maximizing element in this range is $\overline{\rho} = 0$ and the associated rewards are 42. Note that $\overline{\rho}^* = \frac{1}{2}$ precisely for $\kappa = 116$. $\square$

Formulas from Example 3.5. We used the following insights.
(1) The set of Pareto-efficient elements of $conv\ E'$ is nonempty and is a subset of the boundary of $conv\ E'$.
(2) $bd\left(conv\ E'\right) \setminus P^+$ does not contain a Pareto-efficient element.
(3) $M^\kappa = \vartheta^\kappa$ for $\kappa = 118\frac{1}{2}$.
(4) For $\kappa \in \left[0, 118\frac{1}{2}\right]$ it holds that $M^\kappa \geq \vartheta^\kappa$. Hence, if $(x, y) \in P^+$, then $(x, y) \leq (M^\kappa, M^\kappa)$.
(5) For $\kappa \in \left[118\frac{1}{2}, 158\right]$ it holds that $M^\kappa \leq \vartheta^\kappa$. Hence, for any $(x, y) \in P^+ \setminus \{\left(\vartheta^\kappa, v^B\right)\}$ we have that $\varepsilon_1, \varepsilon_2 > 0$ exist such that $(x + \varepsilon_1, y - \varepsilon_2) \in P^+$. (The same argument can be shown in the other direction.) This implies that P^+ does not contain an element that Pareto dominates all other elements of P^+. $\square$

Bibliography

Aumann, R. J. and Maschler, M. (1995). *Repeated Games of Incomplete Information*, (MIT Press, Cambridge, MA).

Bacharach, M. (1993). Variable universe games, in: *Frontiers of Game Theory*, eds.: Binmore, K., Kirman, A. and P Tani, (MIT Press, Cambridge, MA), pp. 255–275.

Billingsley, P. (1986). *Probability and Measure*, (John Wiley & Sons, New York, NY).

Brenner, T. and Witt, U. (2003). Melioration learning in games with constant and frequency-dependent payoffs, *J Econ Behav Organ* **50**, pp. 429–448.

Cellini, R. and Lambertini, L (2003). Advertising with spillover effects in a differential oligopoly game with differentiated goods, *Eur J Oper Res* **11**, pp. 409–423.

Cellini, R, Lambertini, L and Mantovani, A. (2008). Persuasive advertising under Bertrand competition: A differential game, *OR Letters* **36**, pp. 381–384.

Darby, M. and Karni, E. (1973). Free competition and the optimal amount of fraud, *J Law & Econ* **16**, pp. 67-88.

Dockner, E., Jørgensen, S., Van Long, N. and Sorger, G., (2000). *Differential Games in Economics and Management Science*, (Cambridge University Press, Cambridge, UK).

Dorfman, R., and Steiner, P. O. (1954). Optimal advertising and optimal quality, *Amer Econ Rev* **44**, pp. 826–836.

Feichtinger, G., Hartl, R. F. and Sethi, S. P. (1994). Dynamic optimal control models in advertising: recent developments, *Managem Science* **40**, pp. 195–226.

Fershtman, C. (1984). Goodwill and market shares in oligopoly, *Economica* **51**, pp. 271–281.

Forges, F. (1986). An approach to communication equilibria, *Econometrica* **54**, pp. 1375–1385.

Friedman, J. W. (1983). Advertising and oligopolistic equilibrium, *Bell J Econ*, **14**, pp. 464–473.

Gintis, H. (2001). *Game Theory Evolving*, (Princeton University Press, Princeton, NJ).

Hart, S., (1985). Nonzero-sum two-person repeated games with incomplete information, *Math Oper Res* **10**, pp. 117–153.

Herrnstein, R. J. (1997). *The Matching Law: Papers in Psychology and Economics*, (Harvard University Press, Cambridge, MA).

Janssen, M. C. W (1998). Focal points, in: *New Palgrave Dictionary of Economics and Law*, ed. Newman, P., MacMillan, London, Vol II, pp. 150–155.

Janssen, M. C. W (2001). Rationalizing focal points, *Theory and Decision* **50**, pp. 119–148.

Joosten, R. (1996). *Dynamics, Equilibria, and Values*, Ph.D.-thesis, Maastricht University.

Joosten, R. (2004). *Strategic interaction and externalities: FD-games and pollution*, Papers Econ & Evol, **#0417**, Max Planck Institute of Economics, Jena.

Joosten, R. (2005). A note on repeated games with vanishing actions, *Int Game Theory Rev* **7**, pp. 107–115.

Joosten, R. (2007a). Small Fish Wars: A new class of dynamic fishery-management games, *ICFAI J Managerial Economics* **5**, pp. 17–30.

Joosten, R. (2007b). Small Fish Wars and an authority, in: *The Rules of the Game: Institutions, Law, and Economics*, eds.: Prinz, A., Steenge, A. E. and Schmidt, J., LIT, Muenster, pp. 131-162.

Joosten, R. (2008). *Patience, Fish Wars, rarity value and Allee effects*, Papers Econ & Evol, **#0724**, Max Planck Institute of Economics, Jena.

Joosten, R., Brenner, T. and Witt, U. (2003). Games with frequency-dependent stage payoffs, *Int J Game Theory* **31**, pp. 609–620.

Jørgensen, S. and Zaccour, G. (2004). *Differential Games in Marketing*, (Kluwer Academic Publishers, Dordrecht).

Kydland, F. (1977). Equilibrium solutions in dynamic dominant-player models, *J Econ Theory* **15**, pp. 307–324.

Lambertini, L. (2005). Advertising in a dynamic spatial monopoly, *Eur J Oper Res* **16**, pp. 547–556.

Nash, J. (1951). Noncooperative games, *Ann Math* **54**, pp. 284–295.

Nelson, P. (1970). Information and consumer behavior, *J Pol Econ* **78**, pp. 311–329.

Nerlove, M. and Arrow, K. J. (1962). Optimal advertising policy under dynamic conditions *Economica* **29**, pp. 129–142.

Osborne, M. J. and Rubinstein, A. (1994), *A Course in Game Theory*, (MIT Press, Cambridge, MA).

Prescott, E. (1973), Market structure and monopoly profits: a dynamic theory, *J Econ Theory* **6**, pp. 546–557.

Schelling, T. C. (1960). *The Strategy of Conflict*, (Harvard University Press, Cambridge, MA).

Schoenmakers, G. and Flesch, J. and Thuijsman, F. (2002). Coordination games with vanishing actions, *Int Game Theory Rev* **4**, pp. 119–126.

Schmalensee, R. (1978). A model of advertising and product quality, *J Pol Economy* **86**, pp. 485–503.

Sethi, S. P. (1977). Dynamic optimal control models in advertising: a survey, *SIAM Rev* **19**, pp. 685–725.

Sugden, R. (1995). A theory of focal points, *Economic Journal*, **105**, pp. 533–550.

Telser, L. (1962). Advertising and cigarettes, *J Pol Econ* **70**, pp. 471–499.

Van Damme, E. E. C. (1991). *Stability and Perfection of Nash Equilibria*, (Springer Verlag, Berlin).

Vidale, M. L. and Wolfe, H. B. (1957). An operations-research study of sales response to advertising, *Operations Research* **5**, pp. 370-381.

Waugh, F. V. (1959). Needed research on the effectiveness of farm products promotions, *J Farm Econ* **41**, pp. 364-376.

Chapter 4

Connections Between Some Concepts in Reliability and Economics

Subhash Kochar* and Maochao Xu
Department of Mathematics and Statistics
Portland State University, Portland, Oregon 97201, USA
e-mail*: kochar@pdx.edu

Abstract

Unaware of the developments in each other's areas, researchers in the disciplines of Economics and Reliability Theory have been working independently. The objective of this paper is to point out some interesting relationships that exist between some of the notions in these two areas. In particular, we will discuss notions of NBUE (New Better Than Used in Expectation) order, HNBUE order, TTT (Total Time on Test) order from Reliability Theory and those of Lorenz Order, Excess Wealth Order from Economics. Some new results obtained recently about these stochastic orders will also be presented.

Key Words: Reliability, Economics, TTT, NBUE, HNBUE

4.1 Introduction

Both in Life Testing and Reliability Theory, and in Economics we are generally dealing with non-negative random variables with skewed probability distributions. Unaware of the developments in each other's areas, researchers in these two disciplines have been working independently. The objective of this paper is to point out some interesting relationships that exist between some of the notions developed in these two areas. While some of the concepts from one area have direct interpretation in the other, there are many other notions which need further investigation. It should be noted that the list of topics discussed in this paper is by no means exhaustive. It only reflects our personal interests in the area of Reliability Theory.

In the literature, there are many notions introduced and studied in both areas. For example, the well-known stochastic orders defined in reliability literature, like the likelihood ratio order and the hazard rate order, have been applied already in economics and finance (see *Kijima and Ohnishi*, 1999). One may also refer to *Denuit, et al.* (2005) for the applications of stochastic orders in economic and actuarial contexts.

4.2 Transforms from Economics and Reliability

Let the random variable X represent the income of households in a population and let F be its cumulative distribution function. The Lorenz curve of X is defined as

$$L_F(p) = \frac{1}{\mu_F} \int_0^p F^{-1}(x) dx, \quad 0 \leq p \leq 1, \tag{4.1}$$

where μ_F is the mean of X, and $F^{-1}(t) = \inf\{x : F(x) \geq t\}$ is the right continuous inverse of F. Let $x_{(1)} \leq x_{(2)} \leq \cdots \leq x_{(n)}$ denote the orders statistics of a random sample $x_1, x_2, \ldots, x_n$ from F. The sample estimate of L_F is the polygon joining the points $\left(\frac{k}{n}, \hat{L}_F(k)\right)$, where

$$\hat{L}_F(k) = \frac{\sum_{j=1}^k x_{(j)}}{\sum_{j=1}^n x_{(j)}}, \ k = 0, 1, 2, \ldots, n;$$

$\hat{L}_F(0) = 0$. Thus a Lorenz Curve is a graph (x, y) that shows, for the bottom $100x\%$ of the households, the percentage $100y\%$ of the total income which they posses. If every one in the population has the same income, the bottom $p\%$ of the population would always have $p\%$ of the total income and the Lorenz curve would be the diagonal line $y = x$. The more the Lorenz Curve is below the diagonal line, the more is the disparity between the incomes.

The Lorenz curves can also be used to compare the extent of inequality that exists between two distributions F and G. If

$$L_F(p) \leq L_G(p), \quad p \in [0, 1], \tag{4.2}$$

then F shows more inequality than G.

Gini coefficient is also an important concept in economics and it is closely related to Lorenz curve. It measures the extent to which the Lorenz curve is below the diagonal line and is defined by

$$G_X = 1 - 2 \int_0^1 L_X(p) dp.$$

The Gini coefficient is a number between 0 and 1, where perfect equality has a Gini coefficient of zero, and absolute inequality yields a Gini coefficient of 1. Gini coefficients are often expressed as percentages.

Let $X_{1:n} \leq X_{2:n} \leq \cdots \leq X_{n:n}$ be order statistics from the random variable X. The sample Gini statistic can be represented as,

$$G_n = \frac{\sum_{i=1}^{n-1} i(n-i)(X_{i+1:n} - X_{i:n})}{(n-1)\sum_{i=1}^{n} X_{i:n}}.$$

In reliability theory, the ith *total time on test* (TTT) statistic is defined as

$$T_i = \sum_{j=1}^{i}(n-j+1)(X_{j:n} - X_{j-1:n}).$$

T_i can be written as

$$T_i = n \int_0^{F_n^{-1}(i/n)} \bar{F}_n(x)dx,$$

where $\bar{F}_n = 1 - F_n$ is the empirical survival function. It should be noted that, with probability one (cf. *Barlow and Campo*, 1975),

$$\frac{1}{n}T_i \longrightarrow \int_0^{F^{-1}(p)} \bar{F}(x)dx \quad \text{for} \quad n \to \infty, \; i/n \to p,$$

where $\bar{F} = 1 - F$ is the survival function of population.

The function

$$T_X(p) = \int_0^{F^{-1}(p)} \bar{F}(x)dx \quad \text{for} \quad 0 \leq p \leq 1,$$

is called TTT tansform of F.

The scaled TTT to the ith failure time is defined as,

$$W_i = \frac{T_i}{\sum_{j=1}^{n} X_{j:n}}.$$

The cumulative total time on test statistic is,

$$V_n = \frac{1}{n-1}\sum_{i=1}^{n-1} W_i.$$

The corresponding population transforms are respectively,

$$W_X(p) = \frac{T_X(p)}{\mathrm{E}(X)} \quad \text{for} \quad 0 \leq p \leq 1,$$

and

$$V_X = \int_0^1 W_X(u)du.$$

It is well known in the reliability theory literature that the TTT transform can be used to illustrate and characterize different aging properties (cf. *Barlow and Campo*, 1975 and *Klefsjö*, 1983). It can also be used to test whether one distribution has more aging property than the other, like "more NBUE" aging property (cf. *Kochar and Wiens*, 1987 and *Kochar*, 1989). One may also refer to *Barlow* (1979), *Bergman* (1979), *Klefsjö* (1984), *Bergman and Klefsjö* (1984) and *Klefsjö* (1991) for different aspects of the TTT concept.

The relationships between the various transforms could be represented as follows (cf. *Chandra and Singpurwalla*, 1981).

$$G_X = 1 - V_X, \quad W_X(p) = L_X(p) + \frac{1}{E(X)}(1-p)F^{-1}(p).$$

$$V_n = 1 - G_n, \quad W_i = L_n\left(\frac{i}{n}\right) + \frac{(n-i)X_{i:n}}{\sum_{j=1}^n X_{j:n}}.$$

Another transform closely related to TTT transform is the excess wealth (EW) transform, (cf. *Shaked and Shanthikumar*, 1998)

$$\begin{aligned}EW_X(p) &= \int_{F^{-1}(p)}^\infty \bar{F}(x)dx \\ &= E[X - F^{-1}(p)]^+ \\ &= E[X] - T_X(p) \quad \text{for} \quad 0 \le p \le 1.\end{aligned}$$

The excess wealth transform was independently introduced and studied by *Fernández-Ponce, Kochar, and Muñoz-Pérez* (1998). They called it as *right spread* transform. EW transform is of great interest in economics. For example, if F is the distribution of wealth in some community, $EW_X(p)$ can be thought of as the additional wealth (on top of the pth percentile income) of the of the richest $100(1-p)\%$ individuals in that community.

4.3 Partial Orders in Economics and Reliability

Based on TTT transform, *Kochar, Li and Shaked* (2002) proposed a stochastic order called TTT order. For two nonnegative random variables

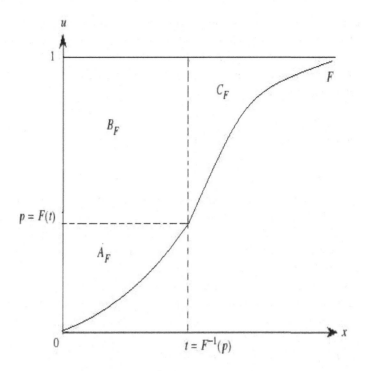

Fig. 4.1 Depiction of $A_F(t)$, $B_F(t)$ and $C_F(t)$

X and Y with distributions F and G, respectively, X is said to be smaller than Y in terms of TTT order (denoted by $X \leq_{\text{TTT}} Y$) if,

$$T_X(t) \leq T_Y(t), \quad 0 \leq t \leq 1.$$

Through Figure 4.1, it is easy to see that

$$T_X(t) \equiv \|A_F(t) \cup B_F(t)\|, \quad t \in (0,1),$$

corresponds to the total time on test (TTT) transform associated with this distribution. Hence, the comparison

$$\|A_F(t) \cup B_F(t)\| \leq \|A_G(t) \cup B_G(t)\|$$

yields the TTT order.

TTT order is closely related to a new variability order known as *excess wealth* (EW) order, written as $X \leq_{ew} Y$ (see *Shaked and Shanthikumar, 1998*). The random variable X is said to be less dispersed than another ranodm variable Y if

$$\int_{F^{-1}(t)}^{\infty} \bar{F}(x)dx \leq \int_{G^{-1}(t)}^{\infty} \bar{G}(x)dx \quad \text{for} \quad 0 \leq t \leq 1.$$

The excess wealth order is also called the *right spread* order (see *Fernández-Ponce, Kochar, and Muñoz-Pérez*, 1998). It gives a second order comparison of variability and $X \leq_{ew} Y$, in particular, implies $var(X) \leq var(Y)$. For properties of this order see the above two references. From Figure 4.1, it follows that the comparison

$$\|C_F(t)\| \leq \|C_G(t)\|, \quad t \in (0,1),$$

yields the EW order. It should be noted that

$$\frac{\|A_F(t)\|}{E(X)} \leq \frac{\|A_G(t)\|}{E(Y)}, \ \forall\, t \in (0,1) \iff X \geq_{\text{Lorenz}} Y,$$

where $\geq_{\text{Lorenz}}$ denotes the Lorenz order (see *Shaked and Shanthikumar* (2007, Section 3.A)) which is used to compare the amount of inequality (in incomes) among different populations. The Lorenz order has been well studied and applied in economics. One may refer to *Arnold* (1987) and *Aaberge* (2001) for more details.

The NBUE (new better than used in expectation) order of *Kochar and Wiens* (1987) (see next section for the definition) can also be characterized by the sets above as follows

$$\frac{\|A_F(t) \cup B_F(t)\|}{E(X)} \leq \frac{\|A_G(t) \cup B_G(t)\|}{E(Y)}, \ \forall\, t \in (0,1) \iff X \geq_{\text{nbue}} Y.$$

When $E(X) = E(Y)$ then the order $\leq_{\text{ttt}}$ is equivalent to the orders $\leq_{\text{ew}}$ and $\geq_{\text{nbue}}$ in the sense

$$X \leq_{\text{ttt}} Y \iff X \geq_{\text{ew}} Y \iff X \geq_{\text{nbue}} Y.$$

However, these orders are distinct when $E(X) \neq E(Y)$. For more details, please refer to *Kochar, Li and Shaked* (2002).

The various stochastic orders mentioned above share some similarities, but they are all distinct, and each is useful in different contexts. For example, the *more* NBUE order compares ageing mechanisms of different items. The EW order is location independent (and thus it can also be used to compare random variables that are not nonnegative) and it compares the variability of the underlying random variables. Similarly, the Lorenz order compares variability among populations after scaling the random variables by their means.

Now, let us give an application of the EW order in reliability theory. Consider a parallel system of n components with i.i.d. NBUE lifetimes. Denote their common mean by μ. Let $Y_1, Y_2, \ldots, Y_n$ be i.i.d. exponential random variables with mean μ. Recall from *Belzunce* (1999) that a random variable X with mean μ is NBUE if and only if

$$X \leq_{\text{ew}} \text{Exp}(\mu),$$

where $\text{Exp}(\mu)$ denotes an exponential random variable with mean μ.

Kochar, Li and Shaked (2002) have shown that $X \leq_{\text{ew}} Y$ implies

$$\max\{X_1, X_2, \ldots, X_n\} \leq_{\text{ew}} \max\{Y_1, Y_2, \ldots, Y_n\}.$$

which in turn implies

$$\text{E}[\max\{X_1, X_2, \ldots, X_n\}] \leq \text{E}[\max\{Y_1, Y_2, \ldots, Y_n\}], \quad \text{as well as}$$
$$\text{Var}[\max\{X_1, X_2, \ldots, X_n\}] \leq \text{Var}[\max\{Y_1, Y_2, \ldots, Y_n\}]$$

Now, computing

$$\text{E}[\max\{Y_1, Y_2, \ldots, Y_n\}] = \int_0^\infty \left[1 - \left(1 - e^{-\frac{x}{\mu}}\right)^n\right] dx = \mu \sum_{k=1}^n \frac{1}{k},$$

and

$$\text{E}[(\max\{Y_1, Y_2, \ldots, Y_n\})^2] = 2\mu^2 \sum_{k=1}^n \frac{(-1)^{k+1}}{k^2} \binom{n}{k},$$

we obtain the following upper bounds on the mean and the variance of the lifetime of the parallel system with NBUE components.

$$\text{E}[\max\{X_1, X_2 \ldots, X_n\}] \leq \mu \sum_{k=1}^n \frac{1}{k}$$

and

$$\text{Var}[\max\{X_1, X_2 \ldots, X_n\}] \leq \mu^2 \left[2 \sum_{k=1}^n \frac{(-1)^{k+1}}{k^2} \binom{n}{k} - \left(\sum_{k=1}^n \frac{1}{k}\right)^2\right].$$

4.4 Relative Aging Orders

A component or system with exponential life distribution does not age with time in the sense that a used component is as good as a new one irrespective of its age. In the reliability literature exponential distribution is taken as bench mark and the relative aging of any component or system is compared with it. Attempts have been made to compare the relative aging of two arbitrary life distributions when none of them is necessarily exponential. First we describe the various partial orders for this purpose.

Let X and Y be two nonnegative random variables with survival functions $\bar{F}$ and $\bar{G}$, and distribution functions F and G, respectively. Let the corresponding density functions be denoted by f and g, and their failure (hazard) rates be denoted by $r_F = f/\bar{F}$ and $r_G = g/\bar{G}$, respectively.

Definition 4.1. F is said to be more IFR than G or F is convex-ordered with respect to G (written as $F \leq_c G$) if $G^{-1}F$ is convex.

It can be seen that F is an increasing failure rate (IFR) distribution if and only if it is convex ordered with respect to exponential distribution which has a constant failure rate.

Definition 4.2. F is said to be more IFRA than G or F is star-ordered with respect to G (written as $F \leq_* G$) if $G^{-1}F(x)/x$ is increasing.

Since the average failure of F at x is
$$\tilde{r}_F(x) = \frac{1}{x}\int_0^x r_F(u)du = \frac{-\ln \bar{F}(x)}{x}.$$
Thus $F \leq_* G$ can be interpreted in terms of average failure rates as
$$\frac{\tilde{r}_F(F^{-1}(u))}{\tilde{r}_G(G^{-1}(u))}$$
is increasing in $u \in (0,1]$. Again F is an IFRA distribution if only if F is star-ordered with respect to exponential distribution.

Definition 4.3. F is said to be more NBU (new better than used) than G or F is super-additive with respect to G (written as $F \leq_{su} G$) if
$$G^{-1}F(x+y) \geq G^{-1}F(x) + G^{-1}F(y).$$
It is easy to see that F is NBU if and only if F is super-additive with respect to exponential distribution.

The mean residual life (MRL) of a random variable X is defined as
$$\mu_F(t) = E[X-t|X>t] = \frac{\int_t^\infty \bar{F}(x)dx}{\bar{F}(t)}.$$

Definition 4.4. F is said to be more DMRL (decreasing mean residual life) than G or F is smaller than G in the DMRL order (written as $F \leq_{DMRL} G$) if
$$\frac{\mu_F(F^{-1}(u))}{\mu_G(G^{-1}(u))} \quad \text{is decreasing in } u \in (0,1].$$
Note that, the above condition is equivalent to
$$\frac{\int_{F^{-1}(u)}^\infty \bar{F}(x)dx}{\int_{G^{-1}(u)}^\infty \bar{G}(x)dx} \quad \text{is decreasing in } u \in (0,1].$$

Since the mean residual life function of an exponential random variable is constant, F is DMRL if and only if it is DMRL ordered with respect to exponential distribution.

Definition 4.5. F is said to be more NBUE (new better than used in expectation) than G or F is smaller than G in the NBUE order (written as $F \leq_{NBUE} G$) if
$$\frac{\mu_F(F^{-1}(u))}{\mu_G(G^{-1}(u))} \leq \frac{\mu_F}{\mu_G}, \quad \text{for all } u \in (0,1],$$
which is equivalent to
$$\frac{1}{\mu_F} \int_{F^{-1}(u)}^{\infty} \bar{F}(x)dx \leq \frac{1}{\mu_G} \int_{G^{-1}(u)}^{\infty} \bar{G}(x)dx, \quad \text{for all } u \in (0,1].$$
Note that F is NBUE if and only if $F \leq_{NBUE} G$, where G is exponential distribution.

Definition 4.6. F is said to be more HNBUE (harmonic new better than used in expectation) than G or F is smaller than G in the HNBUE order (written as $F \leq_{HNBUE} G$) if
$$\int_t^{\infty} \bar{F}(x\mu_F)dx \leq \int_t^{\infty} \bar{G}(x\mu_G)dx, \quad \text{for all } t \geq 0.$$
Note that F is HNBUE if and only if $F \leq_{HNBUE} G$, where G is exponential distribution.

We have the following chain of implications
$$\begin{array}{ccccc} F \leq_c G & \Longrightarrow & F \leq_* G & \Longrightarrow & F \leq_{su} G \\ \Downarrow & & \Downarrow & & \\ F \leq_{DMRL} G & \Longrightarrow & F \leq_{NBUE} G & \Longrightarrow & F \leq_{HNBUE} G \end{array}$$
All these above partial orders are scale invariant. A good discussion of the convex, star and super-additive orderings can be found in *Barlow and Proschan* (1975). The DMRL, NBUE and HNBUE partial orders were introduced in *Kochar and Wiens* (1987) and studied further in *Kochar* (1989) and *Kochar, Li and Shaked* (2002).

Assuming that the two populations have equal means, *Chandra and Singpurwala* (1981) proved that
$$F \leq_* G \Longrightarrow F \leq_{Lorenz} G.$$
Klefsjö (1984) established this connection between star ordering and Lorenz ordering without any restriction on the means of the two distributions. In fact, he proved a stronger result that $F \leq_* G$ implies
$$\frac{L_G(p)}{L_F(p)} \quad \text{increasing in } p \in (0,1].$$

If the above condition holds we will say that F is smaller than G according to $L\star$ order and denote this by $F \leq_{L\star} G$. Since $L_F(1) = L_G(1) = 1$,

$$F \leq_{L\star} G \implies L_G(p) \leq L_F(p) \quad \text{for all } p \in (0,1].$$

Kochar (1989) proved a much stronger result that Lorenz ordering is equivalent to HNBUE ordering, that is,

$$F \leq_{HNBUE} G \iff F \leq_{Lorenz} G.$$

It follows from *Kochar and Carriere* (1997) that

$$F \leq_{Lorenz} G \iff \frac{|X_1 - X_2|}{\mu_F} \leq_{icx} \frac{|Y_1 - Y_2|}{\mu_G}$$

where $\leq_{icx}$ denotes the increasing convex order (cf. *Shaked and Shanthikumar*, 1998), and $X_1, X_2(Y_1, Y_2)$ are two independent copies of $X(Y)$ which in turn implies $cv(F) \leq cv(G)$ as well as $Gini(F) \leq Gini(G)$ where $cv(F) = \sigma_F^2/\mu_F^2$ denote the coefficient of variation of F and $Gini(F) = E[|X_1 - X_2|]/2\mu_F$ denotes the Gini's coefficient of F.

It will be interesting to study whether the other stronger aging orderings like more IFR, more IFRA, more DMRL and more NBUE orderings have any economic interpretations. For example, even when the income inequality in population 1 is more extreme than that in population 2 in the sense of Lorenz ordering, such an ordering may not hold if we compare the poorest, say, 50% of the population. For this purpose we need to consider orders that are stronger than the Lorenz order and study their properties. The TTT and its related transforms have interesting applications in reliability theory and economics. More recently, it is also found that the EW order is of great interest in auction theory (cf. *Kochar, Li and Xu*, 2007). Further research on this topic is needed.

Acknowledgements. The authors are grateful to two referees for carefully reading the manuscript and for their helpful comments and suggestions which have greatly improved the presentation.

Bibliography

Aaberge, R. (2001). Axiomatic Characterization of the Gini Coefficient and Lorenz Curve Orderings, *Journal of Economic Theory* **101**, pp. 115–132.

Arnold, B. C. (1987). *Majorization and the Lorenz Order: A Brief Introduction*, Lecture Notes in Statistics, Vol. 43, (Springer, New York).

Barlow, R. E. (1979). Geometry of the total time on test transform, *Naval Research Logistics Quarterly* **26**, pp. 393–402.

Barlow, R. E. and Campo, R. (1975). Total time on test processes and applications to failure data analysis, In: *Reliability and Fault Tree Analysis* (edited by R. E. Barlow, J. B. Fussell, and N. D. Singpurwalla), SIAM, Philadephia, pp. 451–481.

Barlow, R. E. and Proschan, F. (1975). *Statistical Theory of Reliability and Life Testing: Probability Models*, (Holt, Rinehart and Winston, New York).

Belzunce, F. (1999). On a characterization of right spread order by the increasing convex order, *Statistics and Probability Letters* **45**, pp. 103–110.

Bergman, B. (1979). On age replacement and the total time on test concept, *Scandinavian Journal of Statistics* **6**, pp. 161–168.

Bergman, B. and Klefsjö, B. (1984). Total time on test transforms, In: Kotz, S., Johnson, N.L., (Eds.), *Encyclopedia of Statistical Sciences* **9**, pp. 297–300.

Chandra, M. and Singpurwalla, N. (1981). Relationships between some notions which are common to reliability theory and economics, *Mathematics of Operations Research* **6**, pp. 113–121.

Denuit, M., Dhaene, J. Goovaerts,M. and Kaas, R. (2005). *Actuarial. Theory for Dependent Risks. Measures, Orders and Models*, (John Wiley & Sons, Chichester).

Fernández-Ponce, J. M., Kochar, S. C., and Muñoz-Pérez, J. (1998). Partial orderings of distributions based on right-spread function, *Journal of Applied Probability* **35**, pp. 221–228.

Kijima, M. and Ohnishi, M. (1999). Stochastic orders and their applications in financial optimization, *Mathematical Methods of Operations Research* **50**, pp. 351–372.

Klefsjö, B. (1983). Testing exponentiality against HNBUE, *Scandinavian Journal of Statistics* **10**, 65-75.

Klefsjö, B. (1984). Reliability interpretations of some concepts from economics, *Naval Research Logistics Quarterly* **31**, pp. 301–308.

Klefsjö, B. (1991). TTT-plotting — a tool for both theoretical and practical problems, *Journal of Statistical Planning and Inference* **29**, pp. 111–124.

Kochar, S. (1989). On extensions of DMRL and related partial orderings of life distributions, *Communications in Statistics—Stochastic Models* **5**, pp. 235–245.

Kochar, S. and Carriere, C. (1997). Connections among various variability orderings, *Statistics and Probability Letters* **35**, pp. 327–333.

Kochar, S., Li, X. and Shaked, M. (2002). The total time on test transform and the excess wealth stochastic order of distributions, *Advances in Applied Probability* **34**, pp. 826–845.

Kochar, S., Li, X. and Xu, M. (2007). Excess wealth order and sample spacings, *Statistical Methodolgy* **4**, pp. 385–392.

Kochar, S. and Wiens, D. P. (1987). Partial orderings of life distributions with respect to their aging properties, *Naval Research Logistics* **34**, pp. 823–829.

Shaked, M. and Shanthikumar, J. G. (1998). Two variability orders, *Probability in the Engineering and Informational Sciences* **12**, pp. 1–23.

Shaked, M. and Shanthikumar, J. G. (2007). *Stochastic Orders and Their Applications*, (Springer Series in Statistics, Springer, New York).

Chapter 5

A New Axiomatization of the Shapley Value for TU-games in terms of Semi-null Players Applied to 1-concave Games

Theo S.H. Driessen
Faculty of Electrical Engineering, Mathematics and Computer Science
Department of Applied Mathematics
University of Twente
P.O. Box 217, 7500 AE Enschede, The Netherlands
e-mail: t.s.h.driessen@ewi.utwente.nl

Abstract

In the framework of cooperative games, we introduce the notions of semi-null players and semi-dummy players in order to provide a new axiomatization for the Shapley value. A semi-null player is powerful as a singleton, but powerless by joining another nonempty coalition. According to the axiomatic approach to solutions, semi-null players receive the egalitarian payoff. It is shown that the Shapley value is the unique solution verifying semi-null player property, symmetry, efficiency, and linearity. Shapley's original proof technique is essentially adapted in that the basis of unanimity games is replaced by the basis of the so-called complementary unanimity games being the main representatives of the class of 1-concave games.

Key Words: Cooperative game, Shapley value, semi-null player, semi-dummy player, 1-concave game

5.1 Introductory Notions

Throughout the chapter, let N be a nonempty, finite set and $\mathcal{P}(N) = \{S \mid S \subseteq N\}$ be its power set.

Example 5.1. *The bankruptcy game* ([O'Neill (1982)], [Driessen (1998)]) A *bankruptcy situation* is given by the *estate* $E > 0$ of a bankrupt firm and the non-negative *claims* $d_1, d_2, \ldots, d_n$ of *creditors* $1, 2, \ldots, n$, such that the estate is insufficient to meet all the claims, i.e., $E < \sum_{j \in N} d_j$. With any bankruptcy situation, there is associated the so-called *bankruptcy game* $\langle N, v \rangle$, the finite *player set* N of which consists of all n creditors (but not the bankrupt firm) and its *characteristic function* $v : \mathcal{P}(N) \to \mathbb{R}$ is given, for all $S \subseteq N$, by $v(S) = E - \sum_{k \in N \setminus S} d_k$ whenever $E \geq \sum_{k \in N \setminus S} d_k$, and $v(S) = 0$ otherwise. In words, the *worth* $v(S)$ of *coalition* $S \subseteq N$ equals the (non-negative) remainder of the estate whenever the creditors not belonging to coalition S go to court for the reimbursement of their claims and get their claims. Particularly, the worth $v(N)$ of the grand coalition N equals the estate E, whereas for all $i \in N$ the worth $v(N \setminus \{i\})$ equals either $E - d_i$ or zero, whichever is more. An alternative model, called its *dual game* $\langle N, w \rangle$, is such that the worth $w(S)$ of coalition S equals either the sum $\sum_{j \in S} d_j$ of claims of its members, or the estate E, whichever is less. Once again, $w(N) = E$ due to the surplus of the sum of all claims. Moreover, the individual worth $w(\{i\})$ of creditor $i \in N$ equals either the corresponding claim d_i or the estate, whichever is less.

Example 5.2. *The labour game.*
The utility of the individual *labour* $\ell_1, \ell_2, \ldots, \ell_n$ of the n *inhabitants of a small Indian village* is currently measured through the quadratic function $f(x) = x^2$. Let the finite *player set* N consist of all n inhabitants and the corresponding *characteristic function* $v : \mathcal{P}(N) \to \mathbb{R}$ is given, for all $S \subseteq N$, by the square of the sum of the labour performed by members of coalition S, i.e., $v(S) = \left[\sum_{j \in S} \ell_j \right]^2$, whereas $v(\emptyset) = 0$.

Example 5.3. *The airport cost game* ([Littlechild, and Owen, G. (1973)])
The *planes* are divided into m types ($m \geq 1$). Let N_j denote the set of *landings* by planes of type j, $1 \leq j \leq m$, and so, $N = \cup_{j=1}^{m} N_j$ represents the set of all landings at the airport. Let C_j be the *cost of a runway* adequate for planes of type j. Without loss of generality, these types may be ordered so that $0 = C_0 < C_1 < C_2 < \ldots < C_m$. With this airport situation, there is associated the so-called *airport game* $\langle N, c \rangle$, the *characteristic cost function* $c : \mathcal{P}(N) \to \mathbb{R}$ given, for all $S \subseteq N$, $S \neq \emptyset$, by the cost $c(S)$ of a runway adequate to receive all landings in S, i.e., $c(S) = \max[C_j \mid 1 \leq j \leq m, S \cap N_j \neq \emptyset]$.

Generally speaking, the *player set* N is a nonempty, finite set. Its power set $\mathcal{P}(N) = \{S \mid S \subseteq N\}$ consists of all subsets of N (inclusive the empty set $\emptyset$), called *coalitions*. A *transferable utility (TU) game* $\langle N, v \rangle$ is given by its *characteristic function* $v : \mathcal{P}(N) \to \mathbb{R}$ satisfying $v(\emptyset) = 0$. The real number $v(S)$ is called the *worth of coalition* S. With every game $\langle N, v \rangle$, there is associated the *dual game* $\langle N, v^* \rangle$ defined by $v^*(S) = v(N) - v(N \backslash S)$ for all $S \subseteq N$. In words, the worth of a coalition in the dual game equals its marginal contribution in the original game with respect to the grand coalition. Clearly, $v^*(N) = v(N)$ and $(v^*)^*(S) = v(S)$ for all $S \subseteq N$. Throughout the chapter, n denotes the size (cardinality) of the player set N, whereas the size of coalitions $S, T \subseteq N$ are denoted by s, t, $0 \leq s \leq n$, $0 \leq t \leq n$.

In the setting of a game $\langle N, v \rangle$, it is tacitly supposed that the grand coalition N forms and that its worth $v(N)$, achieved by mutual cooperation, should be allocated among the players, preferably in a fair manner. For that purpose, the solution theory is mainly concerned with payoff vectors $\vec{x} = (x_i)_{i \in N} \in \mathbb{R}^N$ satisfying the so-called *efficiency* constraint $\sum_{j \in N} x_j = v(N)$. The *excess* of coalition S at payoff vector $\vec{x}$ is defined by $e^v(S, \vec{x}) = v(S) - \sum_{j \in S} x_j$. The *core* of the game $\langle N, v \rangle$ is defined to be the set of all efficient payoff vectors satisfying only non-positive excesses, that is, $\sum_{j \in S} x_j \geq v(S)$ for all $\emptyset \neq S \subsetneq N$. In the setting of the bankruptcy game of Example 5.1 almost all core-constraints are redundant because it turns out that an efficient payoff vector $\vec{x}$ belongs to the core of the bankruptcy game if and only if $0 \leq x_i \leq d_i$ for all $i \in N$. In the setting of the labour game of Example 5.2, it turns out that its core contains the efficient payoff vector induced by the *proportionality rule*, i.e., $x_i = \ell_i \cdot \sum_{j \in N} \ell_j$ for all $i \in N$. At the end of this section we show that the latter "proportional" allocation coincides with the so-called Shapley value of the labour game. In this chapter we introduce the Shapley value as the optimal solution of a suitably chosen optimization problem which minimizes the sum of squares of non-trivial excesses. The determination of the (unique) optimal solution is based on the *Method of Lagrange multipliers*. In fact, we deal with a family of optimal solutions, known as *least square values*, since the minimization problem takes into account a collection of weights with reference to the sizes of both player sets and coalitions.

Theorem 5.1. *([Ruiz, Valenciano and Zarzuelo (1998)], Theorem 3, page 113)*
Fix the size $n \geq 2$ of the player set and fix the collection $\mathcal{W} = (w_{n,s})_{s=1}^{n-1}$ of non-negative weights with reference to the non-trivial coalition sizes s, $1 \leq s \leq n - 1$. In the context of n-person games $\langle N, v \rangle$, the objective

function $f^v : \mathbb{R}^N \to \mathbb{R}$ is defined by

$$f^v(\vec{x}) = \sum_{\emptyset \neq S \subsetneq N} w_{n,s} \cdot \left[v(S) - \sum_{j \in S} x_j \right]^2 \quad \text{for all } \vec{x} \in \mathbb{R}^N. \tag{5.1}$$

Thus, $f^v(\vec{x})$ represents the weighted sum of squares of non-trivial excesses at $\vec{x}$ over all the coalitions in the n-person game. The minimum value of the objective function $f^v(\vec{x})$ subject to the efficiency constraint $\sum_{j \in N} x_j = v(N)$ is attained at the optimal solution $\vec{y}$ given, for all $i \in N$, by

$$y_i = \frac{v(N)}{n} + \frac{1}{\beta^{\mathcal{W}}} \cdot \left[\alpha_i^{\mathcal{W},v} - \frac{1}{n} \cdot \sum_{j \in N} \alpha_j^{\mathcal{W},v} \right], \tag{5.2}$$

where $\beta^{\mathcal{W}} = \sum_{s=1}^{n-1} \binom{n-2}{s-1} \cdot w_{n,s}$, and $\alpha_i^{\mathcal{W},v} = \sum_{\substack{S \subsetneq N, \\ S \ni i}} w_{n,s} \cdot v(S)$ for all $i \in N$.

In case $w_{n,s} = \frac{(s-1)! \cdot (n-1-s)!}{(n-1)!}$ for all $1 \leq s \leq n-1$ (yielding $\beta^{\mathcal{W}} = 1$), the optimal solution of (5.2) is called the **Shapley value** and denoted by $Sh(N,v)$.

The optimal solution $\vec{y}$ of (5.2) is called the *least square value* $LS^{\mathcal{W}}(N,v)$ of the n-person game $\langle N, v \rangle$ with respect to the collection $\mathcal{W}$ of non-negative weights. Any least square value is the additive combination of two allocation rules, the first rule being the egalitarian distribution of the worth $v(N)$ of the grand coalition, and the second rule allocates the netto gains (or netto losses, if negative) the amount of the individual profits $\alpha_i^{\mathcal{W},v}$, $i \in N$, in comparison with the averaging amount of these profits. Here the individual profit $\alpha_i^{\mathcal{W},v}$ of player i equals the weighted sum of worths of coalitions containing player i.

In order to rewrite the formula (5.2), notice the following:

$$\sum_{j \in N} \alpha_j^{\mathcal{W},v} = \sum_{j \in N} \sum_{\substack{S \subsetneq N, \\ S \ni j}} w_{n,s} \cdot v(S) = \sum_{\substack{S \subsetneq N, \\ S \neq \emptyset}} s \cdot w_{n,s} \cdot v(S) \quad \text{and so, for all } i \in N,$$

$$\alpha_i^{\mathcal{W},v} - \frac{1}{n} \cdot \sum_{j \in N} \alpha_j^{\mathcal{W},v} = \sum_{\substack{S \subsetneq N, \\ S \ni i}} \frac{n-s}{n} \cdot w_{n,s} \cdot v(S) - \sum_{\substack{S \subseteq N \setminus \{i\}, \\ S \neq \emptyset}} \frac{s}{n} \cdot w_{n,s} \cdot v(S)$$

In summary, another representation of any least square value is given, for all $i \in N$, by

$$LS_i^{\mathcal{W}}(N,v) = \frac{v(N)}{n} + \frac{1}{\beta^{\mathcal{W}}} \cdot \left[\sum_{\substack{S \subsetneq N, \\ S \ni i}} \frac{n-s}{n} \cdot w_{n,s} \cdot v(S) - \sum_{\substack{S \subseteq N \setminus \{i\}, \\ S \neq \emptyset}} \frac{s}{n} \cdot w_{n,s} \cdot v(S) \right]$$

$$\tag{5.3}$$

In case $w_{n,s} = \frac{(s-1)! \cdot (n-1-s)!}{(n-1)!}$ for all $1 \leq s \leq n-1$, the equality $t \cdot w_{n,t} = (n-1-t) \cdot w_{n,t+1}$ holds for all $1 \leq t \leq n-2$ and therefore, this particular least square value, called the Shapley value, can be represented, for all $i \in N$, by

$$Sh_i(N,v) = \sum_{S \subseteq N \setminus \{i\}} \frac{s! \cdot (n-1-s)!}{n!} \cdot \left[v(S \cup \{i\}) - v(S) \right] \quad (5.4)$$

In words, the Shapley value of a player in a game agrees with the expectation of player's marginal returns, based on the probability distribution $p_{n,s} = \frac{s! \cdot (n-1-s)!}{n!} = \frac{1}{n} \cdot \frac{1}{\binom{n-1}{s}}$ that any size s, $0 \leq s \leq n-1$, of the coalition is equally likely, as well as coalitions of the same size are equally likely too. The formula (5.4) of the Shapley value is already well-known.

Remark 5.1. In order to determine the Shapley value of the labour game of Example 5.2, observe that the marginal contribution of player i for joining coalition $S \subseteq N \setminus \{i\}$ equals $v(S \cup \{i\}) - v(S) = \ell_i^2 + 2 \cdot \ell_i \cdot \sum_{j \in S} \ell_j$. Write $p_{n,s} = \frac{s! \cdot (n-1-s)!}{n!}$ for all $0 \leq s \leq n-1$, which refers to a probability distribution over the collection of subsets not containing player i. So, by (5.4), for all $i \in N$,

$$Sh_i(N,v) = \sum_{S \subseteq N \setminus \{i\}} p_{n,s} \cdot \left[\ell_i^2 + 2 \cdot \ell_i \cdot \sum_{j \in S} \ell_j \right] = \ell_i^2 + 2 \cdot \ell_i \cdot \sum_{S \subseteq N \setminus \{i\}} p_{n,s} \sum_{j \in S} \ell_j$$

It remains to determine the latter double sum as follows:

$$\sum_{S \subseteq N \setminus \{i\}} p_{n,s} \cdot \sum_{j \in S} \ell_j = \sum_{j \in N \setminus \{i\}} \ell_j \cdot \sum_{\substack{S \subseteq N \setminus \{i\}, \\ S \ni j}} p_{n,s} = \sum_{j \in N \setminus \{i\}} \ell_j \cdot \sum_{s=1}^{n-1} \binom{n-2}{s-1} \cdot p_{n,s}$$

$$= \sum_{j \in N \setminus \{i\}} \ell_j \cdot \sum_{s=1}^{n-1} \frac{s}{(n-1) \cdot n} = \frac{1}{2} \cdot \sum_{j \in N \setminus \{i\}} \ell_j$$

Hence, the Shapley value is given, for all $i \in N$, by $Sh_i(N,v) = \ell_i^2 + \ell_i \cdot \sum_{j \in N \setminus \{i\}} \ell_j = \ell_i \cdot \sum_{j \in N} \ell_j$. In words, we conclude that the Shapley value of the labour game of Example 5.2 agrees with the proportionality rule applied to the individual labour ℓ_i, $i \in N$.

5.2 The Classical Axiomatization of the Shapley Value Using Null Player Property

Let $\mathcal{G}_N$ denote the linear space of all TU games with fixed player set N. Generally speaking, the solution theory for TU games deals with mappings

$f : \mathcal{G}_N \to \mathbb{R}^N$ that assign to every TU game $\langle N, v \rangle$ a so-called payoff vector $f(N, v) = (f_i(N, v))_{i \in N} \in \mathbb{R}^N$. We say the *solution* f meets the *efficiency* principle if, for every TU game $\langle N, v \rangle$, its corresponding payoff vector $f(N, v)$ represents an allocation of the worth $v(N)$ of the grand coalition N among the players in that it satisfies the efficiency constraint $\sum_{j \in N} f_j(N, v) = v(N)$. Set-valued solution concepts (like the core) are beyond the scope of this chapter. The least square values, as described by (5.2) or (5.3), are appealing examples of efficient solutions for TU games. ¿From the mathematical viewpoint, these linear square values possess the so-called *linearity property* in that the solution of the sum of two games agrees with the sum of the solutions of the separate games. Formally, the solution f is said to be *linear* on $\mathcal{G}_N$ if $f(N, \alpha_1 \cdot v_1 + \alpha_2 \cdot v_2) = \alpha_1 \cdot f(N, v_1) + \alpha_2 \cdot f(N, v_2)$ for all TU games $\langle N, v_k \rangle$, and all scalars α_k, $k = 1, 2$. Here the game $\langle N, \alpha_1 \cdot v_1 + \alpha_2 \cdot v_2 \rangle$ is defined by $(\alpha_1 \cdot v_1 + \alpha_2 \cdot v_2)(S) = \alpha_1 \cdot v_1(S) + \alpha_2 \cdot v_2(S)$ for all $S \subseteq N$. This section surveys a number of traditional fairness properties for solutions that axiomatically characterize the Shapley value on $\mathcal{G}_N$.

Two players i, j are said to be *symmetric* (or *substitutes*) in the game $\langle N, v \rangle$ if the worth for joining any coalition does not differ, i.e., $v(S \cup \{i\}) = v(S \cup \{j\})$ for all $S \subseteq N \backslash \{i, j\}$). A solution f on $\mathcal{G}_N$ is said to verify *symmetry* if symmetric players receive the same payoff, that is $f_i(N, v) = f_j(N, v)$ for symmetric players i, j in the game $\langle N, v \rangle$. It is well-known that the Shapley value fulfils the symmetry property. Concerning the labour game of Example 5.2, two inhabitants of the small Indian village are symmetric if and only if their individual labour is identical. Concerning the bankruptcy game of Example 5.1, two creditors i, j are symmetric if and only if their claims satisfy either $d_i = d_j < E$ or $(d_i \geq E$ and $d_j \geq E)$.

Player i is said to be a *null player* in the game $\langle N, v \rangle$ if, in accordance with the player's zero individual worth, the worth of any coalition does not change when player i joins, i.e., $v(S \cup \{i\}) = v(S)$ for all $S \subseteq N \backslash \{i\}$ (particularly, $v(\{i\}) = 0$). A solution f on $\mathcal{G}_N$ is said to verify *the null player property* if null players receive zero payoff, that is $f_i(N, v) = 0$ for every null player i in the game $\langle N, v \rangle$.

Player i is said to be a *dummy player* in the game $\langle N, v \rangle$ if the increase (or decrease) of the worth of any coalition to which player i joins does not differ from the player's individual worth, i.e., $v(S \cup \{i\}) - v(S) = v(\{i\})$ for all $S \subseteq N \backslash \{i\}$. In words, a null player is a dummy player with zero individual

A New Axiomatization of the Shapley Value for TU-games   63

worth. A solution f on $\mathcal{G}_N$ is said to verify *the dummy player property* if dummy players receive their individual worth, that is $f_i(N,v) = v(\{i\})$ for every dummy player i in the game $\langle N, v \rangle$. Clearly, the dummy player property is the weakest version in that, if f fulfils the dummy player property, then f fulfils the null player property too. For instance, by (5.4), the Shapley value verifies the dummy player property. Concerning the labour game of Example 5.2, an inhabitant of the small Indian village is a dummy or null player if and only if the individual labour is at zero level. Concerning the bankruptcy game of Example 5.1, a creditor is a dummy if and only if his claim vanishes (provided that there are at least two strictly positive claims).

The well-known collection of *unanimity games* $\{\langle N, u_T \rangle \mid \emptyset \neq T \subseteq N\}$ is defined by $u_T(S) = 1$ if $T \subseteq S$ and $u_T(S) = 0$ otherwise. For a fixed nonempty set T, the members of T are considered to be the *friends* of the society and a coalition has a non-zero worth if and only if this coalition contains all friends of T. In fact, any *non-friend* $i \in N \backslash T$ is a *null player in the unanimity game* $\langle N, u_T \rangle$ since the equality $u_T(S \cup \{i\}) = u_T(S)$ holds for all $i \in N \backslash T$ and $S \subseteq N \backslash \{i\}$.

It is well-known ([Shapley (1953)], [Roth (1988)], [Owen (1982)]) that the null player property, symmetry, efficiency, together with linearity, fully characterize the Shapley value on $\mathcal{G}_N$. Shapley's initial proof (1953) exploits the fact that the collection of unanimity games $\{\langle N, u_T \rangle \mid \emptyset \neq T \subseteq N\}$ forms a *basis* of the linear space $\mathcal{G}_N$ such that, for all $\langle N, v \rangle$,

$$v = \sum_{\substack{T \subseteq N, \\ T \neq \emptyset}} \lambda_T^v \cdot u_T \quad \text{where} \quad \lambda_T^v = \sum_{S \subseteq T} (-1)^{t-s} \cdot v(S) \quad \text{for all } T \subseteq N, T \neq \emptyset.$$

(5.5)

For each unanimity game $\langle N, u_T \rangle$, its Shapley value is fully determined by the three properties called null player property, symmetry, efficiency, in that $Sh_i(N, u_T) = 0$ for $i \in N \backslash T$ (non-friends of T receive nothing) and $Sh_i(N, u_T) = \frac{1}{t}$ for $i \in T$ (friends of T distribute one monetary unit equally). Together with the linearity axiom applied to the linear space $\mathcal{G}_N$, ([Shapley (1953)]) succeeded to derive the formula (5.4) for the Shapley value. The goal of this chapter is to replace the role of the collection of unanimity games by a recently discovered basis for a certain class of games called 1-concave games and consequently, to replace both the null player notion and the null player property by slightly adapted versions.

5.3 A New Axiomatization of the Shapley Value Using Semi-null Player Property and 1-concave Games

Player i is said to be a *semi-null player* in the game $\langle N, v \rangle$ if, in spite of the player's non-zero individual worth, the worth of any non-trivial coalition does not change when player i joins, i.e.,

$$v(S \cup \{i\}) = v(S) \quad \text{for all } S \subseteq N\setminus\{i\}, \ S \neq \emptyset, \text{ whereas} \quad v(\{i\}) = 1. \quad (5.6)$$

A solution f on $\mathcal{G}_N$ is said to verify *the semi-null player property* if *semi-null players* receive the *egalitarian payoff*, that is $f_i(N, v) = \frac{1}{n}$ for every semi-null player i in the n-person game $\langle N, v \rangle$.

Player i is said to be a *semi-dummy player* in the game $\langle N, v \rangle$ if the increase (or decrease) of the worth of any non-trivial coalition to which player i joins does not differ from the surplus of the player's individual worth, i.e.,

$$v(S \cup \{i\}) - v(S) = 1 - v(\{i\}) \quad \text{for all } S \subseteq N\setminus\{i\}, \ S \neq \emptyset. \quad (5.7)$$

Obviously, (5.6) implies (5.7). In words, a semi-null player is a semi-dummy player with unitary individual worth. A solution f on $\mathcal{G}_N$ is said to verify *the semi-dummy player property* if semi-dummy players receive some expected payoff of the player's individual worth and its surplus, that is

$$f_i(N, v) = \frac{v(\{i\})}{n} + \frac{n-1}{n} \cdot \left[1 - v(\{i\})\right]$$

for every semi-dummy player i in the n-person game $\langle N, v \rangle$. Clearly, the semi-dummy player property is the weakest version in that, if f fulfils the semi-dummy player property, then f fulfils the semi-null player property too. It follows immediately from its probabilistic representation (5.4) that the Shapley value verifies the semi-dummy player property. Notice that semi-null players and null players exclude each other in any game.

Instead of unanimity games, which possess no semi-null players at all, we are interested in a certain collection of games which forms a basis of the linear space $\mathcal{G}_N$ and of which every game possesses semi-null players as well. We highlight the recently discovered collection of so-called *complementary unanimity games* $\{\langle N, \bar{u}_T \rangle \mid T \subsetneq N\}$, defined by

$$\bar{u}_T(S) = 1 \quad \text{if } S \cap T = \emptyset, \ S \neq \emptyset, \text{ and} \quad \bar{u}_T(S) = 0 \quad \text{otherwise.} \quad (5.8)$$

For a fixed set $T \subsetneq N$, the members of T are considered to be the *enemies* of the society and a coalition has a non-zero worth if and only if this coalition contains no enemies of T at all. In case $T = \emptyset$ (no enemies), each nonempty coalition has a non-zero worth in the corresponding complementary unanimity game $\langle N, \bar{u}_\emptyset \rangle$ and consequently, every player is a semi-null player in the game $\langle N, \bar{u}_\emptyset \rangle$. In fact, any *non-enemy* $i \in N \backslash T$ is a *semi-null player* in the complementary unanimity game $\langle N, \bar{u}_T \rangle$ since the equalities $\bar{u}_T(\{i\}) = 1$ and $\bar{u}_T(S \cup \{i\}) = \bar{u}_T(S)$ hold for all $i \in N \backslash T$ and $S \subseteq N \backslash \{i\}$, $S \neq \emptyset$. Hence, for any solution f verifying the semi-null player property, it holds $f_i(N, \bar{u}_T) = \frac{1}{n}$ for all $i \in N \backslash T$.

Note that two enemies $j, k \in T$ are *symmetric players* in the game $\langle N, \bar{u}_T \rangle$ since $\bar{u}_T(S \cup \{j\}) = 0 = \bar{u}_T(S \cup \{k\})$ for all $S \subseteq N \backslash \{j, k\}$. From this we derive that $f_j(N, \bar{u}_T) = f_k(N, \bar{u}_T)$ for all $j, k \in T$ due to the symmetry property of f, provided it applies. Further, notice that $\bar{u}_T(N) = 0$ whenever $T \neq \emptyset$, whereas $\bar{u}_\emptyset(N) = 1$. To conclude with, if a solution f verifies the three properties called semi-null player property, symmetry, and efficiency, then it holds

$$f_i(N, \bar{u}_T) = \frac{1}{n} \quad \text{for } i \in N \backslash T, \text{ while } \quad f_i(N, \bar{u}_T) = \frac{1}{n} - \frac{1}{t} \quad \text{for } i \in T. \quad (5.9)$$

As shown by ([Driessen, Khmelnitskaya and Sales (2005)], Theorem 3.2), the collection of complementary unanimity games $\{\langle N, \bar{u}_T \rangle \mid T \subsetneq N\}$ forms a *basis* of the linear space $\mathcal{G}_N$ such that, for all $\langle N, v \rangle$,

$$v = \sum_{T \subsetneq N} \bar{\lambda}^v_T \cdot \bar{u}_T \quad \text{where} \quad \bar{\lambda}^v_T = \sum_{S \supseteq N \backslash T} (-1)^{s-n+t} \cdot v(S) \quad \text{for all } T \subsetneq N.$$
(5.10)

Remark 5.2. As proven in [Driessen, Khmelnitskaya and Sales (2005)], the complementary unanimity games of (5.8) are the most important representatives of the class of 1-*concave (cost) games*. A (cost) game $\langle N, c \rangle$ is said to be 1-*concave* if for all $S \subseteq N$, $S \neq \emptyset$,

$$c(S) + \sum_{j \in N \backslash S} \left[c(N) - c(N \backslash \{j\}) \right] \geq c(N) \geq \sum_{j \in N} \left[c(N) - c(N \backslash \{j\}) \right]. \quad (5.11)$$

The 1-concavity condition (5.11) states that the sum of all marginal contributions $c(N) - c(N \backslash \{j\})$, $j \in N$, does not exceed the cost $c(N)$ of the joint enterprise, whereas every partial sum of marginal contributions, together with the cost of the complementary coalition, exceeds the cost $c(N)$. It appears ([Driessen (1988)], page 74) that 1-concave cost games are fully charaterized by some appropriate core structure so that the nucleolus coincides with the centre of gravity of the core, whereas the Shapley value may fail to belong to the core. The library cost game ([Driessen, Khmelnitskaya and Sales (2005)]) is an appealing example of 1-concavity for TU cost games.

Example 5.4. ([Driessen, Khmelnitskaya and Sales (2005)])
In the setting of the airport cost game $\langle N, c \rangle$ of Example 5.3, we consider its corresponding cost savings game with respect to the cost C_m of the largest runway and we claim this cost savings game to be a positive linear combination of complementary unanimity games $\langle N, \bar{u}_T \rangle$, $T \subsetneq N$, as follows:

$$C_m - c = \sum_{k=1}^{m}(C_k - C_{k-1}) \cdot \bar{u}_{M_k} \quad \text{such that } \langle N, C_m \rangle \text{ is a constant game,}$$
(5.12)

and $M_k := \cup_{\ell=k}^{m} N_\ell$ is the set of planes of type k and larger. The proof of (5.12) is left to the reader ([Driessen, Khmelnitskaya and Sales (2005)]). Finally, we determine the Shapley value of the airport cost game by applying the additivity property of the Shapley value to this additive decomposition (5.12) of the cost savings game, using that the Shapley value of any complementary unanimity game $\langle N, \bar{u}_T \rangle$ is given by $Sh_i(N, \bar{u}_T) = \frac{1}{n} - \frac{1}{t}$ for all $i \in T$, and $Sh_i(N, \bar{u}_T) = \frac{1}{n}$ otherwise.

Suppose $i \in N_j$, so $i \in M_k$ for all $1 \leq k \leq j$, whereas $i \notin M_k$ for all $j < k \leq m$. From this, together with (5.12), we determine its Shapley cost allocation by

$$Sh_i(N, c) = Sh_i(N, C_m) - \sum_{k=1}^{m}(C_k - C_{k-1}) \cdot Sh_i(N, \bar{u}_{M_k})$$

$$= \frac{C_m}{n} - \sum_{k=1}^{m} \frac{C_k - C_{k-1}}{n} + \sum_{k=1}^{j} \frac{C_k - C_{k-1}}{m_k} = \sum_{k=1}^{j} \frac{C_k - C_{k-1}}{m_k}$$

where m_k denotes the cardinality of the set M_k, $1 \leq k \leq m$. In words, the Shapley cost allocation to a landing by a plane of type j is based on the equal charge rule applied to the incremental costs of "small runways" taking care of the number of users of these small runways. Both the latter formula and interpretation of the Shapley cost allocation for airport cost games are already known ([Littlechild, and Owen, G. (1973)]), obtained by a much different approach.

Theorem 5.2. *There exists a unique solution on $\mathcal{G}_N$ verifying the semi-null player property, symmetry, efficiency, linearity, and it concerns the Shapley value of the form (5.4).*

Proof. It remains to prove the uniqueness part. Let f be a solution on $\mathcal{G}_N$ verifying the semi-null player property, symmetry, efficiency, and linearity. Our main goal is to show that f agrees with the right hand of formula (5.4).

By the first three properties, (5.9) is valid. Fix the n-person game $\langle N, v \rangle$ and player $i \in N$. By applying the linearity property of f to (5.10), we arrive at

$$f_i(N, v) = f_i\left(N, \sum_{T \subsetneq N} \bar{\lambda}_T^v \cdot \bar{u}_T\right) = \sum_{T \subsetneq N} \bar{\lambda}_T^v \cdot f_i(N, \bar{u}_T). \quad \text{Thus, by (5.9),}$$

$$f_i(N, v) = \sum_{\substack{T \subsetneq N, \\ T \ni i}} \bar{\lambda}_T^v \cdot \left[\frac{1}{n} - \frac{1}{t}\right] + \sum_{T \subseteq N \setminus \{i\}} \frac{\bar{\lambda}_T^v}{n} = \frac{1}{n} \cdot \sum_{T \subsetneq N} \bar{\lambda}_T^v - \sum_{\substack{T \subsetneq N, \\ T \ni i}} \frac{\bar{\lambda}_T^v}{t}. \quad (5.13)$$

Part 1. On the one hand, by reversing the order of a first double sum of (5.13) concerning the complementary dividends $\bar{\lambda}_T^v$ as given by (5.10), we obtain

$$\sum_{T \subsetneq N} \bar{\lambda}_T^v = \sum_{T \subsetneq N} \sum_{S \supseteq N \setminus T} (-1)^{s-n+t} \cdot v(S) = \sum_{R \neq \emptyset} \left[\sum_{\substack{T \supseteq N \setminus R, \\ T \neq N}} (-1)^{r-n+t}\right] \cdot v(R)$$

$$= \sum_{R \neq \emptyset} \left[\sum_{t=n-r}^{n-1} \binom{r}{t-n+r} \cdot (-1)^{r-n+t}\right] \cdot v(R) = \sum_{R \neq \emptyset} \left[\sum_{k=0}^{r-1} \binom{r}{k} \cdot (-1)^k\right] \cdot v(R)$$

$$= \sum_{R \neq \emptyset} \left[((-1) + 1)^r - (-1)^r\right] \cdot v(R) = \sum_{R \neq \emptyset} (-1)^{r+1} \cdot v(R)$$

where the last but one equality is due to the binomial theorem. So far, we conclude

$$\frac{1}{n} \cdot \sum_{T \subsetneq N} \bar{\lambda}_T^v = \frac{1}{n} \cdot \sum_{R \neq \emptyset} (-1)^{r+1} \cdot v(R) = \frac{1}{n} \cdot \sum_{T \subseteq N \setminus \{i\}} (-1)^t \cdot \left[v(T \cup \{i\}) - v(T)\right].$$

Part 2. On the other hand, by reversing the order of a second double sum of (5.13) concerning the averaging of the complementary dividends $\bar{\lambda}_T^v$ as given by (5.10), we obtain

$$\sum_{\substack{T\subseteq N,\\ T\ni i}} \frac{\bar{\lambda}_T^v}{t}$$

$$= \sum_{\substack{T\subseteq N,\\ T\ni i}} \frac{1}{t} \cdot \sum_{S\supseteq N\setminus T} (-1)^{s-n+t} \cdot v(S) = \sum_{\substack{R\subseteq N,\\ R\neq\emptyset}} \left[\sum_{\substack{T\supseteq N\setminus R,\\ T\neq N, T\ni i}} \frac{(-1)^{r-n+t}}{t} \right] \cdot v(R)$$

$$= \sum_{\substack{R\subseteq N\setminus\{i\},\\ R\neq\emptyset}} \left[\sum_{\substack{T\supseteq N\setminus R,\\ T\neq N}} \frac{(-1)^{r-n+t}}{t} \right] \cdot v(R) + \sum_{\substack{R\subseteq N,\\ i\in R, R\neq\{i\}}} \left[\sum_{\substack{T\supseteq (N\setminus R)\cup\{i\},\\ T\neq N}} \frac{(-1)^{r-n+t}}{t} \right] \cdot v(R)$$

$$= \sum_{\substack{R\subseteq N\setminus\{i\},\\ R\neq\emptyset}} \left[\sum_{t=n-r}^{n-1} \binom{r}{t-n+r} \cdot \frac{(-1)^{r-n+t}}{t} \right] \cdot v(R)$$

$$+ \sum_{\substack{R\subseteq N,\\ i\in R, R\neq\{i\}}} \left[\sum_{t=n-r+1}^{n-1} \binom{r-1}{t-n+r-1} \cdot \frac{(-1)^{r-n+t}}{t} \right] \cdot v(R)$$

$$= \sum_{\substack{R\subseteq N\setminus\{i\},\\ R\neq\emptyset}} \left[\sum_{k=0}^{r-1} \binom{r}{k} \cdot \frac{(-1)^k}{n-r+k} \right] \cdot v(R) + \sum_{\substack{R\subseteq N,\\ i\in R, R\neq\{i\}}} \left[\sum_{\ell=0}^{r-2} \binom{r-1}{\ell} \cdot \frac{(-1)^{\ell+1}}{n-r+\ell+1} \right] \cdot v(R)$$

$$= \sum_{\substack{T\subseteq N\setminus\{i\},\\ T\neq\emptyset}} \left[\sum_{k=0}^{t-1} \binom{t}{k} \cdot \frac{(-1)^k}{n-t+k} \right] \cdot v(T) + \sum_{\substack{T\subseteq N\setminus\{i\},\\ T\neq\emptyset}} \left[\sum_{k=0}^{t-1} \binom{t}{k} \cdot \frac{(-1)^{k+1}}{n-t+k} \right] \cdot v(T\cup\{i\})$$

$$= -\sum_{\substack{T\subseteq N\setminus\{i\},\\ T\neq\emptyset}} \left[\sum_{k=0}^{t-1} \binom{t}{k} \cdot \frac{(-1)^k}{n-t+k} \right] \cdot \left[v(T\cup\{i\}) - v(T) \right].$$

Part 3. From Part 1 and Part 2, we derive that (5.13) reduces to the following equality:

$$f_i(N,v) = \frac{1}{n} \cdot \sum_{T\subsetneq N} \bar{\lambda}_T^v - \sum_{\substack{T\subseteq N,\\ T\ni i}} \frac{\bar{\lambda}_T^v}{t} = \frac{1}{n} \cdot \sum_{T\subseteq N\setminus\{i\}} (-1)^t \cdot \left[v(T\cup\{i\}) - v(T) \right]$$

$$+ \sum_{\substack{T\subseteq N\setminus\{i\},\\ T\neq\emptyset}} \left[\sum_{k=0}^{t-1} \binom{t}{k} \cdot \frac{(-1)^k}{n-t+k} \right] \cdot \left[v(T\cup\{i\}) - v(T) \right]$$

$$= \frac{v(\{i\})}{n} + \sum_{\substack{T\subseteq N\setminus\{i\},\\ T\neq\emptyset}} \left[\frac{(-1)^t}{n} + \sum_{k=0}^{t-1} \binom{t}{k} \cdot \frac{(-1)^k}{n-t+k} \right] \cdot \left[v(T\cup\{i\}) - v(T) \right]. \quad (5.14)$$

Part 4. The formula (5.14) in Part 3 agrees with the right hand of formula (5.4) if and only if the following claim holds:

$$\sum_{k=0}^{t} \binom{t}{k} \cdot \frac{(-1)^k}{n-t+k} = \frac{t! \cdot (n-t-1)!}{n!} \qquad \text{for all } 1 \leq t \leq n-1 \qquad (5.15)$$

To prove (5.15), recall that $\int_0^1 x^{p-1} dx = \frac{1}{p}$ for all $p \geq 1$. Let $1 \leq t \leq n-1$. We obtain, by reversing the order of summation and integration, and using again the binomial theorem,

$$\sum_{k=0}^{t} \binom{t}{k} \cdot \frac{(-1)^k}{n-t+k} = \sum_{k=0}^{t} \binom{t}{k} \cdot (-1)^k \cdot \left[\int_0^1 x^{n-t+k-1} dx\right]$$

$$= \int_0^1 \left[\sum_{k=0}^{t} \binom{t}{k} \cdot (-1)^k \cdot x^{n-t+k-1}\right] dx$$

$$= \int_0^1 x^{n-t-1} \cdot \left[\sum_{k=0}^{t} \binom{t}{k} \cdot (-x)^k\right] dx$$

$$= \int_0^1 x^{n-t-1} \cdot (1-x)^t dx = \frac{t! \cdot (n-t-1)!}{n!}$$

where the last equality is a well-known result from integration by parts. So, the claim (5.15) holds. Thus, $f_i(N,v) = Sh_i(N,v)$ for all $i \in N$ and all $\langle N, v \rangle$, as was to be shown. □

Remark 5.3. In [van den Brink (2007)], the egalitarian rule (i.e., $f_i(N,v) = \frac{v(N)}{n}$ for all $i \in N$) is axiomatized through the nullifying player property, symmetry, efficiency, and linearity. Its uniqueness part is proven by using the standard basis (instead of the unanimity basis). The nullifying player property states that a nullifying player should receive nothing. Here a nullifying player i is identified by $v(S) = 0$ for all coalitions $S \subseteq N$ with $i \in S$.

In [van den Brink (2001)], the Shapley value is axiomatized through the null player property, efficiency and fairness. A solution f on $\mathcal{G}_N$ satisfies *fairness* if for all symmetric players i, j in the game $\langle N, w \rangle$, it holds that $f_i(N, v+w) - f_i(N, v) = f_j(N, v+w) - f_j(N, v)$ for all games $\langle N, v \rangle$. That is, the payoffs of two players change by the same amount when adding a game, in which both players are symmetric.

5.4 Appendix: the Proof of Theorem 5.1

Fix the size $n \geq 2$ of the player set and fix the collection $\mathcal{W} = (w_{n,s})_{s=1}^{n-1}$ of non-negative weights with reference to the coalition sizes s, $1 \leq s \leq n-1$. Let $\vec{x}$ denote the variable (payoff) vector $\vec{x} = (x_j)_{j \in N} \in \mathbb{R}^N$. Define the function $g : \mathbb{R}^N \to \mathbb{R}$ by $g(\vec{x}) = \sum_{j \in N} x_j$. In the context of n-person games $\langle N, v \rangle$, the objective function $f^v : \mathbb{R}^N \to \mathbb{R}$ is defined by

$$f^v(\vec{x}) = \sum_{\emptyset \neq S \subsetneq N} w_{n,s} \cdot \left[v(S) - \sum_{j \in S} x_j\right]^2 \quad \text{for all } \vec{x} \in \mathbb{R}^N$$

To find the minimum value of the objective function $f^v(\vec{x})$ subject to the efficiency constraint $g(\vec{x}) = v(N)$, we use *the Method of Lagrange multipliers* in that the gradient vectors ∇f^v and ∇g must be parallel, that is $\nabla f^v = \lambda \cdot \nabla g$ for a certain scalar $\lambda \in \mathbb{R}$. For all $i \in N$, the corresponding partial derivatives $(f^v)_{x_i}$ and g_{x_i} are given by

$$g_{x_i}(\vec{x}) = 1 \quad \text{and} \quad (f^v)_{x_i}(\vec{x}) = -2 \cdot \sum_{\substack{S \subsetneq N, \\ S \ni i}} w_{n,s} \cdot \left[v(S) - \sum_{j \in S} x_j\right]$$

So, the gradient vector equation $\nabla f^v = \lambda \cdot \nabla g$ reduces to the following system of equations:

$$\sum_{\substack{S \subsetneq N, \\ S \ni i}} w_{n,s} \cdot \left[v(S) - \sum_{j \in S} x_j\right] = \frac{-\lambda}{2} \quad \text{for all } i \in N, \text{ or equivalently,}$$

$$\sum_{\substack{S \subsetneq N, \\ S \ni i}} w_{n,s} \cdot v(S) - x_i \cdot \sum_{\substack{S \subsetneq N, \\ S \ni i}} w_{n,s} - \sum_{\substack{S \subsetneq N, \\ S \ni i}} w_{n,s} \cdot \sum_{j \in S \setminus \{i\}} x_j = \mu \quad \text{for all } i \in N,$$

where $\mu = \frac{-\lambda}{2}$. By some combinatorial calculations, for all $i \in N$, we derive the following chain of equations:

$$\sum_{\substack{S \subsetneq N, \\ S \ni i}} w_{n,s} \cdot \sum_{j \in S \setminus \{i\}} x_j = \sum_{\substack{S \subsetneq N, \\ S \ni i}} w_{n,s} \cdot \sum_{j \in N \setminus \{i\}} x_j \cdot 1_S(j)$$

$$= \sum_{j \in N \setminus \{i\}} x_j \cdot \sum_{\substack{S \subsetneq N, \\ S \ni i}} w_{n,s} \cdot 1_S(j) = \sum_{j \in N \setminus \{i\}} x_j \cdot \sum_{\substack{S \subsetneq N, \\ S \ni i,j}} w_{n,s}$$

$$= \sum_{j \in N \setminus \{i\}} x_j \cdot \sum_{s=2}^{n-1} \binom{n-2}{s-2} \cdot w_{n,s}$$

$$= \sum_{j \in N \setminus \{i\}} x_j \cdot \beta_1^{\mathcal{W}} = \beta_1^{\mathcal{W}} \cdot (v(N) - x_i)$$

where $\beta_1^{\mathcal{W}} = \sum_{s=2}^{n-1} \binom{n-2}{s-2} \cdot w_{n,s}$. Put $\beta_2^{\mathcal{W}} = \sum_{s=1}^{n-1} \binom{n-1}{s-1} \cdot w_{n,s}$ and $\alpha_i^{\mathcal{W},v} = \sum_{\substack{S \subseteq N, \\ S \ni i}} w_{n,s} \cdot v(S)$ for all $i \in N$. In summary, the gradient vector equation $\nabla f^v = \lambda \cdot \nabla g$ reduces to the following system of equations: for all $i \in N$

$$\alpha_i^{\mathcal{W},v} - \beta_2^{\mathcal{W}} \cdot x_i - \beta_1^{\mathcal{W}} \cdot (v(N) - x_i) = \mu \qquad \text{or equivalently,}$$

$$(\beta_1^{\mathcal{W}} - \beta_2^{\mathcal{W}}) \cdot x_i = -\alpha_i^{\mathcal{W},v} + \mu + \beta_1^{\mathcal{W}} \cdot v(N).$$

To eliminate μ, summing up over all $i \in N$ yields

$$(\beta_1^{\mathcal{W}} - \beta_2^{\mathcal{W}}) \cdot v(N) = -\sum_{j \in N} \alpha_j^{\mathcal{W},v} + n \cdot (\mu + \beta_1^{\mathcal{W}} \cdot v(N)).$$

Finally, the substitution of μ in the former expression yields the optimal payoff vector $\vec{x}$ given, for all $i \in N$, by

$$x_i = \frac{v(N)}{n} + \frac{1}{\beta_2^{\mathcal{W}} - \beta_1^{\mathcal{W}}} \cdot \left[\alpha_i^{\mathcal{W},v} - \frac{1}{n} \cdot \sum_{j \in N} \alpha_j^{\mathcal{W},v} \right].$$

Notice that $\beta_2^{\mathcal{W}} - \beta_1^{\mathcal{W}} = \sum_{s=1}^{n-1} \binom{n-1}{s-1} \cdot w_{n,s} - \sum_{s=2}^{n-1} \binom{n-2}{s-2} \cdot w_{n,s} = \sum_{s=1}^{n-1} \binom{n-2}{s-1} \cdot w_{n,s} = \beta^{\mathcal{W}}$. $\square$

Bibliography

van den Brink, R. (2001). An axiomatization of the Shapley value using a fairness property, *International Journal of Game Theory* **30**, pp. 309–319.

van den Brink, R. (2007). Null or nullifying players: The difference between the Shapley value and the equal division solutions, *Journal of Economic Theory*, in press.

Driessen, T. S. H. (1988). *Cooperative Games, Solutions, and Applications*, (Kluwer Academic Publishers, Dordrecht, The Netherlands).

Driessen, T. S. H. (1998). The greedy bankruptcy game: an alternative game theoretic analysis of a bankruptcy problem, in *Game Theory and Applications IV*, eds: L.A. Petrosjan and V.V. Mazalov, Nova Science Publishers, Inc. New York, USA, pp. 45–61.

Driessen, T. S. H., Khmelnitskaya, A. B. and Sales, J. (2005). 1-*concave basis for TU games*. Memorandum no. 1777, Department of Applied Mathematics, University of Twente, Enschede, The Netherlands.

Littlechild, S. C. and Owen, G.(1973). A simple expression for the Shapley value in a special case, *Management Science* **20**, pp. 370–372.

O'Neill, B. (1982). A problem of rights arbitration from the Talmud, *Mathematical Social Sciences* **2**, pp. 345–371.

Owen, G. (1982). *Game Theory. Second Edition*, (Academic Press, New York, U.S.A).

Roth, A. E.(editor) (1988). *The Shapley value: Essays in honor of Lloyd S. Shapley*, (Cambridge University Press, Cambridge, U.S.A).

Ruiz, L. M., Valenciano, F. and Zarzuelo, J. M. (1998). The family of least square values for transferable utility games, *Games and Economic Behavior* **24**, pp. 109–130.

Shapley, L. S. (1953). A value for n-person games, *Annals of Mathematics Study* **28**, pp. 307–317 (Princeton University Press). Also in [Roth (1988)], (1988), pp. 31–40.

Chapter 6

Coextrema Additive Operators

Atsushi Kajii
Institute of Economic Research, Kyoto University, Japan
e-mail: kajii@kier.kyoto-u.ac.jp
Hiroyuki Kojima
Department of Economics, Teikyo University, Japan.
e-mail: hkojima@main.teikyo-u.ac.jp
Takashi Ui
Faculty of Economics, Yokohama National University, Japan
e-mail: oui@ynu.ac.jp

Abstract

This chapter proposes a class of weak additivity concepts for an operator on the set of real valued functions on a finite state space Ω. Let $\mathcal{E} \subseteq 2^\Omega$ be a set of subsets of Ω. Two functions x and y on Ω are $\mathcal{E}$-coextrema if, for each $E \in \mathcal{E}$, the set of minimizers of x restricted on E and that of y have a common element, and the set of maximizers of x restricted on E and that of y have a common element as well. An operator I on the set of functions on Ω is $\mathcal{E}$-coextrema additive if $I(x+y) = I(x) + I(y)$ whenever x and y are $\mathcal{E}$-coextrema. So an additive operator is $\mathcal{E}$-coextrema additive, and an $\mathcal{E}$-coextrema additive operator is comonotonic additive. The main result characterizes homogeneous $\mathcal{E}$-coextrema additive operators.

Key Words: Coextrema additive games, Choquet integral, comonotonicity, non-additive probabilities, capacities

6.1 Introduction

The purpose of this chapter is to characterize operators on the set of real valued functions on a finite set which is *coextrema additive*: let Ω be a finite set and let $\mathcal{E} \subseteq 2^\Omega$ be a set of subsets of Ω. Two functions x and y on Ω are said to be $\mathcal{E}$-coextrema if, for each $E \in \mathcal{E}$, the set of minimizers of function x restricted on E and that of function y have a common element, and the set of maximizers of x restricted on E and that of y have a common element as well. An operator I on the set of functions on Ω is $\mathcal{E}$-coextrema additive if $I(x+y) = I(x) + I(y)$ whenever x and y are $\mathcal{E}$-coextrema. Note that if two functions are comonotonic, then they are $\mathcal{E}$-coextrema, *a fortiori*.

The main result shows that a homogeneous coextrema additive operator I can be represented as $I(x) = \sum_{\omega \in \Omega} p(\omega) x(\omega) + \sum_{E \in \mathcal{E} \setminus \mathcal{F}_1} \{\lambda_E \max_{\omega \in E} x(\omega) + \mu_E \min_{\omega \in E} x(\omega)\}$, where λ_E and μ_E are unique constants, $\mathcal{F}_1$ is the set of singleton subsets of Ω, and $p \in \mathbb{R}^\Omega$, when $\mathcal{E}$ satisfies a certain regularity condition. This expression can also be written as the Choquet integral with respect to a certain non-additive (signed) measure. Therefore, a homogeneous coextrema additive operator corresponds to a special class of the Choquet integral, which is expressed as a weighted sum of "optimistic evaluation" $\max_{\omega \in E} x(\omega)$ and "pessimistic evaluation" $\min_{\omega \in E} x(\omega)$. For the case where $I(1) = 1$, we have $\sum_{\omega \in \Omega} p(\omega) + \sum_{E \in \mathcal{E}} (\lambda_E + \mu_E) = 1$, and then these weights can be interpreted as beliefs on events if these are non-negative numbers.

As a corollary, our result shows that for the special case where $\mathcal{E}$ consists of singletons and the whole set Ω, a homogeneous $\mathcal{E}$-coextrema operator is exactly the Choquet integral of a NEO-additive capacity, which is axiomatized by Chateauneuf, Eichberger, and Grant (2002). Eichberger, Kelsey, and Schipper (2006) applied a NEO-additive capacity model to the Bertrand and Cournot competition models to study combined effects of optimism and pessimism in economic environments.

While in the NEO-additive capacity, optimism and pessimism are about the whole states of the world, our model can accommodate more delicate combinations of optimism and pessimism measured in a family of events. Thus our $\mathcal{E}$-coextrema additivity model provides a rich framework for analyzing effects of optimism and pessimism in economic problems. Consequently, we contend that our result provides a natural, and important generalization of NEO-additive capacity models.

Kajii, Kojima, and Ui (2007) considered the class of $\mathcal{E}$-cominimum additive operators, and it is shown that an $\mathcal{E}$-cominimum additive operator is a weighted sum of minimums. The class of $\mathcal{E}$-comaximum operators is defined and characterized similarly. If an operator is both $\mathcal{E}$-cominimum and $\mathcal{E}$-comaximum additive, it is $\mathcal{E}$-coextrema additive, but the converse is not necessarily the case: for instance, there exists an $\mathcal{E}$-coextrema additive operator which is neither $\mathcal{E}$-cominimum additive nor $\mathcal{E}$-comaximum additive. In this sense, the class of $\mathcal{E}$-coextrema additive operators is *not* the intersection of the two classes, and the characterization result reported in this chapter cannot be done by any simple application of these results. In fact, the reader will see that the issue of characterization is far more technically involved.

Ghirardato, Maccheroni, and Marinacci (2004) proposed the following class of operators called the α-MEU functional: $I(x) = \alpha \min_{q \in C} \int x dq + (1 - \alpha) \max_{q \in C} \int x dq$ where C is a convex set of additive measures.[1] In the special case of $\mathcal{E}$ consisting of singletons and the whole set Ω, it can be readily verified that an $\mathcal{E}$-coextrema additive operator is an α-MEU functional, but for general $\mathcal{E}$, there is no direct connection as far as we can tell.

The organization of this chapter is as follows. After a summary of basic concepts and preliminary results in Section 6.2, a formal definition of the $\mathcal{E}$-coextrema operator is given in Section 6.3. Section 6.3 also contains some discussions on the operator, including potential applications to economics and social sciences. The main result is stated in Section 6.4 and a proof is provided in Section 6.5.

6.2 The Model and Preliminary Results

Let Ω be a finite set, whose generic element is denoted by ω. Denote by $\mathcal{F}$ the set of all non-empty subsets of Ω, and by $\mathcal{F}_1$ the set of singleton subsets of Ω. A typical interpretation is that Ω is the set of the states of the world and a subset $E \subseteq \Omega$ is an event.

We shall fix $\mathcal{E} \subseteq \mathcal{F}$ with $\mathcal{E} \neq \emptyset$ throughout the analysis. Write $\sigma(\mathcal{E})$ for the algebra of Ω generated by $\mathcal{E}$, i.e., the smallest algebra containing each element of $\mathcal{E}$. Let $\Pi(\mathcal{E}) \subseteq \mathcal{F}$ be the set of minimal elements of $\sigma(\mathcal{E})$, which constitutes a well defined *partition* of Ω, since Ω is a finite set. A generic element of partition $\Pi(\mathcal{E})$ will be denoted by S. For each $F \in \mathcal{F}$, let $\kappa(F) \in \sigma(\mathcal{E})$ denote the minimal $\sigma(\mathcal{E})$-measurable set containing F; that is, $\kappa(F) := \cap \{E \in \sigma(\mathcal{E}) : F \subseteq E\}$.

Remark 6.1. Note that every element of $\Pi(\mathcal{E})$ belongs to $\sigma(\mathcal{E})$ and that any element $E \in \sigma(\mathcal{E})$ is the union of some elements of $\Pi(\mathcal{E})$. So in particular, for every $E \in \mathcal{E}$ and every $S \in \Pi(\mathcal{E})$, either $S \subseteq E$ or $S \subseteq E^c$ holds. By construction, $\kappa(F) = \cup \{S \in \Pi(\mathcal{E}) : S \cap F \neq \emptyset\}$, i.e., $\kappa(F)$ is the

---

[1] Ghirardato, Maccheroni, and Marinacci (2004) axiomatized the α-MEU functional on infinite state spaces.

union of elements in partition $\Pi(\mathcal{E})$ intersecting F. It is readily verified that if $E \in \sigma(\mathcal{E})$, then $\kappa(E \cap F) = E \cap \kappa(F)$ holds for any $F \in \mathcal{F}$, and so in particular $\kappa(E) = E$.

Example 6.1. Let $\Omega = \{1, 2, \cdots, 8\}$ and $\mathcal{E} = \{E_1, E_2, E_3, E_4\}$ where $E_1 = \{1,2,3,4\}$, $E_2 = \{3,4,5,6\}$, $E_3 = \{1,2,5,6\}$, $E_4 = \{5,6,7,8\}$. Then, $\Pi(\mathcal{E}) = \{S_1, \ldots, S_4\}$, where $S_1 = \{1,2\}$, $S_2 = \{3,4\}$, $S_3 = \{5,6\}$, $S_4 = \{7,8\}$. In this case, $E_1 = S_1 \cup S_2$, $E_2 = S_2 \cup S_3$, $E_3 = S_1 \cup S_3$, $E_4 = S_3 \cup S_4$. For instance, for $R = \{1,3,5,7\}$, we have $\kappa(R) = \Omega$, because every $S \in \Pi(\mathcal{E})$ intersects R.

A set function $v : 2^\Omega \to \mathbb{R}$ with $v(\emptyset) = 0$ is called a *game* or a *non-additive signed measure*. Since each game is identified with a point in $\mathbb{R}^\mathcal{F}$, we denote by $\mathbb{R}^\mathcal{F}$ the set of all games. For a game $v \in \mathbb{R}^\mathcal{F}$, we use the following definitions:

- v is *non-negative* if $v(E) \geq 0$ for all $E \in 2^\Omega$.
- v is *monotone* if $E \subseteq F$ implies $v(E) \leq v(F)$ for all $E, F \in 2^\Omega$. A monotone game is non-negative.
- v is *additive* if $v(E \cup F) = v(E) + v(F)$ for all $E, F \in 2^\Omega$ with $E \cap F = \emptyset$, which is equivalent to $v(E) + v(F) = v(E \cup F) + v(E \cap F)$ for all $E, F \in 2^\Omega$.
- v is *convex* (or *supermodular*) if $v(E) + v(F) \leq v(E \cup F) + v(E \cap F)$ for all $E, F \in 2^\Omega$.
- v is *normalized* if $v(\Omega) = 1$.
- v is a *non-additive measure* if it is monotone. A normalized non-additive measure is called a *capacity*.
- v is a *measure* if it is non-negative and additive. A normalized measure is called a *probability measure*.
- The *conjugate* of v, denoted by v', is defined as $v'(E) = v(\Omega) - v(\Omega \backslash E)$ for all $E \in 2^\Omega$. Note that $(v')' = v$ and $(v + w)' = v' + w'$ for $v, w \in \mathbb{R}^\mathcal{F}$.

For $T \in \mathcal{F}$, let $u_T \in \mathbb{R}^\mathcal{F}$ be the *unanimity game* on T defined by the rule: $u_T(E) = 1$ if $T \subseteq E$ and $u_T(E) = 0$ otherwise. Let w_T be the conjugate of u_T. Then $w_T(E) = 1$ if $T \cap E \neq \emptyset$ and $w_T(E) = 0$ otherwise. Note that when $T = \{\omega\}$, i.e., T is a singleton set, $u_T = w_T$ and they are additive. The following result is well known as the Möbius inversion in discrete and combinatorial mathematics (cf. Shapley, 1953).

Lemma 6.1. *The set $\{u_T\}_{T \in \mathcal{F}}$ is a linear base for $\mathbb{R}^{\mathcal{F}}$, so is the set $\{w_T\}_{T \in \mathcal{F}}$. The unique set of coefficients $\{\beta_T\}_{T \in \mathcal{F}}$ satisfying $v = \sum_{T \in \mathcal{F}} \beta_T u_T$ is given by $\beta_T = \sum_{E \subseteq T, E \neq \emptyset} (-1)^{|T|-|E|} v(E)$.*

By convention, we shall omit the empty set in the summation indexed by subsets of Ω. By the definition of u_T, we have $v(E) = \sum_{T \subseteq E} \beta_T$ for all $E \in \mathcal{F}$. The set of coefficients $\{\beta_T\}_{T \in \mathcal{F}}$ is referred to as the Möbius transform of v. If $v = \sum_{T \in \mathcal{F}} \beta_T u_T$, then the conjugate v' is given by $v' = \sum_{T \in \mathcal{F}} \beta_T w_T$. Using the formula in Lemma 6.1, by direct computation, one can show that for each $E \in \mathcal{F}$:

$$w_E = \sum_{T \subseteq E} (-1)^{|T|-1} u_T. \tag{6.1}$$

Remark 6.2. If $v = \sum_{T \in \mathcal{F}} \beta_T u_T$, the game v is additive if and only if $\beta_T = 0$ unless $|T| = 1$. Obviously, $\sum_{\omega \in \Omega} \beta_{\{\omega\}} u_{\{\omega\}}$ is an additive game. So, we can also write $v = p + \sum_{T \in \mathcal{F}, |T| > 1} \beta_T u_T$ where p is an additive game.

By convention, a function $x : \Omega \to \mathbb{R}$ is identified with an element of $\mathbb{R}^{\Omega}$, and we denote by 1_E the indicator function of event $E \in \mathcal{F}$. For a function $x \in \mathbb{R}^{\Omega}$, and an event E, we write $\min_E x := \min_{\omega \in E} x(\omega)$ and $\arg\min_E x := \arg\min_{\omega \in E} x(\omega)$. Similarly, we write $\max_E x := \max_{\omega \in E} x(\omega)$ and $\arg\max_E x := \arg\max_{\omega \in E} x(\omega)$.

Definition 6.1. For $x \in \mathbb{R}^{\Omega}$ and $v \in \mathbb{R}^{\mathcal{F}}$, the Choquet integral of x with respect to v is defined as

$$\int_\Omega x \, dv = \int_0^\infty v(x \geq \alpha) d\alpha + \int_{-\infty}^0 (v(x \geq \alpha) - 1) d\alpha \tag{6.2}$$

where $v(x \geq \alpha) = v(\{\omega \in \Omega : x(\omega) \geq \alpha\})$.

By definition, $\int 1_E dv = v(E)$. A direct computation reveals that, for any two sets E and F in $\mathcal{F}$,

$$\int (1_E + 1_F) \, dv = v(E \cup F) + v(E \cap F). \tag{6.3}$$

Then for each event T, we see from (6.2) that $\int x \, du_T = \min_T x$ and $\int x \, dw_T = \max_T x$. Also it can be readily verified that the Choquet integral is additive in games. Recall that for a game v, there is a unique set of coefficients $\{\beta_T : T \in \mathcal{F}\}$ such that $v = \sum_T \beta_T u_T$ by Lemma 6.1. Using additivity, therefore, we have $\int x \, dv = \sum_T \beta_T \min_T x$, as is pointed out in Gilboa and Schmeidler (1994).

Note that the additivity implies the following property: for any $T \in \mathcal{F}$ and real numbers λ and μ, $\int xd(\lambda w_T + \mu u_T) = \int xd(\lambda w_T) + \int xd(\mu u_T) = \lambda \max_T x + \mu \min_T x$, and so

$$\int xd\left(\sum_{E \in \mathcal{F}'} \lambda_E w_E + \mu_E u_E\right) = \sum_{E \in \mathcal{F}'} \{\lambda_E \max_E x + \mu_E \min_E x\}, \quad (6.4)$$

for any set of events $\mathcal{F}' \subseteq \mathcal{F}$ and sets of real numbers $\{\lambda_E : E \in \mathcal{F}'\}$ and $\{\mu_E : E \in \mathcal{F}'\}$.

Definition 6.2. Let $\mathcal{E} \subseteq \mathcal{F}$ be a set of events. Two functions $x, y \in \mathbb{R}^\Omega$ are said to be $\mathcal{E}$-cominimum, provided $\arg \min_E x \cap \arg \min_E y \neq \emptyset$ for all $E \in \mathcal{E}$. Two functions $x, y \in \mathbb{R}^\Omega$ are said to be $\mathcal{E}$-comaximum, provided $\arg \max_E x \cap \arg \max_E y \neq \emptyset$ for all $E \in \mathcal{E}$.

Remark 6.3. Clearly, x and y are $\mathcal{E}$-cominimum if and only if $-x$ and $-y$ are $\mathcal{E}$-comaximum. Also, the $\mathcal{E}$-cominimum and the $\mathcal{E}$-comaximum relations are invariant of adding a constant. In particular, if two indicator functions 1_A and 1_B are $\mathcal{E}$-cominimum, $1_{\Omega \setminus A}$ ($=1 - 1_A$) and $1_{\Omega \setminus B}$ ($=1 - 1_B$) are $\mathcal{E}$-comaximum, and vice versa.

A function $I : \mathbb{R}^\Omega \to \mathbb{R}$ is referred to as an operator.

Definition 6.3. An operator I is said to be homogeneous if $I(\alpha x) = \alpha I(x)$ for any $\alpha > 0$.

Kajii, Kojima, and Ui (2007) studied $\mathcal{E}$-cominimum and $\mathcal{E}$-comaximum operators defined as follows:

Definition 6.4. An operator $I : \mathbb{R}^\Omega \to \mathbb{R}$ is $\mathcal{E}$-cominimum (resp. comaximum) additive provided $I(x + y) = I(x) + I(y)$ whenever x and y are $\mathcal{E}$-cominimum (resp. comaximum).

A pair of functions x and y are said to be *comonotonic* if $(x(\omega) - x(\omega'))(y(\omega) - y(\omega')) \geq 0$ for any $\omega, \omega' \in \Omega$. Notice that if $\mathcal{E} = \mathcal{F}$, a pair of functions x and y are comonotonic if and only if they are $\mathcal{E}$-cominimum, as well as $\mathcal{E}$-comaximum. So when $\mathcal{E} = \mathcal{F}$, the $\mathcal{E}$-cominimum additivity, as well as the $\mathcal{E}$-comaximum additivity, is equivalent to the comonotonic additivity which Schmeidler (1986) characterized. Then in general both the $\mathcal{E}$-cominimum and the $\mathcal{E}$-comaximum additivity imply the comonotonic additivity. Therefore, the following can be obtained from Schmeidler's theorem in a straightforward manner.[2]

---

[2]Schmeidler (1986) assumes monotonicity instead of homogeneity of the operator, but the method of his proof can be adopted for this result with little modification.

Theorem 6.1. *If an operator* $I : \mathbb{R}^\Omega \to \mathbb{R}$ *is homogenous and satisfies* $\mathcal{E}$-*cominimum additivity (or* $\mathcal{E}$-*comaximum additivity), then there exists a unique game* $v \in \mathbb{R}^{\mathcal{F}}$ *such that* $I(x) = \int x dv$ *for all* $x \in \mathbb{R}^\Omega$. *Moreover, game* v *is defined by the rule* $v(E) = I(1_E)$.

We say that a game v is $\mathcal{E}$-cominimum additive (resp. $\mathcal{E}$-comaximum additive) if the operator $I(x) := \int x dv$ is $\mathcal{E}$-cominimum additive (resp. $\mathcal{E}$-comaximum additive). Since $\mathcal{E}$-cominimum additivity as well as $\mathcal{E}$-comaximum additivity implies comonotonic additivity, Theorem 6.1 assures that this is a consistent terminology.

Obviously, the properties of $\mathcal{E}$-cominimum additive or $\mathcal{E}$-comaximum additive operators depend on the structure of $\mathcal{E}$.

Definition 6.5. Let $\mathcal{E} \subseteq \mathcal{F}$ be a set of events. An event $T \in \mathcal{F}$ is $\mathcal{E}$-complete provided, for any two distinct points ω_1 and ω_2 in T, there is $E \in \mathcal{E}$ such that $\{\omega_1, \omega_2\} \subseteq E \subseteq T$. The set of all $\mathcal{E}$-complete events is denoted by $\Upsilon(\mathcal{E})$. A set $\mathcal{E}$ is said to be complete if $\mathcal{E} = \Upsilon(\mathcal{E})$.

Note that a singleton set is automatically $\mathcal{E}$-complete, so is any $E \in \mathcal{E}$. For each T, consider the graph where the set of vertices is T and the set of edges consists of the pairs of vertices $\{\omega_1, \omega_2\}$ with $\{\omega_1, \omega_2\} \subseteq E \subseteq T$ for some $E \in \mathcal{E}$. This graph is a complete graph if and only if T is $\mathcal{E}$-complete.

Remark 6.4. For $E, E' \in \mathcal{E}$, $E \cup E'$ is not necessarily $\mathcal{E}$-complete. However, by definition, for any $T \in \Upsilon(\mathcal{E})$ with $|T| > 1$, T coincides with the union of sets in $\mathcal{E}$ which are included in T, thus T is the union of (partition) elements in $\Pi(\mathcal{E})$ which are included in T. In particular, T must contain at least one element of $\Pi(\mathcal{E})$.

It can be shown that for any $\mathcal{E} \subseteq \mathcal{F}$, $\Upsilon(\mathcal{E})$ is complete, i.e., $\Upsilon(\mathcal{E}) = \Upsilon(\Upsilon(\mathcal{E}))$. See Kajii, Kojima, and Ui (2007) for further discussions on this concept, as well as for the proofs of the results shown in the rest of this section.

Example 6.2. In Example 6.1, $S = S_1 \cup S_2 \cup S_3 = \{1, 2, 3, 4, 5, 6\}$ is $\mathcal{E}$-complete, but $S_2 \cup S_3 \cup S_4 = \{3, 4, 5, 6, 7, 8\}$ is not $\mathcal{E}$-complete since there is no $E \in \mathcal{E}$ with $\{3, 7\} \subseteq E \subseteq S_2 \cup S_3 \cup S_4$.

The completeness plays a crucial role in our analysis, as is indicated in the next result:

Lemma 6.2. *Two functions x and y are $\mathcal{E}$-cominimum (resp. $\mathcal{E}$-comaximum) if and only if they are $\Upsilon(\mathcal{E})$-cominimum (resp. $\Upsilon(\mathcal{E})$-comaximum).*

The idea of "cominimum" can be stated in terms of sets by looking at the indicator functions. Say that a pair of sets A and B is an $\mathcal{E}$-decomposition pair if for any $E \in \mathcal{E}$, $E \subseteq A \cup B$ implies that $E \subseteq A$ or $E \subseteq B$ or both. Then the following can be shown:

Lemma 6.3. *Two indicator functions 1_A and 1_B are $\mathcal{E}$-cominimum if and only if the pair of sets A and B constitutes an $\mathcal{E}$-decomposition pair.*

Remark 6.5. From Lemma 6.3 and Remark 6.3, we see that two indicator functions 1_A and 1_B are $\mathcal{E}$-comaximum if and only if for any $E \in \mathcal{E}$, $E \subseteq \Omega \setminus (A \cap B)$ implies that $E \subseteq \Omega \setminus A$ or $E \subseteq \Omega \setminus B$ or both.

Finally, a characterization of cominimum additive and comaximum additive operators is given below (Kajii, Kojima, and Ui, 2007).

Theorem 6.2. *Let $v \in \mathbb{R}^{\mathcal{F}}$ be a game, and let $I(x) = \int x dv$. Write $v = \sum_{T \in \mathcal{F}} \beta_T u_T = \sum_{T \in \mathcal{F}} \eta_T w_T$. Then,*
(1) the following three statements are equivalent: (i) operator I is $\mathcal{E}$-cominimum additive; (ii) $v(A) + v(B) = v(A \cup B) + v(A \cap B)$ for any $\mathcal{E}$-decomposition pair A and B; (iii) $\beta_T = 0$ for any $T \notin \Upsilon(\mathcal{E})$, and
(2) the following three statements are equivalent: (i) operator I is $\mathcal{E}$-comaximum additive; (ii) $v(A^c) + v(B^c) = v(A^c \cup B^c) + v(A^c \cap B^c)$ for any $\mathcal{E}$-decomposition pair A and B; (iii) $\eta_T = 0$ for any $T \notin \Upsilon(\mathcal{E})$.

Note that since $\int 1_E dv = v(E)$, condition (ii) above implies that it suffices to check indicator functions to study $\mathcal{E}$-cominimum and/or $\mathcal{E}$-comaximum additivity.

6.3 Coextrema Additive Operators

In this chapter we study pairs of functions which share *both* a minimizer and a maximizer for events in $\mathcal{E}$, which is fixed throughout.

Definition 6.6. Two functions $x, y \in \mathbb{R}^{\Omega}$ are said to be $\mathcal{E}$-coextrema, provided they are both $\mathcal{E}$-cominimum and $\mathcal{E}$-comaximum; that is, $\arg\min_E x \cap \arg\min_E y \neq \emptyset$ and $\arg\max_E x \cap \arg\max_E y \neq \emptyset$ for all $E \in \mathcal{E}$.

Analogous to the cases of cominimum and comaximum functions, the notion of $\mathcal{E}$-coextrema functions induces the following additivity property of an operator $I : \mathbb{R}^\Omega \to \mathbb{R}$.

Definition 6.7. An operator $I : \mathbb{R}^\Omega \to \mathbb{R}$ is $\mathcal{E}$-coextrema additive provided $I(x+y) = I(x) + I(y)$ whenever x and y are $\mathcal{E}$-coextrema.

The completion $\Upsilon(\mathcal{E})$ plays an important role here again: the following is an immediate consequence of the definition and Lemma 6.2.

Lemma 6.4. *Two functions x and y are $\mathcal{E}$-coextrema if and only if they are $\Upsilon(\mathcal{E})$-coextrema.*

By definition, the $\mathcal{E}$-coextrema additivity implies the comonotonic additivity. So by Theorem 6.1, we obtain the following result.

Lemma 6.5. *If an operator $I : \mathbb{R}^\Omega \to R$ is homogeneous and $\mathcal{E}$-coextrema additive for some $\mathcal{E} \subseteq \mathcal{F}$, then there exists a unique game v such that $I(x) = \int x dv$ for any $x \in \mathbb{R}^\Omega$. Moreover, v is defined by the rule $v(E) = I(1_E)$.*

Thus the following definition is justified:

Definition 6.8. A game v is said to be $\mathcal{E}$-coextrema additive provided $\int (x+y) dv = \int x dv + \int y dv$ whenever x and y are $\mathcal{E}$-coextrema.

Our goal is to establish that a game v is $\mathcal{E}$-coextrema additive if and only if v can be expressed in the form

$$v = \sum_{E \in \Upsilon(\mathcal{E})} \{\lambda_E w_E + \mu_E u_E\}, \tag{6.5}$$

under some condition on $\mathcal{E}$. Note that from (6.4), this is equivalent to say that the original operator I can be written as

$$I(x) = \sum_{E \in \Upsilon(\mathcal{E})} \{\lambda_E \max_E x + \mu_E \min_E x\}. \tag{6.6}$$

Remark 6.6. Note that by definition $u_{\{\omega\}} = w_{\{\omega\}}$, and they are the probability measure δ_ω which assigns probability one to $\{\omega\}$. Since $\Upsilon(\mathcal{E})$ contains all the singleton subsets of Ω, (6.5) has a trivial redundancy for E with $|E| = 1$. Taking this into account, (6.5) can be written as:

$$v = p + \sum_{E \in \Upsilon(\mathcal{E}) \setminus \mathcal{F}_1} \{\lambda_E w_E + \mu_E u_E\}, \tag{6.7}$$

where p is an additive measure given by $p := \sum_{\omega \in \Omega}(\lambda_{\{\omega\}} + \mu_{\{\omega\}})\delta_\omega$. Similarly, (6.6) can be written as

$$I(x) = \int x \, dp + \sum_{E \in \Upsilon(\mathcal{E}) \setminus \mathcal{F}_1} \{\lambda_E \max_E x + \mu_E \min_E x\}. \qquad (6.8)$$

We will also show that these expressions are unique under some conditions.

As we mentioned before, a leading case for our set up is to interpret Ω as the set of states describing uncertainty and function x as a random variable over Ω. Then the class of operators which can be written as in (6.8) with underlying capacity of the form (6.7) has a natural interpretation that the value of x is the sum of its expected value $\int x \, dp$ and a weighted average of the most optimistic outcome and the most pessimistic outcomes on events in $\Upsilon(\mathcal{E})$. That is, $I(x)$ is the expectation biased by optimism and pessimism conditional on various events in $\Upsilon(\mathcal{E})$.

Alternatively, interpret Ω as a collection of individuals (i.e., a society), and $x(\omega)$ as the wealth allocated to individual ω. Then $\int x \, dp$ can be seen as the (weighted) average income of the society, and $\max_E x$ and $\min_E x$ correspond to the wealthiest and the poorest in group E, respectively. In particular, when p is the uniform distribution and $\lambda_E = -1$ and $\mu_E = 1$, then the problem of maximizing (6.8) subject to $\int x \, dp$ being held constant means that that of reducing the sum of wealth differences in various groups in $\Upsilon(\mathcal{E})$.

An interesting special subclass of (6.8) is the class of NEO-additive capacities obtained by Chateaunuf, Eichberger, and Grant (2002): a NEO-additive capacity is a capacity of the form $v = (1 - \lambda - \mu)q + \lambda w_\Omega + \mu u_\Omega$, i.e., $\mathcal{E} = \{\Omega\}$ in (6.8) and $I(1_\Omega) = 1$.[3] More generally, let $\mathcal{E}$ be a partition of Ω, and write $\mathcal{E} = \{E_1, ..., E_K\}$. Then (6.8) is essentially $v = p + \sum \lambda_k w_{E_k} + \mu_k u_{E_k}$, where p is an additive game. Not only this is a generalization of the NEO-additive capacity, but also it is a generalization of the E-capacities of Eichberger and Kelsey (1999), which correspond to the case where $\lambda_k = 0$ for all k.

6.4 Main Characterization Result

One direction of the characterization can be readily established, as is shown below.

---

[3]When $\lambda = 0$, i.e., there is no part for optimism, this type of capacity is also referred to as an ε-contamination. See Kajii, Kojima, and Ui (2007) for more discussions.

Lemma 6.6. Let $v = \sum_{E \in \Upsilon(\mathcal{E})} \{\lambda_E w_E + \mu_E u_E\}$. Then v is $\mathcal{E}$-coextrema additive.

Proof. Let x and y be $\mathcal{E}$-coextrema functions. Then by Lemma 6.4, x and y are $\Upsilon(\mathcal{E})$-coextrema. For every $E \in \Upsilon(\mathcal{E})$, let $\bar{\omega} \in \arg\max_E x \cap \arg\max_E y$ and $\underline{\omega} \in \arg\min_E x \cap \arg\min_E y$. Then, $\max_E(x+y) = (x+y)(\bar{\omega}) = x(\bar{\omega}) + y(\bar{\omega}) = \max_E x + \max_E y$, and $\min_E(x+y) = (x+y)(\underline{\omega}) = x(\underline{\omega}) + y(\underline{\omega}) = \min_E x + \min_E y$.

Using these relations, since the Choquet integral is additive in games (see (6.4)), we have

$$\int (x+y) dv = \int (x+y) d[\sum_{E \in \Upsilon(\mathcal{E})} \{\lambda_E w_E + \mu_E u_E\}],$$

$$= \sum_{E \in \Upsilon(\mathcal{E})} \{\lambda_E \max_E(x+y) + \mu_E \min_E(x+y)\},$$

$$= \sum_{E \in \Upsilon(\mathcal{E})} \{\lambda_E (\max_E x + \max_E y) + \mu_E (\min_E x + \min_E y)\},$$

$$= \sum_{E \in \Upsilon(\mathcal{E})} \{\lambda_E \max_E x + \mu_E \min_E x\}$$

$$+ \sum_{E \in \Upsilon(\mathcal{E})} \{\lambda_E \max_E y + \mu_E \min_E y\},$$

$$= \int x \, dv + \int y \, dv,$$

which completes the proof. □

The other direction is far more complicated. Observe first that since both $\{u_T : T \in \mathcal{F}\}$ and $\{w_T : T \in \mathcal{F}\}$ constitute linear bases, if the set of events $\Upsilon(\mathcal{E})$ contains a sufficient variety of events, not only coextrema additive games but also many other games can be expressed as in (6.5) or (6.6). In other words, for these expressions to be interesting, it is important to establish the uniqueness, and one can easily expect that the set $\Upsilon(\mathcal{E})$ should not contain too many elements for this purpose. On the other hand, each element of $\mathcal{E}$ must be large enough as the following example shows.

Example 6.3. Let $|\Omega| \geq 3$ and $\mathcal{E} = \{\{1, 2, 3\}\}$. Then $\Upsilon(\mathcal{E}) \setminus \mathcal{F}_1 = \mathcal{E}$. Notice that in general when $|E| = 3$, if x and y are coextrema on E, then x and y are automatically comonotonic on E. So any game v of the form $v = \sum_{T \subseteq \{1,2,3\}} \beta_T u_T$ is $\mathcal{E}$-coextrema additive, in particular $u_{\{1,2\}}$ is $\mathcal{E}$-coextrema additive. But it can be shown that $u_{\{1,2\}}$ cannot be written in the form (6.5).

To exclude cases like Example 6.3, we also need to guarantee that each element of $\mathcal{E}$ contains a large enough number of points. The key condition formally stated below roughly says that the elements of $\mathcal{E}$, as well as their intersections, are not too small, i.e., the set $\mathcal{E}$ is "coarse" enough:

Coarseness Condition $|E| \geq 4$ for every $E \in \mathcal{E}$ and $|S| \geq 2$ for every $S \in \Pi(\mathcal{E})$.

The Coarseness Condition is satisfied in Example 6.1, but it is violated in Example 6.3. It will turn out that $\Upsilon(\mathcal{E})$ cannot contain too many elements if the Coarseness Condition is satisfied.

Remark 6.7. Obviously, if $\mathcal{E}$ is coarse, it contains no singleton set. However, as far as the representation result stated below is concerned, singletons are inessential since $\Upsilon(\mathcal{E})$ automatically contains all the singletons anyway. Put it differently, we could state the condition by first excluding singletons from $\mathcal{E}$ and then construct the relevant field and partition.

We are now ready to state the main result of this chapter.

Theorem 6.3. *Let $\mathcal{E}$ be a set of events which satisfies the coarseness condition. Let v be a game. Then the following two conditions are equivalent: (i) v is $\mathcal{E}$-coextrema additive; (ii) there exist an additive game p and two sets of real numbers, $\{\lambda_E : E \in \Upsilon(\mathcal{E}) \backslash \mathcal{F}_1\}$ and $\{\mu_E : E \in \Upsilon(\mathcal{E}) \backslash \mathcal{F}_1\}$, such that*

$$v = p + \sum_{E \in \Upsilon(\mathcal{E}) \backslash \mathcal{F}_1} \{\lambda_E w_E + \mu_E u_E\}. \tag{6.9}$$

Moreover, (6.9) is unique; that is, if $v = p' + \sum_{E \in \Upsilon(\mathcal{E}) \backslash \mathcal{F}_1} \{\lambda'_E w_E + \mu'_E u_E\}$ where p' is additive, then $p = p'$, and $\lambda'_E = \lambda_E$ and $\mu'_E = \mu_E$ hold for every $E \in \Upsilon(\mathcal{E}) \backslash \mathcal{F}_1$.

We shall prove this result in the next section, but we note here that the coarseness condition is indispensable for Theorem 6.3. Recall that in Example 6.3 the coarseness condition is violated and there is a coextrema additive game which cannot be expressed in the form (6.9). The next example is also instructive for this point.[4]

---

[4] Admittedly, this example is extreme and the reader may wonder if the coarseness condition is tight for the characterization result. In fact, we could relax the coarseness condition slightly but the weaker condition is extremely complicated and hardly instructive. Also we could not tell if the weaker condition itself is tight. So we choose not to present it here.

Example 6.4. Let $\Omega = \{1,2,3,4\}$, $\mathcal{E} = \{\{1,2,3\},\{1,2,4\},\{3,4\}\}$. In this case, it is $\Upsilon(\mathcal{E})\backslash\mathcal{F}_1 = \mathcal{E} \cup \{\Omega\}$. But if x and y are $\mathcal{E}$-coextrema, then it is comonotonic on both $\{1,2,3\}$ and $\{1,2,4\}$, and hence it is comonotonic on Ω. So any non-additive measure v is $\mathcal{E}$-coextrema additive.

Let us conclude this section with a couple of applications of Theorem 6.3. The first concerns a characterization of the generalized NEO-additive, E-capacities outlined before. Let $\mathcal{E}$ be a partition of Ω, and write $\mathcal{E} = \{E_1, ..., E_K\}$ as before. It can be readily verified that $\Upsilon(\mathcal{E}) = \mathcal{E} \cup \mathcal{F}_1$. Trivially, $\Pi(\mathcal{E}) = \mathcal{E}$. So if $|E_k| \geq 4$ for every $k = 1,...,K$, by Theorem 6.3, $\mathcal{E}$ satisfies the coarseness condition and then v is $\mathcal{E}$-coextrema additive if and only if v can be written as $v = p + \sum (\lambda_k w_{E_k} + \mu_k u_{E_k})$, where p is an additive game.

The second is a generalization of the variation averse operator proposed in Gilboa (1989). Let $T > 1$ and $M \geq 2$ be integers and set $\Omega = \{(m,t) : m = 1,...,2M, t = 1,...,T\}$. The intended interpretation is that t is the time and at each time t there are m states representing some uncertainty. Let $\mathcal{E}$ be the set of all sets of the following forms: $\{(m,t) : m = 1,...,2M\}$; $\{(m,t) : m = 1,...,M\} \cup \{(m,t+1) : m = 1,...,M\}$; and $\{(m,t) : m = M+1,...,2M\} \cup \{(m,t+1) : m = M+1,...,2M\}$. It can be readily verified that $\Upsilon(\mathcal{E}) = \mathcal{E} \cup \mathcal{F}_1$, and every set in $\Pi(\mathcal{E})$ contains M points. So the coarseness condition is met, and by Theorem 6.3, an $\mathcal{E}$-extrema additive capacity has the form in (6.8). Arguing analogously as in Kajii, Kojima, and Ui (2007), the coefficients for the $\mathcal{E}$-events of the form $\{(m,t) : m = 1,...,2M\}$ represent measurements of optimism and pessimism about the uncertainty, whereas the coefficients for the $\mathcal{E}$-events of the other forms represent measurements of (conditional) degrees of variation loving and variation aversion.

6.5 The Proof

This section is devoted to the proof of Theorem 6.3. Since Lemma 6.5 has already shown that (ii) implies (i), it suffices to establish the other direction. The proof consists of several steps: basically, starting with an $\mathcal{E}$-coextrema game v, we shall first show that a restriction of v is $\mathcal{E}$-comaximum. Then we show that this construction is invariant of the way the restriction is chosen as long as a certain condition is satisfied, which then implies the existence of a well-defined $\mathcal{E}$-comaximum additive game v_1. We then show that the game $v_2 := v - v_1$ is $\mathcal{E}$-cominimum additive. Theorem 6.2 can be applied to v_1 and v_2 to obtain the desired expression.

Let v be an $\mathcal{E}$-coextrema additive game with $v = \sum_{T \in \mathcal{F}} \beta_T u_T$. For any $R \in \mathcal{F}$, let $v_{|R}$ be the game defined by the rule $v_{|R}(E) = v(E \cap R)$ for all $E \in \mathcal{F}$, i.e., $v_{|R} = \sum_{T \subseteq R} \beta_T u_T$. Define $\mathcal{E}_{\cap R} = \{E \cap R \mid E \in \mathcal{E}, E \cap R \neq \emptyset\}$, which is the set of intersections of elements of $\mathcal{E}$ and R, and also define $\mathcal{E}_{\subseteq R} = \{E \mid E \in \mathcal{E}, E \subseteq R\}$, which is the set of elements of $\mathcal{E}$ contained in R. Note that $\mathcal{E}_{\subseteq R} \subseteq \mathcal{E}_{\cap R}$.

To construct the desired $\mathcal{E}$-comaximum additive game v_1, we first observe the following property.

Lemma 6.7. *Let v be $\mathcal{E}$-coextrema additive. Let $R \in \mathcal{F}$ be such that $\mathcal{E}_{\subseteq R} = \emptyset$ and $\mathcal{E}_{\cap R} \neq \emptyset$. Then, $v_{|R}$ is $\mathcal{E}_{\cap R}$-comaximum additive.*

Proof. Let 1_S and 1_T be $\mathcal{E}_{\cap R}$-comaximum. It is enough to show that $v_{|R}(S \cup T) + v_{|R}(S \cap T) = v_{|R}(S) + v_{|R}(T)$, which is rewritten as $v((S \cap R) \cup (T \cap R)) + v((S \cap R) \cap (T \cap R)) = v(S \cap R) + v(T \cap R)$. Therefore, it suffices to show that $1_{S \cap R}$ and $1_{T \cap R}$ are $\mathcal{E}$-coextrema because v is $\mathcal{E}$-coextrema additive.

Fix any $E \in \mathcal{E}$. Since $\mathcal{E}_{\subseteq R} = \emptyset$, either $E \cap R = \emptyset$, or $E \cap R \neq \emptyset$ and $E \setminus R \neq \emptyset$. If $E \cap R = \emptyset$, then $1_{S \cap R}$ and $1_{T \cap R}$ are 0 on E and thus have a common minimizer and maximizer on E. If $E \cap R \neq \emptyset$ and $E \setminus R \neq \emptyset$, then $1_{S \cap R}$ and $1_{T \cap R}$ have a common maximizer in $E \cap R \subseteq E$ since 1_S and 1_T are $\mathcal{E}_{\cap R}$-comaximum, and $1_{S \cap R}$ and $1_{T \cap R}$ have a common minimizer in $E \setminus R \subseteq E$ since $1_{S \cap R}$ and $1_{T \cap R}$ are 0 on R^c. Therefore, $1_{S \cap R}$ and $1_{T \cap R}$ are $\mathcal{E}$-coextrema. $\square$

By this lemma and Theorem 6.2, $v_{|R}$ has a unique expression

$$v_{|R} = \sum_{\omega \in R} \nu^R_{\{\omega\}} w_{\{\omega\}} + \sum_{E' \in \Upsilon(\mathcal{E}_{\cap R}) \setminus \mathcal{F}_1} \nu^R_{E'} w_{E'}. \qquad (6.10)$$

To obtain the desired game v_1 which will constitute a part of the expression (6.9), we want the second part of the right hand side of (6.10) in the following form: $\sum_{E \in \Upsilon(\mathcal{E}) \setminus \mathcal{F}_1} \nu^R_{E \cap R} w_{E \cap R}$. Since each $E' \in \mathcal{E}_{\cap R} \setminus \mathcal{F}_1$ is written as $E' = E \cap R$ for some $E \in \mathcal{E}$, one way to proceed is to associate each E' with the corresponding E. Of course, this procedure is not well defined in general, since there may be many such E for candidates. So our next step is to find a condition on the set R so that this procedure in fact unambiguously works. It turns out that the following property is suitable for this purpose.

Definition 6.9. A set $R \in \mathcal{F}$ is a representation of $\mathcal{E}$ if $\mathcal{E}_{\subseteq R} = \emptyset$, $\kappa(R) = \Omega$, and $|R \cap E| \geq 2$ for all $E \in \mathcal{E}$. Moreover we say that $R \in \mathcal{F}$ is a minimal representation of $\mathcal{E}$ if R is a representation of $E \in \mathcal{E}$ and any proper subset of R is not a representation.

In Example 6.1, the set R is a representation for $\mathcal{E}$. Another example follows below.

Example 6.5. Let $\Omega = \{1,2,3,4,5,6\}$ and set $\mathcal{E} = \{\{1,2,3,4\}, \{3,4,5,6\}.\}$ Then $\Pi(\mathcal{E}) = \{\{1,2\}, \{3,4\}, \{5,6\}\}$. $R = \{3,4\}$ is not a representation, since $\kappa(R) = \{3,4\} \neq \Omega$. $R = \{2,3,4,6\}$ is a representation but not minimal. $R = \{2,4,6\}$ is a minimal representation.

Lemma 6.8. *Assume that $\mathcal{E}$ is coarse. Then if $T \in \mathcal{F}$ satisfies $\mathcal{E}_{\subseteq T} = \emptyset$, there is a representation R such that $T \subseteq R$.*

Proof. Construct R by the following procedure which adds points to T: first set $R = T$ and then for each $S \in \Pi(\mathcal{E})$; if $S \in \mathcal{E}$ and $|T \cap S| \leq 1$, then add a point or two to R from $S \setminus T$ (recall that $|S| \geq 4$ if $S \in \mathcal{E}$ by the coarseness) so that exactly two points from S are contained in R; if $S \in \mathcal{E}$ and $|T \cap S| \geq 2$, do nothing; if $S \notin \mathcal{E}$ and $T \cap S = \emptyset$, then add a point to R (note $S \setminus R \neq \emptyset$ by the coarseness); if $S \notin \mathcal{E}$ and $T \cap S \neq \emptyset$, do nothing. Then by construction, $\kappa(R) = \Omega$, and $|R \cap E| \geq 2$ for all $E \in \mathcal{E}$. Notice also that since $\mathcal{E}_{\subseteq T} = \emptyset$, for any $E \in \mathcal{E}$, there is some point in E which is not added to R, so $\mathcal{E}_{\subseteq R} = \emptyset$ follows. □

Note that if R is a representation of $\mathcal{E}$, $\kappa(R) = \Omega$ holds by definition and so every $S \in \Pi(\mathcal{E})$ must necessarily intersect R. Roughly speaking, a representation is obtained by choosing some representative elements from each S in $\Pi(\mathcal{E})$ when $\mathcal{E}$ is coarse. Formally, we have the following result:

Lemma 6.9. *Assume that $\mathcal{E}$ is coarse. Then there exists a minimal representation, which can be constructed by the following rule: for each $S \in \Pi(\mathcal{E})$, choose two distinct elements from S if $S \in \mathcal{E}$, and one element if $S \notin \mathcal{E}$, and let R be the set of chosen elements. Moreover, every minimal representation can be constructed in this way, and so in particular minimal representations contain exactly the same number of points.*

Proof. The set R constructed as above is well defined since by the coarseness condition every S has at least two elements. We claim that R is a representation. For all $E \in \mathcal{E}$, there is $S \in \Pi(\mathcal{E})$ with $S \subseteq E$. If $S = E$, R contains exactly two points belonging to E. If $S \subsetneq E$, then there is another $S' \neq S$ with $S' \subseteq E$ because E is the union of some elements in $\Pi(\mathcal{E})$. Since R contains one element of S and S', it contains at least two points belonging to E. Therefore, $|R \cap E| \geq 2$ for all $E \in \mathcal{E}$. Also, every $S \in \Pi(\mathcal{E})$ intersects with R and so $\kappa(R) = \Omega$. Finally, notice that $E \subseteq R$ is possible only if $E \in \Pi(\mathcal{E})$. But by the coarseness condition, $|E| \geq 4$ and so this case cannot occur in the construction, thus $\mathcal{E}_{\subseteq R} = \emptyset$.

Next we claim that R is minimal. Let R' be a proper subset of R and pick any $\omega \in R \backslash R'$. Let $S \in \Pi(\mathcal{E})$ be the set where ω is chosen from. If $S \in \mathcal{E}$, then R contains exactly two elements of S by construction. Then $|R' \cap S| = 1$, and so R' is not a representation. If $S \notin \mathcal{E}$, then ω is the only one element from S. Then $S \cap R' = \emptyset$ which implies $\kappa(R') \subseteq \Omega \backslash S$, and so R' is not a representation.

Finally, let R be a minimal representation. Then $S \cap R \neq \emptyset$ for every $S \in \Pi(\mathcal{E})$ so R contains at least one point from each S. If $S \in \Pi(\mathcal{E})$ and $S \in \mathcal{E}$, then $|R \cap S| \geq 2$ so at least two points from such S must be contained in R. Let R' be the set of all these points in the intersections, which is a minimum representation as we have shown above. Since $R' \subseteq R$, we conclude $R' = R$, which completes the proof. □

Example 6.6. In Example 6.1, none of elements in $\Pi(\mathcal{E})$ belongs to $\mathcal{E}$. So to obtain a minimal representation one can choose exactly one point from each $S \in \Pi(\mathcal{E})$. For instance, $R = \{1, 3, 5, 7\}$ is a minimal representation.

When R constitutes a representation of $\mathcal{E}$, we can associate each $E \in \Upsilon(\mathcal{E}_{\cap R}) \backslash \mathcal{F}_1$ to some *unique* element in $\Upsilon(\mathcal{E}) \backslash \mathcal{F}_1$, as is shown in the next result.

Lemma 6.10. *Assume that $\mathcal{E}$ is coarse, and let $R \in \mathcal{F}$ be a representation of $\mathcal{E}$. Then $\kappa(F) \in \Upsilon(\mathcal{E}) \backslash \mathcal{F}_1$ for any $F \in \Upsilon(\mathcal{E}_{\cap R}) \backslash \mathcal{F}_1$. Conversely, if $E \in \Upsilon(\mathcal{E}) \backslash \mathcal{F}_1$, then $E \cap R$ is a unique element of $\Upsilon(\mathcal{E}_{\cap R}) \backslash \mathcal{F}_1$ such that $\kappa(E \cap R) = E$. In short, given representation R, the restriction of κ, denoted by κ_R, constitutes a bijection between $\Upsilon(\mathcal{E}_{\cap R}) \backslash \mathcal{F}_1$ and $\Upsilon(\mathcal{E}) \backslash \mathcal{F}_1$ by the rule $\kappa_R(F) = \kappa(F)$ for all $F \in \Upsilon(\mathcal{E}_{\subseteq R}) \backslash \mathcal{F}_1$, and $\kappa_R^{-1}(E) = E \cap R$ for all $E \in \Upsilon(\mathcal{E}) \backslash \mathcal{F}_1$.*

Proof. Note that $\mathcal{E}$ and $\mathcal{E}_{\cap R}$ contain no singleton since $\mathcal{E}$ is coarse and R is a representation of $\mathcal{E}$. Also note that from the basic property of κ and $\kappa(R) = \Omega$ by the definition of representation, we have for each $E \in \mathcal{E}$, $E \cap R \in \mathcal{E}_{\cap R}$ and $\kappa(E \cap R) = \kappa(E) \cap \kappa(R) = \kappa(E) \cap \Omega = \kappa(E) = E$.

We first show that $\kappa(F) \in \Upsilon(\mathcal{E}) \backslash \mathcal{F}_1$ for all $F \in \Upsilon(\mathcal{E}_{\cap R}) \backslash \mathcal{F}_1$. Fix any $F \in \Upsilon(\mathcal{E}_{\cap R}) \backslash \mathcal{F}_1$. Choose two distinct points $\omega_1, \omega_2 \in \kappa(F)$ arbitrarily, and we shall show that there is an $E \in \mathcal{E}$ such that $\{\omega_1, \omega_2\} \subseteq E \subseteq \kappa(F)$. By the construction of $\kappa(F)$, there are $S_1, S_2 \in \Pi(\mathcal{E})$ (possibly $S_1 = S_2$) such that $\omega_1 \in S_1$, $\omega_2 \in S_2$, and both $S_1 \cap F$ and $S_2 \cap F$ are non-empty. Suppose first that $S_1 \neq S_2$. Then we can select two distinct points $\omega_1' \in S_1 \cap F$ and $\omega_2' \in S_2 \cap F$. Since $F \in \Upsilon(\mathcal{E}_{\cap R}) \backslash \mathcal{F}_1$, there exists $F' \in \mathcal{E}_{\cap R}$ such that $\omega_1', \omega_2' \in F' \subseteq F$ by the definition of completeness. By the definition of $\mathcal{E}_{\cap R}$, there is $E \in \mathcal{E}$ with $F' = E \cap R$. Using the property of κ (see Remark 6.1), and the definition of a representation, $\kappa(F') = E \cap \kappa(R) = E$ and $\kappa(F') \subseteq \kappa(F)$. So we have $\{\omega_1, \omega_2\} \subseteq F' \subseteq \kappa(F') = E \subseteq \kappa(F)$, as we wanted. Suppose then $S_1 = S_2 \, (= \hat{S})$. Recall that $F \in \Upsilon(\mathcal{E}_{\cap R}) \backslash \mathcal{F}_1$ implies that F is the union of some elements in $\mathcal{E}_{\cap R}$. Since $\hat{S} \in \Pi(\mathcal{E})$, this means that there is at least one $E \in \mathcal{E}$ such that $\hat{S} \subseteq E$ and $E \cap R \subseteq F$. Then again by the definition of representation, $E = \kappa(E \cap R) \subseteq \kappa(F)$, and so this E has the desired property.

Next, we show that the restriction κ_R is a map from $\Upsilon(\mathcal{E}_{\cap R}) \backslash \mathcal{F}_1$ onto $\Upsilon(\mathcal{E}) \backslash \mathcal{F}_1$. Fix any $E \in \Upsilon(\mathcal{E}) \backslash \mathcal{F}_1$. Since $E \in \sigma(\mathcal{E})$, $\kappa_R(E \cap R) = \kappa(E) \cap \kappa(R) = \kappa(E) \cap \Omega = \kappa(E) = E$; that is, $E \cap R$ is in the inverse image of κ_R. Thus, it is enough to show that $E \cap R \in \Upsilon(\mathcal{E}_{\cap R}) \backslash \mathcal{F}_1$. By the definition of completeness, there exist $E_1, \ldots, E_K \in \mathcal{E}$ such that $E = \bigcup_{k=1}^K E_k$ and that for any pair of points $\omega, \omega' \in E$, $\omega, \omega' \in E_k$ holds for some k. So in particular, for any distinct points $\omega, \omega' \in E \cap R \subseteq E$, there exists k with $\omega, \omega' \in E_k$ and thus $\omega, \omega' \in E_k \cap R \in \mathcal{E}_{\cap R}$ since $\omega, \omega' \in R$. Therefore, κ_R is onto.

Finally we show that κ_R is one to one, i.e., $\kappa_R(F) = E$ occurs for $F \in \Upsilon(\mathcal{E}_{\cap R}) \backslash \mathcal{F}_1$ only if $F = E \cap R$. Note that $F \in \Upsilon(\mathcal{E}_{\cap R}) \backslash \mathcal{F}_1$ implies that there exist $E_1, \ldots, E_K \in \mathcal{E}$ such that $F = \bigcup_{k=1}^K (E_k \cap R) = (\bigcup_{k=1}^K E_k) \cap R$. Since R is a representation, R must intersect any $\Pi(\mathcal{E})$-component of E_k for all k, and so $\kappa(F) = \kappa((\bigcup_{k=1}^K E_k) \cap R) = \bigcup_{k=1}^K E_k$. So $\kappa_R(F) = E$ implies $\bigcup_{k=1}^K E_k = E$ and so $F = E \cap R$ must hold. This completes the proof. $\square$

By Lemma 6.10, if R be a representation of $\mathcal{E}$, then, by rewriting (6.10), we have

$$v_{|R} = \sum_{\omega \in R} \lambda_{\{\omega\}}^R w_{\{\omega\}} + \sum_{E \in \Upsilon(\mathcal{E}) \backslash \mathcal{F}_1} \lambda_E^R w_{E \cap R} \qquad (6.11)$$

where $\lambda_E^R = \nu_{E \cap R}^R$ for each $E \in \Upsilon(\mathcal{E})$. By construction, the coefficients $\{\lambda_E^R : E \in \Upsilon(\mathcal{E})\}$ are uniquely determined with respect to a representation R except for singletons. It turns out that these do not depend upon the choice of representation R, which we shall demonstrate in the following in a few lemmas. Let $R, R' \in \mathcal{F}$ be representations of $\mathcal{E}$, and so there are corresponding expressions of the form (6.11). We write $R \stackrel{\circ}{=} R'$ if $\lambda^R_{\{\omega\}} = \lambda^{R'}_{\{\omega\}}$ for all $\omega \in R \cap R'$ and $\lambda^R_E = \lambda^{R'}_E$ for all $E \in \Upsilon(\mathcal{E})\backslash\mathcal{F}_1$. Note that the first part holds vacuously if $R \cap R' = \emptyset$.

Lemma 6.11. *Assume that $\mathcal{E}$ is coarse and let v be $\mathcal{E}$-coextrema additive. Let $R, R', R'' \in \mathcal{F}$ be representations of $\mathcal{E}$. Suppose that $R \stackrel{\circ}{=} R'$ and $R' \stackrel{\circ}{=} R''$. Then, $R \stackrel{\circ}{=} R''$ holds if $R \cap R'' \subseteq R'$.*

Proof. By definition, $\lambda^R_{\{\omega\}} = \lambda^{R''}_{\{\omega\}}$ for all $\omega \in R \cap R' \cap R''$ ($= R \cap R''$) and $\lambda^R_E = \lambda^{R''}_E$ for all $E \in \Upsilon(\mathcal{E})\backslash\mathcal{F}_1$. □

Lemma 6.12. *Assume that $\mathcal{E}$ is coarse and let v be $\mathcal{E}$-coextrema additive. Let $R, R' \in \mathcal{F}$ be representations of $\mathcal{E}$. Then $R \stackrel{\circ}{=} R'$ holds if $R \cap R'$ is a representation. In particular, if $R \subseteq R'$, $R \stackrel{\circ}{=} R'$ holds.*

Proof. Set $R^* = R \cap R'$. Note that by construction, for all $T \in \mathcal{F}$, $v_{|R^*}(T) = v_{|R}(T \cap R^*) = v_{|R'}(T \cap R^*)$. Using (6.11) on the other hand, we have

$$v_{|R}(T \cap R^*) = \sum_{\omega \in R} \lambda^R_{\{\omega\}} w_{\{\omega\}}(T \cap R^*) + \sum_{E \in \Upsilon(\mathcal{E})\backslash\mathcal{F}_1} \lambda^R_E w_{E \cap R}(T \cap R^*)$$

$$= \sum_{\omega \in R^*} \lambda^R_{\{\omega\}} w_{\{\omega\}}(T) + \sum_{E \in \Upsilon(\mathcal{E})\backslash\mathcal{F}_1} \lambda^R_E w_{E \cap R^*}(T)$$

and

$$v_{|R'}(T \cap R^*) = \sum_{\omega \in R^*} \lambda^{R'}_{\{\omega\}} w_{\{\omega\}}(T) + \sum_{E \in \Upsilon(\mathcal{E})\backslash\mathcal{F}_1} \lambda^{R'}_E w_{E \cap R^*}(T).$$

Thus, for all $T \in \mathcal{F}$,

$$\sum_{\omega \in R^*} \lambda^R_{\{\omega\}} w_{\{\omega\}}(T) + \sum_{E \in \Upsilon(\mathcal{E})\backslash\mathcal{F}_1} \lambda^R_E w_{E \cap R^*}(T)$$

$$= \sum_{\omega \in R^*} \lambda^{R'}_{\{\omega\}} w_{\{\omega\}}(T) + \sum_{E \in \Upsilon(\mathcal{E})\backslash\mathcal{F}_1} \lambda^{R'}_E w_{E \cap R^*}(T).$$

Since R^* is also a representation by assumption, by Lemma 6.10, $\Upsilon(\mathcal{E}_{|R^*})\backslash\mathcal{F}_1$ and $\Upsilon(\mathcal{E})\backslash\mathcal{F}_1$ are isomorphic. Since $\{w_T : T \in \mathcal{F}\}$ are linearly independent, this means that the games in $\{w_{\{\omega\}}\}_{\omega \in R^*} \cup \{w_{E \cap R^*}\}_{E \in \Upsilon(\mathcal{E})\backslash\mathcal{F}_1}$ are linearly independent. Therefore, the respective coefficients on the both sides of the above equation must coincide each other, which completes the proof. □

Lemma 6.13. *Assume that $\mathcal{E}$ is coarse and let v be $\mathcal{E}$-coextrema additive. Let $R, R' \in \mathcal{F}$ be minimal representations of $\mathcal{E}$. Then $R \stackrel{\circ}{=} R'$.*

Proof. If $R = R'$, then obviously $R \stackrel{\circ}{=} R'$, and so let $R \neq R'$. By Lemma 6.9, $|R| = |R'|$ and so there is $\omega' \in R' \backslash R$. Let $S \in \Pi(\mathcal{E})$ be the unique element with $\omega' \in S$. Recall that a representation intersects every elements of $\Pi(\mathcal{E})$, and hence we can pick an $\omega \in R \cap S$. By construction $\omega \neq \omega'$. Set $R^1 = (R \backslash \{\omega\}) \cup \{\omega'\}$, i.e., R^1 is obtained by substituting ω with ω' both of which belong to S. So R^1 is also a minimal representation by Lemma 6.9.

We shall show that $R \stackrel{\circ}{=} R^1$. For this, consider first $\hat{R} = R \cup \{\omega'\}$. Notice that $\hat{R}$ is a representation; since $R \subseteq \hat{R}$ and R is a representation, it is clear that $\kappa(\hat{R}) = \Omega$, and $\left|\hat{R} \cap E\right| \geq 2$ for all $E \in \mathcal{E}$. Since $\mathcal{E}$ is coarse and R is minimal, for all $E \in \mathcal{E}$, we have $|E \backslash R| \geq 2$ and so $\left|E \backslash \hat{R}\right| \geq 1$. Hence $\mathcal{E}_{\subseteq \hat{R}} = \emptyset$, which proves that $\hat{R}$ is a representation. By construction, both $R \cap \hat{R} = R$ and $R^1 \cap \hat{R} = R^1$ are representations, so by Lemma 6.12, $R \stackrel{\circ}{=} \hat{R}$ and $\hat{R} \stackrel{\circ}{=} R^1$. Note that $R \cap R^1 \subseteq \hat{R}$, which implies that $R \stackrel{\circ}{=} R^1$ by Lemma 6.11.

Recall that both R and R' are finite and they can be obtained by the method described in Lemma 6.9, so repeating the argument above, i.e., replacing one ω in R with another $\omega' \in R' \backslash R$, we can construct a sequence of minimal representations $R^0 (= R)$, R^1, R^2, $\cdots$, $R^k = R'$ such that $R^{m-1} \stackrel{\circ}{=} R^m$ for each $m = 1, .., k$. By definition, $\lambda_E^{R^{m-1}} = \lambda_E^{R^m}$ holds for all $E \in \Upsilon(\mathcal{E}) \backslash \mathcal{F}_1$ for every $m = 1, ..., k$, hence $\lambda_E^R = \lambda_E^{R'}$ holds for all $E \in \Upsilon(\mathcal{E}) \backslash \mathcal{F}_1$. For any $\omega \in R \cap R'$, since such ω is never replaced along the sequence above, we have $\lambda_{\{\omega\}}^{R^{m-1}} = \lambda_{\{\omega\}}^{R^m}$ for every $m = 1, ..., k$, and hence $\lambda_{\{\omega\}}^R = \lambda_{\{\omega\}}^{R'}$. Therefore, we conclude that $R \stackrel{\circ}{=} R'$. □

Lemma 6.14. *Assume that $\mathcal{E}$ is coarse and let v be $\mathcal{E}$-coextrema additive. Let $R, R' \in \mathcal{F}$ be representations of $\mathcal{E}$. Then $R \stackrel{\circ}{=} R'$.*

Proof. Choose any two minimal representations Γ and Γ' such that $\Gamma \subseteq R$, $\Gamma' \subseteq R'$, and $\Gamma \cap \Gamma' \subseteq R \cap R'$. Notice that by Lemma 6.9 such minimal representations always exist and can be constructed as follows: for any $\omega \in R \cap R'$, then select this ω from $S \in \Pi(\mathcal{E})$ which contains ω. Now by Lemma 6.12, $R \stackrel{\circ}{=} \Gamma$ and $\Gamma' \stackrel{\circ}{=} R'$ hold. Also, by Lemma 6.13, $\Gamma \stackrel{\circ}{=} \Gamma'$ holds. These imply that $\lambda_E^R = \lambda_E^{R'}$ for all $E \in \Upsilon(\mathcal{E}) \backslash \mathcal{F}_1$ and that $\lambda_{\{\omega\}}^R = \lambda_{\{\omega\}}^{R'}$ for all $\omega \in \Gamma \cap \Gamma'$. Since the choice of $\Gamma \cap \Gamma' \subseteq R \cap R'$ is arbitrary as is pointed out above, we must have $\lambda_{\{\omega\}}^R = \lambda_{\{\omega\}}^{R'}$ for all $\omega \in R \cap R'$. Therefore, we conclude that $R \stackrel{\circ}{=} R'$. □

Since there is a representation containing any $\omega \in \Omega$, Lemma 6.14 implies that there exists a unique set of constants $\{\lambda_E\}_{E \in \Upsilon(\mathcal{E})}$ such that, for any representation R of $\mathcal{E}$, $v_{|R} = \sum_{\omega \in R} \lambda_{\{\omega\}} w_{\{\omega\}} + \sum_{E \in \Upsilon(\mathcal{E}) \setminus \mathcal{F}_1} \lambda_E w_{E \cap R}$. Using this set, define two games v_1 and v_2 by the following rule:

$$v_1 = \sum_{E \in \Upsilon(\mathcal{E})} \lambda_E w_E \text{ and } v_2 = v - v_1. \quad (6.12)$$

By Theorem 6.2, v_1 is $\mathcal{E}$-comaximum additive. To show that v_2 is $\mathcal{E}$-cominimum additive, we use the following property of v_2.

Lemma 6.15. *Assume that $\mathcal{E}$ is coarse and let v be $\mathcal{E}$-coextrema additive. Then for any $T \in \mathcal{F}$,*

$$v_2(T) = v_2(\bigcup_{E \in \mathcal{E}_{\subseteq T}} E). \quad (6.13)$$

Proof. Case 1: $\mathcal{E}_{\subseteq T} = \emptyset$, i.e., no element in $\mathcal{E}$ is contained in T. Then, $v_2(\bigcup_{E \in \mathcal{E}_{\subseteq T}} E) = v(\emptyset) - v_1(\emptyset) = 0$, so we need to show that $v_2(T) = v(T) - v_1(T) = 0$. Note that there exists a representation R of $\mathcal{E}$ such that $T \subseteq R$ (see Lemma 6.8). Then, $v(T) = v(T \cap R) = v_{|R}(T) = \sum_{E \in \Upsilon(\mathcal{E}), E \cap T \neq \emptyset} \lambda_E = v_1(T)$, as claimed.

Case 2: $\mathcal{E}_{\subseteq T} \neq \emptyset$. Let $E^* = \bigcup_{E \in \mathcal{E}_{\subseteq T}} E$ and $T^* = T \setminus E^*$. We want to show that $v_2(T) = v_2(E^*)$. By construction, $E^* \in \sigma(\mathcal{E})$ is the union of some elements in $\Pi(\mathcal{E})$, choose one point from each of these elements and let A be the set of these points. Note that $\kappa(A) = E^*$, and that $\mathcal{E}_{\subseteq A} = \mathcal{E}_{\subseteq T^* \cup A} = \emptyset$ follows from the coarseness. Thus, Case 1 applies to A and $T^* \cup A$, and we have

$$v_2(A) = v_2(T^* \cup A) = 0. \quad (6.14)$$

Now we claim that 1_{E^*} and $1_{T^* \cup A}$ are $\mathcal{E}$-coextrema. Note first that $E^* \cap (T^* \cup A) = A$ by construction. To see that they are $\mathcal{E}$-comaximum, recall Remark 6.5, and pick $F \in \mathcal{E}$ with $F \subseteq \Omega \setminus A$. Then $F \cap E^* = \emptyset$ must follow, since both F and E^* are in $\sigma(\mathcal{E})$ and so for any $S \in \Pi(\mathcal{E})$ with $S \subseteq F$, $A \cap S \neq \emptyset$ would hold if $S \subseteq E^*$. Then $F \subseteq \Omega \setminus E^*$ as desired. To see that they are $\mathcal{E}$-cominimum as well, notice that if $F \in \mathcal{E}$ and $F \subseteq E^* \cup (T^* \cup A) = T$, then $F \subseteq E^*$ by construction. Thus E^* and $(T^* \cup A)$ are an $\mathcal{E}$-decomposition pair, and so apply Lemma 6.3.

By the coextrema additivity of v, $v(E^* \cup (T^* \cup A)) + v(E^* \cap (T^* \cup A)) = v(E^*) + v(T^* \cup A)$, which can be re-written as

$$v(E^* \cup T^*) + v(A) = v(E^*) + v(T^* \cup A). \quad (6.15)$$

On the other hand, since 1_{E^*} and $1_{T^* \cup A}$ are $\mathcal{E}$-comaximum and v_1 is $\mathcal{E}$-comaximum additive,

$$v_1(E^* \cup T^*) + v_1(A) = v_1(E^*) + v_1(T^* \cup A). \tag{6.16}$$

Subtracting (6.16) from (6.15), and using the definition of v_2, and the fact $T = E^* \cup T^*$, we have

$$v_2(T) + v_2(A) = v_2(E^*) + v_2(T^* \cup A).$$

Applying (6.14) here, we obtain the desired equation. $\square$

Now we are ready to show that v_2 is $\mathcal{E}$-cominimum additive.

Lemma 6.16. *Assume that $\mathcal{E}$ is coarse and let v be $\mathcal{E}$-coextrema additive. Then, v_2 is $\mathcal{E}$-cominimum additive and thus it has a unique expression*

$$v_2 = \sum_{E \in \Upsilon(\mathcal{E})} \mu_E u_E.$$

Proof. Let 1_A and 1_B be $\mathcal{E}$-cominimum, i.e., A and B constitute an $\mathcal{E}$-decomposition pair by Lemma 6.3. We need to show that $v_2(A \cup B) + v_2(A \cap B) = v_2(A) + v_2(B)$.

Note that for each $S \in \Pi(\mathcal{E})$ such that there is an $E \in \mathcal{E}$ with $S \subseteq E \subseteq A \cup B$, if $S \not\subseteq A \cap B$, then either $S \cap (B \backslash A) \neq \emptyset$ or $S \cap (A \backslash B) \neq \emptyset$, but not both; if both hold then A and B would not be an $\mathcal{E}$-decomposition pair.

For each $S \in \Pi(\mathcal{E})$ with $S \not\subseteq A \cap B$, choose a point ω_S from $S \cap (A \backslash B)$ if $S \cap (A \backslash B) \neq \emptyset$, or from $S \cap (B \backslash A)$ if $S \cap (B \backslash A) \neq \emptyset$. Let Ω^* be the set of chosen points. Finally, set $A^* = A \cup (B \backslash \Omega^*)$ and $B^* = B \cup (A \backslash \Omega^*)$. Notice that $A^* \cup B^* = A \cup B$ by construction.

We claim that if $E \in \mathcal{E}$ satisfies $E \subseteq A^*$, then $E \subseteq A$. Indeed, suppose that there is a point $\omega \in E \cap (A^* \backslash A)$. Since $E \in \mathcal{E}$, we can find (a unique) $S \in \Pi(\mathcal{E})$ with $\omega \in S \subseteq E$, and $\omega \in S \cap (B \backslash A)$. By the construction of Ω^*, this means that $S \cap ((B \backslash A) \cap \Omega^*) \neq \emptyset$ so $E \cap ((B \backslash A) \cap \Omega^*) \neq \emptyset$, which is impossible since $E \subseteq A^* = A \cup (B \backslash \Omega^*)$.

Similarly, if $E \in \mathcal{E}$ satisfies $E \subseteq B^*$, then $E \subseteq B$. To sum up, the set of $\mathcal{E}$-elements contained in A^*, B^*, $A^* \cup B^*$ and $A^* \cap B^*$ coincide with those of A, B, $A \cup B$ and $A \cap B$, respectively. Therefore, by Lemma 6.15, we are done if $v_2(A^* \cup B^*) + v_2(A^* \cap B^*) = v_2(A^*) + v_2(B^*)$. For this, it suffices to show that 1_{A^*} and 1_{B^*} are $\mathcal{E}$-coextrema. Indeed, since v is $\mathcal{E}$-coextrema additive, we have $v(A^* \cup B^*) + v(A^* \cap B^*) = v(A^*) + v(B^*)$, and since v_1 is $\mathcal{E}$-comaximum additive, we have $v_1(A^* \cup B^*) + v_1(A^* \cap B^*) = v_1(A^*) + v_1(B^*)$. Since $v_2 = v - v_1$, the desired equation is established from these two equations.

To see 1_{A^*} and 1_{B^*} are $\mathcal{E}$-cominimum, notice that A and B constitutes a decomposition pair by assumption, and so do A^* and B^*; if $E \subseteq A^* \cup B^*$ with $E \in \mathcal{E}$, then $E \subseteq A \cup B$, which implies $E \subseteq A$ or $E \subseteq B$ and hence $E \subseteq A^*$ or $E \subseteq B^*$ as we have shown above. Thus 1_{A^*} and 1_{B^*} are $\mathcal{E}$-cominimum by Lemma 6.3.

It remains to show that 1_{A^*} and 1_{B^*} are $\mathcal{E}$-comaximum. Pick any $E \in \mathcal{E}$ with $E \subseteq \Omega \backslash (A^* \cap B^*)$. We need to show that $E \subseteq \Omega \backslash A^*$ or $E \subseteq \Omega \backslash B^*$ or both (see Remark 6.5). Suppose $E \cap (A^* \cup B^*) \neq \emptyset$ or else the implication holds trivially, and so it suffices to show that $E \cap (A^* \backslash B^*) = \emptyset$ or $E \cap (B^* \backslash A^*) = \emptyset$. If neither of these holds, then pick $\omega_A \in E \cap (A^* \backslash B^*)$ and $\omega_B \in E \cap (B^* \backslash A^*)$. Note that $A^* \backslash B^* = A \backslash B^*$ and $B^* \backslash A^* = B \backslash A^*$ holds, and thus ω_A and ω_B must belong to Ω^* by the construction of A^* and B^*. Since $E \in \mathcal{E}$, there must be $S_A \in \Pi(\mathcal{E})$ and $S_B \in \Pi(\mathcal{E})$ and $E_A \in \mathcal{E}$ and $E_B \in \mathcal{E}$ such that $\omega_A \in S_A \subseteq E_A \cap E \subseteq A \cup B$ and $\omega_B \in S_B \subseteq E_B \cap E \subseteq A \cup B$. But then, by the coarseness, both $S_A \cap B^*$ and $S_B \cap A^*$ are non-empty, which implies $E \cap (A^* \cap B^*) \neq \emptyset$, a contradiction. This completes the proof. $\square$

Since v_1 is $\mathcal{E}$-comaximum additive and v_2 is $\mathcal{E}$-cominimum additive, we have the desired expression $v = v_1 + v_2 = \sum_{E \in \Upsilon(\mathcal{E})} \lambda_E w_E + \sum_{E \in \Upsilon(\mathcal{E})} \mu_E u_E = p + \sum_{E \in \Upsilon(\mathcal{E}) \backslash \mathcal{F}_1} (\lambda_E w_E + \mu_E u_E)$ where $p = \sum_{\omega \in \Omega} p_{\{\omega\}} u_{\{\omega\}}$ and $p_{\{\omega\}} = \lambda_{\{\omega\}} + \mu_{\{\omega\}}$. It remains to show that this is a unique representation.

Lemma 6.17. *Assume that $\mathcal{E}$ is coarse and let v be $\mathcal{E}$-coextrema additive. Then, the expression $v = \sum_{\omega \in \Omega} p_{\{\omega\}} u_{\{\omega\}} + \sum_{E \in \Upsilon(\mathcal{E}) \backslash \mathcal{F}_1} (\lambda_E w_E + \mu_E u_E)$ is unique; that is, if $v = \sum_{\omega \in \Omega} p'_{\{\omega\}} u_{\{\omega\}} + \sum_{E \in \Upsilon(\mathcal{E}) \backslash \mathcal{F}_1} (\lambda'_E w_E + \mu'_E u_E)$ then $p_{\{\omega\}} = p'_{\{\omega\}}$ for all $\omega \in \Omega$, $\lambda_E = \lambda'_E$, and $\mu_E = \mu'_E$ for all $E \in \Upsilon(\mathcal{E}) \backslash \mathcal{F}_1$.*

Proof. Let $R \in \mathcal{F}$ be a representation of $\mathcal{E}$. Then,

$$v|_R = \sum_{\omega \in R} p_{\{\omega\}} u_{\{\omega\}} + \sum_{E \in \Upsilon(\mathcal{E}) \backslash \mathcal{F}_1} \lambda_E w_{E \cap R}$$

$$= \sum_{\omega \in R} p'_{\{\omega\}} u_{\{\omega\}} + \sum_{E \in \Upsilon(\mathcal{E}) \backslash \mathcal{F}_1} \lambda'_E w_{E \cap R}.$$

By Lemma 6.10, $\Upsilon(\mathcal{E}_{\cap R}) \backslash \mathcal{F}_1$ and $\Upsilon(\mathcal{E}) \backslash \mathcal{F}_1$ are isomorphic and thus for any $E, E' \in \Upsilon(\mathcal{E}) \backslash \mathcal{F}_1$, $E \cap R = E' \cap R$ if and only if $E = E'$. In particular, $\{w_{\{\omega\}}\}_{\omega \in R} \cup \{w_{E \cap R}\}_{E \in \Upsilon(\mathcal{E}) \backslash \mathcal{F}_1}$ are linearly independent. Therefore, $p_{\{\omega\}} = p'_{\{\omega\}}$ for all $\omega \in R$ and $\lambda_E = \lambda'_E$ for all $E \in \Upsilon(\mathcal{E}) \backslash \mathcal{F}_1$. Since the choice of R was arbitrary, $p_{\{\omega\}} = p'_{\{\omega\}}$ for all $\omega \in \Omega$. The linear independence also guarantees that the expression $v - \sum_{\omega \in \Omega} p_{\{\omega\}} w_{\{\omega\}} - \sum_{E \in \Upsilon(\mathcal{E}) \setminus \mathcal{F}_1} \lambda_E w_E = \sum_{E \in \Upsilon(\mathcal{E}) \setminus \mathcal{F}_1} \mu_E u_E$ must also be unique. □

The proof of Theorem 6.3 is now complete.

Acknowledgements. Kajii acknowledges financial support by MEXT, Grant-in-Aid for the 21st Century COE Program. Ui acknowledges financial support by MEXT, Grant-in-Aid for Scientific Research.

Bibliography

Chateaunuf, A., Eichberger, J., and Grant, S. (2002). Choice under uncertainty with the best and worst in mind: neo-additive capacities, *Journal of Economic Theory* **137**, pp. 538-567.

Eichberger, J., and Kelsey, D. (1999). E-capacities and the Ellsberg paradox, *Theory and Decision* **46**, pp. 107–140.

Eichberger, J., Kelsey, D., and Schipper, B. (2006). *Ambiguity and social interaction*, available at SSRN: http://ssrn.com/abstract=464242 or DOI: 10.2139/ssrn.464242.

Ghirardato, P., Maccheroni, F., and Marinacci, M. (2004). Differentiating ambiguity and ambiguity attitude, *Journal of Economic Theory* **118**, pp. 133-173.

Gilboa, I. (1989). Expectation and variation in multi-period decisions, *Econometrica* **57**, pp. 1153–1169.

Gilboa, I., and Schmeidler, D. (1994). Additive representation of non-additive measures and the Choquet integral, *Annals of Operations Research* **52**, pp. 43–65.

Kajii, A., Kojima, H., and Ui, T. (2007). Cominimum additive operators, *Journal of Mathematical Economics* **43**, pp. 218-230.

Schmeidler, D. (1986). Integral representation without additivity, *Proceedings of the American Mathematical Society* **97**, pp. 255–261.

Schmeidler, D. (1989). Subjective probability and expected utility without additivity, *Econometrica* **57**, pp. 571–587.

Shapley, L. S. (1953). *A value for n-person games*, in: H. Kuhn and A. Tucker, eds., *Contributions to the Theory of Games II* (Princeton Univ. Press, NJ) pp. 307–317.

Chapter 7

Models without Main Players

T. S. Arthanari
Department of Information Systems & Operations Management
University of Auckland
12, Grafton Road, Auckland, New Zealand
e-mail: t.arthanari@auckland.ac.nz

Abstract

In this chapter we are looking critically at our capitalist market system, or at least at the models we might use to predict market behaviour and make market decisions, and wondering what happened to the main player — the consumer — in the system. Indeed, the consumer, in most models is marginalized or dehumanized if considered at all. A revisit to both the logic and the ideals of capitalism of Adam Smith and others is undertaken. Creatively reengineering the market system is the task addressed. This requires an understanding of what constitutes a creative process.

Two kinds of applications of game theory were outlined in [Aumann (2008)]. In one we get insight into an interactive situation and, in the other, game theory tells us what to do. Aumann calls the second kind of applications, game engineering. Organisational systems gurus, Russell Ackoff and Sheldon Rovin explain how to outsmart bureaucracies in their celebrated work 'Beating the System' [Ackoff and Rovin (2005)]. What is common between game engineering and beating the system is creativity. We apply this understanding to games played in the market place. This leads us to consider thought experiments, social dialogues, and models that stipulate the rightful place for the consumers as main players.

Key Words: Game Theory, Game Engineering, Beating the System, Systems Thinking, Bait-ad, Marketing models, creative capitalism

7.1 Introduction

Is there, perhaps, a flaw or two in our economic modelling? When we look at the "market" do we see a rational decision-making consumer and reasonable choices? Or has the "invisible" hand become a manipulating hand that moulds the consumer's thinking? Is the consumer a consumption addict? Is the market serving needs or creating wants? These are not necessarily rhetorical questions. We are looking critically at our capitalist market system, or at least at the models we might use to predict market behaviour and make market decisions, and wondering what happened to the main player — the consumer — in the system. Something has gone wrong. Some might give emphasis to brainwashing, some might talk about addiction (in the broad sense of consumerism), but whatever the cause, the result is that those of our economic models that ostensibly are based upon the notion of the rational decision-making consumer result in systems that don't serve that consumer well, because she is no longer rational (if she ever was). Indeed, the consumer, in most models is marginalized or dehumanized if considered at all.

If, as these questions suggest, we have moved far away from the logical ideal of capitalism of Adam Smith and others, do we need to revisit both the logic and the ideals? Is capitalism of today a distortion or merely appropriate evolution?

No less a modern day capitalist than Bill Gates, in his Harvard commencement speech, has urged the need for reengineering of the market system as it exists today. He has coined the phrase 'creative capitalism'. " We need to rewrite the rules of the capitalist system to solve inequity in the world. [Grossman (2007)]." Where do we start?

Often, in today's world, problems created by our economic decisions are left to be addressed or solved as social, political or environmental issues. So it pays to have a closer look at the way we formulate our economic problems and the solution concepts we advocate to resolve those problems. In this chapter, we suggest, after considering the distortions that are now so prevalent in the marketplace, beginning with the prototypes — with the models and the mind-maps that we use to predict outcomes and to aid our decision-making.

This chapter discusses some key drivers of economic growth and related theories and models to appreciate the capitalist system, especially the market, as it exists today. We offer a discussion of many of the indicators of a flawed system, alluded to by Gates. Reengineering capitalism also requires the understanding of how to be creative. We consider examples from 'game engineering' by Nobel Laureate Robert Aumann and 'beating the system' by systems gurus, Russell Ackoff and Sheldon Rovin, to draw a pattern for creative problem solving. Finally we apply this understanding of creativity to the capitalist market systems.

7.2 The Key Drivers of Economic Growth

The key drivers of world economic growth are population, employment, consumption, and environment. Among them consumption plays the major role as other activities are enabling consumption, and consumption accelerates the growth of other economic and social activities. This in turn affects the environment. "The global market and the purchasing power of an increasingly wealthy and urban population are driving the homogenization of lifestyles and popular culture. The late 20th century 'consumer society' can be characterized by a growing emphasis on the individual, a search for wider opportunities and experiences, a desire for comfort and autonomy, and personal material accumulation. The advent of international advertising, electronic communications and wide access to the mass media have fed a worldwide public appetite for new and more products, and for travel. Rising affluence has fueled the 'Western' model of consumption, and its emulation all over the world."[Clarke (2000)] The marketing efforts of large corporations engaged in production and distribution of goods and services are directed towards maximising their revenue through increasing consumption. Through competition the market could ensure improvement in quality and reduction in price, resulting in increased consumption. We consider this key driver for further discussion and note how consumption is modeled in the literature and in practice.

While discussing the economic system proposed by Adam Smith, Heilbroner reminds us, "Do not forget that the great beneficiary of the system is the consumer-not the producer. First time in the philosophy of everyday life, the consumer is king." (page 69, [Heilbroner (1999)])

7.3 Elimination of Main Player in Language and in Action

In the marketplace there are many players. Producers, distributors, advertisers, communicators, logistics providers, insurers, brokers, lawyers, bankers, financiers and regulators. There is competition between them.

There is cooperation among them. They form supply chains by coalition and agreement. There is competition between supply chains. Principles of game theory are applied to advance the interest of these players. They are all interested in the consumers and their behaviour. The psychology of buying intentions is studied to develop products, promotions, packagings, exposures and displays, and to convert these intentions to actual buying. How do consumers think, feel, reason, and select between different brands, and products? How are consumers influenced by culture, family, signs, and media? How do consumers behave while shopping or making other marketing decisions? How does a consumer's limited knowledge or interpersonal ability influence decisions and marketing outcomes? [Piana (2001)]

But somehow in all this the real answers being sought have more to do with how to manipulate or brainwash than with seeking the rational decision-making consumer who is prepared to act in his best interests. It becomes the producer, distributor, and other interests that are being maximized. And the otherwise main player — the consumer — is made to serve those other interests, not her.

The language used in the market and the literature to describe the consumers will throw some light on this elimination of real people in our mind maps. The models refer to market share. Like occupying a territory, the models mention capturing markets. Like milk production from a cow, the models speak about consumer spending and revenue generation. And in television advertising the metaphor alludes to rendering one blind or capturing eyeballs. Similarly how many pairs of ears are plugged may matter in the audio products' market. The shopper in a supermarket is like a sleep walker(see [Underhill (1999)] for other observations about shopper's behaviour).

And so much of our actions are to render the consumers ineffective as rational decision-makers able to act in their own best interests . Bait-and-switch advertisement 'offers retail merchandise at an alluringly low price to entice buyers to visit a store; also called bait-ad. In actuality, the customer may find it difficult, if not impossible, to purchase the advertised merchandise because it is the retailer's intention (1) to get the buyer to the store and then (2) to switch him or her to something more costly' [Downes (2006)]. This practice is considered deceptive advertising. Legal action by the Federal Trade Commission on retailers practicing bait-ad is possible. But [Gerstner and Hess (1990)] argued that bait-ad benefits customers and so the legislation against bait-ad should be repealed. However, this was challenged by [Wilkie et. al. (1998a)] and the discussion finally concluded bait-ad is not in the interest of the consumer [Hess and Gerstner (1998)], [Wilkie et. al. (1998b)].

Another kind of bait is to invite the consumer to gamble. These advertisements provide the consumer an opportunity to win a fabulous prize or holiday if they enter a competition, thereby enticing the customer. And through the images of the holiday or the expensive car, the seller is able to switch the consumer's attention from the product to the prize. Irrespective of the chances of winning the prize, the consumer enters the competition.

Retailers, in markets where impulse goods are sold, use loss leader pricing (pricing some product low or below cost) and rain check policy (a promise or commitment by a seller to a buyer that an item currently out of stock can be purchased at a later date for today's sale price.). Loss leader pricing draws customers to the store, where they also buy highly profitable impulse goods. Rain checks are introduced to enhance the effect of the loss leader. By running out of the leader item and offering rain checks, sellers bring the customers to their store a second time. Hess and Gerstner conclude that 'an entrepreneur who introduces rain checks to a market with loss leader pricing initially can earn higher profit. However, should other sellers follow suit, they all would prefer a law that prohibits use of rain checks.

Catching them young to form lifelong influence is attempted through vigourous exposure to young minds, especially regarding food brands. 'At six months of age, babies are forming mental images of corporate logos and mascots. A person's "brand loyalty" may begin as early as age two' (as quoted in [Consumers International (2004)]). Marketing science and practices demonstrate many other ways in which the consumer is hypnotised or misled or tricked or cajoled or enticed or moulded into a habit of consuming; here only a glimpse of that is provided. Total US advertising expenditures in 2006 increased 4.1 percent to $149.6 billion as compared to 2005. During fourth quarter of 2006, the combined load of Brand Appearances and paid advertisement messages through TV, in prime time and other shows, were respectively 39 percent and 55 percent of total content time [TNS media intelligence report (2007)]. A kind of 'addiction without awareness' is what is attempted when advertisers speak of catering to the emotions of the consumer.

Using credit/debit cards in place of cash for payment has made it difficult to kick the consumerism habit. If the consumer runs out of resources to feed his insatiable desire to consume, no problem, there are financial institutions waiting to lend them money or a seller offering interest free loans and repayment holidays. Again the consumer is uninformed about the actual repayment burden from such offers. The law is unable to come to the help of the consumer who is ignorant of the legalese, and can not see the small print that protects the seller. Last but not least, one of the predictions by the futurists on the digital revolution in the twenty-first century, is (page 429,[Burstein and Kline (1995)]): "'... a second 'killer application' in the future is just allowing people to convert an emotion called "I want" —which is usually incoherent, irrational, impulsive, and incomplete — into something that hops through an interface into a digital environment where something can be done with it. ... If you chase that through, it turns today's economics — certainly retail-level economics — on its head." The role of emotions in consumer response to advertisements is a well-studied topic [Du Plessis (2005)]. Advertising firms are already launching new constructs that emotionally connect to consumers. CEO of Saatchi and Saatchi, [Roberts (2007)], in his talk, entitled, 'Loyalty beyond reason in Milan', claims, "Power has at last returned to the people, back where it belongs." It may sound similar to what we heard from Adam Smith, making consumer the king.

But listen to the language! 'Lovemark' is the newest beyond-the-brands mantra. It is about the emotional connectivity with the people, that is in the high love and high respect quadrant; loyalty beyond reason. Roberts uses emotional words like inspiration, dreams (several times), love, mystery, sensuality, intimacy, empathy, and passion. Roberts gives the example of iPod and the iPod halo which have created $70 billion in shareholder value in just three years. Thus, capturing consumers through a web of emotions, images, and digital electronics has already been achieved as predicted by the futurists more than a decade ago.

7.4 Revelation— Models without Main Players

On the other hand the economic models of the consumer portray a rational agent. Modeling rational agents in neoclassical economics and game theory has been studied thoroughly by [Giocoli (2003)]. [Mirowski (2002)] has critically exposed the surrounding history of rationality in economic modelling. The argument put forward by Giocoli is that the evolution of modern decision theory exhibits the profound transformation that reduced the notion of rationality to a formal property of consistency. Also this has transformed a discipline (like economics) dealing with real economic processes to one investigating issues of logical consistency between mathematical relationships. While Giocoli is pointing out how human psychology was overlooked twice in this evolution of modeling rational agents, the psychology of consumer behaviour is never sidetracked in marketing practice. Leaving aside such debates on rationality, we adopt Aumann's simple explanation of rationality and proceed. "What do I mean by 'rationality'? It is this: A person's behaviour is rational if it is in his best interests, given his information"[Aumann (2005)].

One can see from the reality of the consumer markets, the action by the consumer is without informed or conscious decision-making, while the action by the market is premeditated and guided at times by academic findings and experiments on consumers. The consumer is made to behave not in the consumer's best interest. The consumer is provided with distorted information, so his decision is in the interest of the market system, which has many other players, but the consumer is not one of them.

The revelation : Consumers are not the players—They are the Game [1]
What needs to be done?

7.4.1 Need for the Paradigm Shift

In the 7th edition of The Worldly Philosophers published in 1999, Heilbroner addresses, in the last chapter entitled 'The end of the worldly philosophy', his views on Economics as a science. Since "there was no such process in hunting and gathering or command systems: the provision of the very stuff of life by competitive buying and selling is an arrangement that has no parallel in any other social order," capitalism distinguishes itself from other social systems that preceded it. The thesis that depicts economics as a science like physics is faulty as "human behaviour cannot be understood without the concept of volition–the unpredictable capacity to change our minds up to the very last moment." "Hence a careless use of the word "behaviour" can easily conflate two utterly different things, one of them the quintessential element of conscious existence, the other having nothing whatsoever to do with it." Heilbroner objects to the notion that economists can have 'scientific objectivity' as 'social life of humankind is by its very nature political.' This resonates with what Wiener wrote about science half a century ago in The Human Use of Human Beings: 'Science is a way of life which can only flourish when men are free to have faith. A

---

[1] Wild animals, birds, or fish hunted for food or sport. The flesh of these animals, eaten as food.

faith which we follow upon orders imposed from outside is no faith, and a community which puts its dependence upon such a pseudo-faith is ultimately bound to ruin itself because of the paralysis which the lack of a healthily growing science imposes upon it'[Wiener (1954)]. If economics is not to be a science of society, what is to be its ultimate social usefulness? Heilbroner's answer for this question is: " it's purpose is to help us better understand the capitalist setting in which we will most likely have to shape our collective destiny for the foreseeable future."

While discussing guided evolution of society [Banathy (2000)] observes, "When it comes to the design of social and societal systems of all kinds, it is the users, the people in the system, who are the experts. Nobody has the right to design social systems for someone else. It is unethical to do so. Design cannot be legislated, it should not be bought from the expert, and it should not be copied from the design of others. If the privilege of and responsibility for design is "given away," others will take charge of designing our lives and our systems. They will shape our future."

Though this sounds reasonable, unfortunately the consumer is not in a position to take control of her decisions, leave alone redesigning the market systems. Consumption and provocations to consume as seen in the marketplace are clearly not orchestrated by the consumers. As it is impossible to educate an addict to become responsible for her action, before being de-addicted, a phase of 'habit-breaking' is required before the consumer can be made to make decisions in her interest. Fortunately, this is possible as no human is alone. Which leads us to the discussion of being human. [Maturana and Varela (1980)] define life using the concept of autopoiesis, which means self-organising. An autopoietic system, or 'autopoietic machine,' is formally defined by Maturana and Varela (op. cit., p. 79). In this sense, a single cell, or a human is autopoietic. The structural coupling of a living organism to it's environment is an important concept put forth by them. 'Indeed you cannot consider an organism outside the circumstances which made it possible' [Maturana (2001)]. Inside a living organism we see its internal dynamics, its phisiology, that conserves the living. This way we are having a dual existence as a living system and an organism. Being human means, in addition, using 'languaging' as a process to create systems which use humans as components, helping mutual self-organisation (see [Fell et. al. (1994)], [Boden (2000)]).

A community, for instance, consists of coordinating humans in their physical relations through communications made possible by language which itself is a system being similarly created by humans to coordinate such coordinations [Kravchenko (2007)]. Society, government, market, business organisation and university are such autopoietic systems created by humans to survive in this planet and ensure a conscious existence. Market economic activities arise from the biological, psychological and existential needs of human beings and the autopoietic needs of the systems they have created. These needs are intricately mixed in everyday living.

'To be human is to live humanly. We are human only through our relationship with other humans. We find ourselves in structural congruence in our relational space. Our structure changes as our experience changes'. 'Human existence takes place in the relational space of conversation. This means that, even though from a biological perspective we are Homo sapiens, our way of living —that is to say, our human condition —takes place in our form of relating to each other and the world we bring forth in our daily living through conversation' [Maturana and Varela (1998)]. We can summarise this understanding by:

a. the consumer, as a living organism, is structurally coupled with her environment, the physical space;
b. the consumer, as a human, in addition coordinates through language with other humans, who are similarly structurally coupled with their environment;
c. the market like any other system which uses humans as components is self-organising — has a *life* of its own, though not in the same domain as a human;
d. but the human life or survival of the human race is not the concern of the market, though paradoxically it can not survive without the human components.

So who can emancipate the consumer, the human component in the self organising market web? Not the oblivious advertisers whose only concern is how to find newer ways to keep the consumption going, in the name of loyalty, love, sustainability or any other alternative construct. Not the businesses, being autopoietic systems themselves, which like to survive at any cost. Not the governments, as succinctly said by [Roberts (2007)],'Government is largely a disappointing institution, today less trusted than business, too slow, too compromised, too contained, too distant from everyday sentiment'. It is the consumer, herself. It appears, the game is one against many.

Perhaps we who challenge and criticize and we who develop models can be an instrument for rehabilitation. Taking the lead from gurus in the field might be helpful

7.5 On the Ubiquitous Applications of Game Theory

Since the publication of the celebrated book, The Theory of Games and Economic Behavior [von Neumann and Morgenstern (1947)], the influence of game theoretic concepts can be seen in many areas of military, business and politics. Throughout the history of neoclassical economics, the concept of equilibrium has played a central role, as pointed out by [Aumann (1987)]: 'Nash equilibrium is the embodiment of the idea that economic agents are rational; that they simultaneously act to maximize their utility. If there is any idea that can be considered the driving force of economic theory, that is it.' However game theory or resolving conflicts through equilibrium analysis is not without its share of criticisms from different angles: behavioural aspects [Camerer (2003)],[Simon (1997)],[Kahneman et. al. (1982)], robotising the decision process [Mirowski (2002)], replacing economics of the market place by mathematical formalism devoid of psychology or pragmatism [Giocoli (2003)], problems in measurement and representation of utility [Barzilai (2005)],[Barzilai (2006)], and improper arithmetic operations with the value of a game [Barzilai (2007)].

Notwithstanding these criticisms, game theory is applied not only in military and defence but also in the policies of governments, in strategic, managerial and tactical decisions made by corporations, in negotiations between groups, in auctions, in market design, in marketing and advertising, and even in prenuptial agreements or marriages. In addition game theory has provided explanations in evolutionary biology, neuroscience, sociology and philosophy. As noticed by behavioural game theorists, it pays to examine actual behaviour of human decision-makers and compare it with that predicted by the theory, either to change the rules of the game or to postulate other factors that moderate the prediction.

Aumann in his Nobel Prize acceptance lecture at the Royal Swedish Academy, entitled, "War and Peace", outlines the development of non cooperative repetitive games and how repetition enables cooperation. In essence instead of judging war as good or bad he goes into studying the process that keeps war alive [Aumann (2005)]. Through applications of this kind, we get insight into conflicting situations.

7.6 Game Engineering

[Aumann (2008)] describes another kind of application, in which game theory tells us what to do, in his inaugural address at the International Symposium on Mathematical Programming for Decision Making, at Delhi, 2007. Aumann says if you want a different result, that is if you want to change behaviour, you have to play a different game. Aumann explains that existing game theory can provide insight into what the system is doing. He offers in addition what he calls 'game engineering', which tells us what to do.

The first game engineering example given by Aumann is the second price auction by William Vickrey. Second price is defined as an auction in which the bidder who submitted the highest bid is awarded the object being sold and pays a price equal to the second highest amount bid. Similarly, in a procurement auction, the winner is the bidder who submits the lowest bid, and is paid an amount equal to the next lowest submitted bid [Shor (2005)]. The charm of second price auctions, as pointed out by William Vickrey, is that bidding one's true value is a dominant strategy or honesty is the best policy. Aumann gives this as an example for game engineering, as the alternative closed bid auction, in which the highest bidder gets the object and pays the price she quoted, motivates a bidder to quote lower than what one is willing to pay for the object.

Aumann gives the final offer arbitration as the second example for game engineering. Here again the parties involved in a conflict are forced to behave in an honest fashion in providing their offer as the arbitrator is not going for a compromise. The other two examples of game engineering from [Aumann (2008)] relate to traffic. "Everybody was warned by the TV and the radio to stay in. So Bob Aumann goes in. That's game theory, that's game engineering" says Aumann, retelling what he did on February 26, 1993, when the world trade center was attacked by terrorists. The last example, the tale of two cities, points out how the solution to reduce travel time between city A and city B (from 8 hours), by providing a shortcut, resulted in increasing the travel time for all (to $8\frac{1}{2}$ hours). All the four examples can be summarised as in Table 7.1.

Table 7.1 Comparison of Game Engineering Examples

Situation	Common Sense	Game Engineering
Auction	Highest price	Second highest price
Arbitration	Compromise	Final offer
Traffic	Expect jam so avoid travel	Since most avoid, there will be no jam, travel.
Road Design	Built short cut to save time	If all use the short cut expect long travel time

7.7 Beating the System According to Ackoff and Rovin

Russell Ackoff, operations researcher and systems guru and Sheldon Rovin, a leader in healthcare management, exemplify the importance of defeating bureaucracies in today's society in [Ackoff and Rovin (2005)]. Their "system beaters" (called by the generic biblical name, David) display fundamental creative behaviours. Beating the system is structured around a brief introduction on the authors' theory of system-beating, presentation of the episodes, the creative behaviours exhibited in these stories, and possible applications. Not all cases cited demand exceptional creativity though.

7.7.1 *Beating The System - Examples*

The first story from [Ackoff and Rovin (2005)], page 85: David had developed plans for building a new house in a rural area near a large city. He sent them to four contractors for bids. One was lower than the rest. A contractor, whom David knew personally and thought was retired but wasn't, approached David and said he was tired of retirement and offered to match the lowest bid. David reflected on the fact that almost everyone he knew who had built a house wound up paying more than the contracted amount. He then decided to offer his friend the job if he would accept the following conditions: The contractor would tell David how much profit he had figured in his estimate of the cost of building the house. David would pay him that amount before he began the work. Then David would receive and pay all the bills. The contractor would pay every dollar spent over his estimated cost, and David and the contractor would divide equally every dollar below the estimate. The contractor accepted these conditions. David's and the contractor's objectives were now the same. The house was built for several thousand dollars less than estimated and each received half the savings. Ackoff and Rovin distill the moral of the story as: The objective of a contract should always be cooperation and not conflict.

The second story from (ibid. page 61), to illustrate that it can be dangerous to assume that the objectives of those who serve us are compatible with our objectives, is about the CEO who asked a professor, "Why is it that just about every time I ask our corporate lawyers if I can do something, they say no?". The professor replied that he deserved such an answer. The CEO was taken aback and asked, "Why?" "You shouldn't ask that question." replied the professor. "A major responsibility of your lawyers is to keep you out of jail. So they say no because any change in what you're doing has some possibility of putting you there. There're just doing their job." The CEO then asked what he should ask his lawyers. The professor replied, "Ask nothing. Tell them what you are going to do and instruct them to keep you out of jail."

The stories in [Ackoff and Rovin (2005)] cover different systems including airlines, bureaucracy, business, education, employment, health care, hospitality, and military.

7.8 Nature of Creativity

Creativity is defined differently in arts and sciences. A common definition from Webster's goes as follows: Creativity is marked by the ability or power to create, to bring into existence, to invest with a new form, to produce through imaginative skill, to make or bring into existence something new. But this definition does not enable one to be creative, though it can help recognise when creativity is present.

While explaining the nature of creativity Ackoff and Rovin bring out the importance of examining the assumptions made by us and the system about the system and us. The set of identified assumptions or beliefs thus entertained are: assumptions about us, by us; assumptions about the system, by us; assumptions about us, by the system; assumptions about the system, by the system. The constraints imposed by these assumptions restrict our behaviour. Taking Aumann together with Ackoff we are led to an algorithm for creativity — IDEA.

The creative process starts when one of the underlying assumptions is denied. Then the consequences of that denial are explored. New options are exposed in this way. The creative solution process ends in implementation of a new option and advancing and reaping the benefits. These steps can be presented as the IDEA process:

- Identify an assumption
- Deny it
- Explore consequences
- Advance (accrue benefits)

Comparing the six examples cited for game engineering and beating the system, we see the IDEA process in operation in all these examples. However the examples differ in how the consequences are explored.

With this understanding of creativity, we approach the task at hand, reengineering capitalist economic systems. (A quick review of the history of economics [2], as it has evolved in the west, can be found in The Worldly Philosophers, by [Heilbroner (1999)].) Bernard Mandeville, several centuries ago, commented, "To make the Society Happy ..., it is requisite that great numbers should be Ignorant as well as Poor."(op. cit., page 37) As noted by Bill Gates great numbers are poor still. Adam Smith's suggestion: Don't try to do good. Let good emerge as the byproduct of selfishness. As Heilbroner notes, this suggestion has not worked in the capitalist system of a market economy. Though the system in its globalised version has attracted socialist regimes like China and Russia and democratic socialistic India, among other nations. "... Practical men, who believe themselves to be quite exempt from any intellectual influences, are usually the slaves of some defunct economist. Madmen in authority, who hear voices in the air, are distilling their frenzy from some academic scribbler of a few years back. I am sure the power of vested interests is vastly exaggerated compared with the gradual encroachment of ideas." (John Maynard Keynes as quoted in the introduction of [Heilbroner (1999)].)

In this sense, Bill Gates's call to the academics/business graduates, to address the real problem of the capitalist system, is relevant.

Fortunately, even the environment of a consumer who is not in her senses includes other humans who uphold their own autonomy, who can converse with her about regaining her autonomy and dignity. To achieve this creatively, we could resort to the IDEA process outlined in this section.

I: The market assumes the consumer is an entity to be manipulated into slavery or addiction to consumption.
D: Uphold the autonomy and dignity of the consumer as a decision-maker. Consumer has the right to her information. Consumer has the freedom to

---

[2] Not surprisingly, game theory has found application in history of economics as well. [Grew (1994)]

act in her interest. Her life, time, and well being are the different aspects of her interest. This ensures the coordination with other humans in her environment and the structural coupling with the environment.

E: The consequences could be explored using different methods, like: [1] Thought experiments, [2] Dialogues in the community, [3] Simulation of intelligent agents and interpretation of strategies through human participants, [4] Experimental economics [5] certain adaptation of n-person cooperative games and so on.

A: Apply the solutions obtained from the above exploration to the real market systems, and theory. Change the behaviour of the market and the consumer and reap the benefits, individually and collectively.

Though we should not hurry into the discussion of the solution(s) here, the nature of solution lies in creating adjustments to the market system that do not rob or disrespect the autonomy of the consumer.

A dilemma: If the consumer goes after her autonomy, is she not guilty of not saving the good earth that is ailing from global warming or deserting the market system that has provided her with the material welfare?

We could answer this through a thought experiment from [Cohen (2005)]: "Zenon recounts how he was driving along a remote road in the Yorkshire Dales on his motorbike on a wild, stormy night, when he passed by a desolate bus stop, and saw three people huddling in it, waiting in vain for the bus. (He actually saw it pass by ahead earlier.) Zenon stops, and asks who would like a lift. It turned out they all have a good claim for his help. The first to speak is an old lady who looks as if she is about to die and says she needs to get to the hospital urgently. The second is a healthy looking middle-aged man who it turns out is an old friend. He once saved Zenon's life, and now needs to get to town or he will lose his job. The last is a beautiful woman whom Zenon immediately and intuitively realizes is the perfect partner of his dreams–and what's more it is obvious that the woman thinks the same thing. Yet he can only carry one passenger on his motorbike." Zenon needs to make a choice between saving a life, showing gratitude to his friend, and acting in his own interest. Whatever be his choice he is bound to regret. That is the dilemma. Or could we apply the IDEA process to creatively resolve the problem? I: Zenon rides the bike. D: No, his friend could. E: The old lady can be taken by his friend to the hospital and then attend his job. A: Without any ethical dilemma, Zenon could wait with the perfect partner of his dreams on the desolate road, planning for their future. Cohen reports that this experiment was part of a job application, and a smart candidate who came up with this creative answer, got the job.

By applying the analogy: Zenon – the consumer, the old lady – environment, the old friend – market system, and the perfect partner – the consumer's autonomy, thus we have an answer for our market dilemma too.

7.9 Conclusions

Two kinds of applications of game theory were outlined in [Aumann (2008)]. In one we get insight into an interactive situation and, in the other, game theory tells us what to do. Aumann calls the second kind of applications, game engineering. Organisational systems gurus, Russell Ackoff and Sheldon Rovin explain how to outsmart bureaucracies in their celebrated work 'Beating the System'[Ackoff and Rovin (2005)]. What is common between game engineering and beating the system is creativity. Creativity lies in questioning an assumption either made by the system or by us about us or the system. We apply this understanding to some games played in the market place.

Consumption is a key driver of the economy and assumptions are made about consumers in the models and the market place. We find consumers are not the players in these games. In fact they are the game, meaning hunted animals. This chapter identifies this very important gap in economic/marketing modelling. This leads us to consider thought experiments, social dialogues, and models that stipulate the rightful place for the consumers as main players. Simulation of intelligent agents is also possible to evolve desirable solution concepts. The future research is directed towards building models involving consumers also as players in specific marketing situations and using both theory and simulation to gain experience and insight that could suggest which strategies are in the best interest of consumers.

Acknowledgements. The author appreciates the referees' suggestions gratefully. Without the untiring efforts from Bill English the revision of the chapter incorporating the referees' suggestions would have been impossible. The author thanks his wife Jaya for her patience and support during the preparation and revision of this chapter.

Bibliography

Ackoff, R. L. and Rovin, S. (2005). *Beating the System– Using Creativity to Outsmart Bureaucracies*, (Berrett-Koehler Publishers, Inc., San Francisco).

Aumann, R. J. (1987). *What is Game Theory Trying to Accomplish?*, in Arrow, K. J. and Honkapohja, S. (eds), Frontiers of Economics, Basil Blackwell, Oxford, pp. 28–76.

Aumann, R. J. (2005). *War and Peace*, edited version of the lecture delivered at the Royal Swedish academy of Sciences in Stockholm, http://www.pnas.org/cgi/content/full/103/46/17075.

Aumamm, R. J. (2008). *Game Engineering*, in S. K. Neogy, R. B. Bapat, A. K. Das and Parthasarathy, T. (eds), Mathematical Programming and Game Theory for Decision Making, Statistical Science and Interdisciplinary Research, vol. 1, World Scientific, New Jersey.

Banathy, B. H. (2000). *Guided Evolution of Society*, pp. 288–291, (Springer).

Barzilai, J. (2005). Measurement and Preference Function Modelling, *International Transactions in Operational Research* **12**, pp. 173–183.

Barzilai, J. (2006). *On the Mathematical Modelling of Measurement*, http://arxiv.org/, math.GM/0609555, pp. 1–4.

Barzilai, J. (2007). *Game Theory Foundational Errors - Part I*, Technical Report, Dept. of Industrial Engineering, Dalhousie University, pp. 1-2.

Boden, M. A. (2000). Autopoiesis and Life, *Cognitive Science Quarterly*, **1**, pp. 117–145.

Burstein, D. and Kline, D. (1995). *Road Warriors– Dreams and Nightmares along the Information Highway*, (Dutton, New York).

Camerer, C. F. (2003). *Behavioral Game Theory*, (Princeton University Press, Princeton).

Clarke, R. (edited) (2000), *Global Environmental Outlook, GEO-2000*, Earthscan Publications, UK, on behalf of UNEP, Nairobi, http://www.grida.no/geo2000/english

Cohen, M. (2005). *Wittgenstein's Beetle, and Other Classic Thought Experiments*, (Blackwell, Oxford).

Consumers International, (2004). *The Junk Food Generation*, A multi-country survey of the influence of television advertisements on children, http://www.economicswebinstitute.org/essays/junkfoodasia.pdf

Downes, J. and Goodman, J. E. (2006). *Dictionary Of Financial & Investment Terms*, 7th Edition John and Jordan Elliot, Barron's Business Dictionaries.

Du Plessis, E. (2005). *The Advertised Mind : groundbreaking insights into how our brains respond to advertising*, (Sterling, London).

Fell, L., Russell, D. and Stewart, A. (1994). *Seized by Agreement Swamped by Understanding*, University of Western Sydney, Hawkesbury printing, Hawkesbury.

Giocoli, N. (2003). *Modeling Rational Agents : from interwar economics to early modern game theory*, Edward Elgar, Cheltenham, UK ; Northampton, MA.

Gerstner, E. and Hess, J. D. (1990). Can bait-switch benefit customers?, *Marketing Science* **9**, pp. 114–124.

Grew, A. (1994). *Economic History And Game Theory*, Chapter 52, in ,*Handbook of Game Theory with Economic Applications*, Volume 2, Aumann, R. J. and Hart, S. eds., North-Holland, Amsterdam.

Grossman, L. (2007). *Bill Gates Goes Back to School*, Time, June, 18, pp.40-42.

Heilbroner, R. L. (1999). *The Worldly Philosophers : the lives, times, and ideas of the great economic thinkers*, 7th ed., (Simon & Schuster, New York).

Hess, J. D. and Gerstner, E. (1987). Loss leader pricing and raincheck policy, *Marketing Science* **6**, pp. 358–374.

Hess, J. D. and Gerstner, E. (1998). Yes, "bait and switch" really benefits consumers, *Marketing Science* **17**, pp. 283–289.

Kahneman D., Slovic, P. and Tversky, A. (1982). *Judgment Under Uncertainty: heuristics and biases*, (Cambridge University Press, NY).

Kravchenko, A. V. (2007). Essential properties of language, or, why language is not a code, *Language Sciences* **29**, pp. 656–671.

Maturana, H. R. (2001). *Our Genome Does Not Determine Us*, 'Remaining Human' Forum Webpage, http://www.asc-cybernetics.org/2001/RH-Maturana.htm.

Maturana, H. R. and Varela, F. (1980). *Autopoiesis and Cognition: the Realization of the Living*, (Reidel, Dordrecht).

Maturana, H. R. and Varela, F. (1998). *The Tree of Knowledge– The Biology of Human Understanding*, Revised edition, (Shambhala, Boston).

Mirowski, P. (2002). *Machine Dreams : economics becomes a cyborg science*, (Cambridge University Press, Cambridge; New York).

Piana, V. (2001). The Economics Web Institute, *Consumption*, http://www.economicswebinstitute.org/glossary/cons.htm

Roberts, K. (2007). *Loyalty beyond Reason in Milan*, http://www.saatchikevin.com/Loyalty_Beyond_Reason_in_Milan

Shor, M. (2005). *Second Price Auction*, Dictionary of Game Theory Terms, Game Theory.net, http://www.gametheory.net /dictionary

Simon, H. A. (1997). *Models of Bounded Rationality*, Vol. III, (MIT Press, Massachusetts).

Underhill, P. (1999). *Why we buy : the science of shopping*, (Simon & Schuster, New York).

von Neumann, J. and Morgenstern, O. (1947). *The Theory of Games and Economic Behavior*, 2nd edition, (Princeton University Press, Princeton).

TNS Media Intelligence Report. (2007). *U.S. Advertising Expenditures in 2006*, http://www.tns-mi.com/news/03132007.htm

Wiener, N. (1954). *The Human Use of Human Beings*, Doubleday Anchor.

Wilkie, W. L., Mela, C. F., and Gundlach, G. T. (1998). Does "bait and switch" really benefit consumers?, *Marketing Science* **17**, pp. 273–282.

Wilkie, W. L., Mela, C. F., and Gundlach, G. T. (1998). Does "bait and switch" really benefit consumers? Advancing the discussion, *Marketing Science* **17**, pp. 290–293.

Chapter 8

Dynamic Optimal Advertising Expenditure Strategies for Two Successive Generations of High Technology Products

Udayan Chanda
Department of Operational Research, University of Delhi,
Delhi 110 007, India
e-mail: uchanda@or.du.ac.in

A. K. Bardhan
Faculty of Management Studies, University of Delhi,
Delhi 110 007, India
e-mail: amit-bardhan@fms.edu

Abstract

Advertising plays a very important role in drawing consumer attention to new-products and in encouraging early adoptions. The problem of drawing optimal advertising plans for innovations have therefore has received a lot of attention from marketing managers and researchers. But there is need for further research on problems pertaining to new-products that are part of technological generations. This chapter deals with the determination of optimal advertising expenditure for two-generation consumer durables. The model considers intergenerational diffusion effect and also introduces a framework for modeling innovation diffusion for two competing generations.

Key Words: Optimal advertising expenditure, high Technology Products, strategies, dynamic optimal control models

8.1 Introduction

The primary objective of advertising is to give visibility to a product/service and to draw consumer attention towards it. It is one of the key marketing tools managers use to influence their potential customers into acquiring products and also to create new markets. For innovative new products, advertisements play a pivotal role in generating initial sales. They can influence innovators who take independent purchase decisions. Research, as reported by Rogers (1962), have shown that advertising works best on innovators and early adopters. With a launch of a new product a promotional plan is also set in motion. Sometimes promotion starts even before the product hits the market. Though these activities are done under a budget, the actual advertising expenditures are dependent on many factors including the behavior of the sales growth curve. The promotional plan usually includes identification of the target markets, what message company wants to convey, how it will be conveyed and how much money is budgeted for this effort. Successful advertising depends very much on knowing the preferred methods and styles of communications of the target markets.

Thus, drawing up an optimal advertising plan over time for a new product is an important field of study in marketing. There are several approaches propagated by researchers as reviewed by Sethi (1977) and Feichtinger *et al.* (1994). The problem becomes more complex for new products that are part of the technological generations. As a result they have received less attention. Generally technological innovations come in generations and often an organization introduces new innovations without withdrawing the earlier ones from the market. In such situation a consumer can map the new generational product on the merits of earlier generations. But for the firm this further complicates the marketing decision making.

The purpose of this chapter is to suggest optimal advertising expenditure policies for a technology product having two generations in the market. This is done by evaluating intergenerational diffusion effect within a finite time-horizon. Optimal control theory has been used in deriving advertising expenditure strategies. Similar problems in other marketing contexts have been studied (for example, Horsky and Simone 1983). In this chapter the demand function of Horsky and Simone model has been extended for two generational products. The structure of this article is as follows. The following sections of this chapter are the theoretical background, analysis of advertising policies for a general diffusion model, model development, analysis of advertising policies for the proposed model. Finally, the article concludes with discussions on possible extensions, limitations and future avenues of research.

8.2 Literature Review

In the last three decades, dynamic optimal control models involving advertising has become one of the most focused area of research in marketing. Since the pioneering work of Bass (1969) in the field of diffusion theory, many authors have used the Bass model to analyse the impact of marketing variables on new product diffusion (Robinson and Lakhani 1975, Bass and Bultez 1982, Kalish 1983, 1985, Kamakura and Balasubramanium 1988, Horsky 1990, Sethi and Bass 2003). These models incorporate the pricing effects on diffusion. Other models incorporate the effects of advertising on diffusion (Horsky and Simon 1983, Simon and Sebastian 1987, Dockner and Jorgensen 1988). Horsky and Simon (1983) modified the Bass model by incorporating the effects of advertising in Bass' innovation coefficient. Thompson and Teng (1984) in their dynamic oligopoly price-advertising model has incorporated learning curve production cost. Dockner and Jorgensen discussed the optimal advertising policies for diffusion model under monopolistic product situation. Bass et al. (1994) include both price and advertising in their Generalized Bass Model (GBM). The authors found that when the rate of changes in pricing and advertising were kept constant the proposed GBM no longer gives better fit than the Bass model (1969).

While the empirical marketing research pays increasingly growing attention to the single-generation market aspects of the advertising strategy (Sethi 1977, Feichtinger et al. 1994), in comparison the modeling support for the studies of advertising policies distributed over multiple generation products have been lacking. In the present chapter, a model has been proposed which deals with both the dynamic as well as the substitution effects of advertising for multiple-generation products. As discussed, many models have already been developed to study the optimal policies for different marketing variables. But there is very few which take into consideration the collision of market expectations of new products (i.e. generational product diffusion) into the policy making decisions (Bayus 1992). Norton and Bass model (1987) was one of the earliest models to capture the intergeneration diffusion process for multiple generation products, which is built upon the Bass model. During the model development Norton and Bass assumed that the coefficients of innovation and imitation remain unchanged from generation to generation of technology. Islam and Meade (1997) relax the assumption of constant-coefficient. Soon after many researchers tried to extend the Bass model to capture the substitution and adoption patterns of technology generations (Mahajan and Muller 1996, Danaher et al. 2001, Chanda and Bardhan 2008). But none of the models have discussed the normative behavior of the marketing variables.

This chapter deals with the determination of optimal advertising expenditure policies for two generational products under different marketing conditions. Here advertising expenditure is a control variables whose optimal values are to be determined.

8.3 Model Formulation

A new technology aims to provide better utility to the customer than the technology it replaces. The model proposed in this chapter and later used in the optimization problem, assumes that for a sequence of innovations within a product category each new generations is an improvement over the immediate predecessor. But the consumer is free to choose from any generation depending on his perception about utility, price etc. As information about new products spread in the market, consumers who might be committed towards an earlier generation can change their minds and instead adopt the latest one (Mahajan and Muller 1996, Lilly and Walters 1997, Danaher et al. 2001, Chanda and Bardhan 2008). In such situation a marketing-variable, which has positive effect on one generation can have conflicting influence on the other generation (Bardhan and Chanda 2008). For multi-generation products, in addition to leapfroggers there are repeat purchasers also, who upgrade their older technologies with the latest one, lifting up the potential market size. For two-generation situation, diffusion of product generations (due to leapfrogging and upgradation) can be depicted by the following figure.

In this section a sales model is chosen for two generations of a technology product. The analysis is restricted to a monopolistic firm, which controls its advertising expenditure in finite planning horizon. The model is capable of capturing both diffusion and substitution processes for each generation. It is further assumed that there are two groups of buyers: (a) new purchasers, who are first-time adopters of the product generation (b) repeat buyers, who had also adopted the first generation product and now look to upgrade. The potential adopters of both categories can be influenced by advertising and also by word-of-mouth effect.

Based on the assumption stated above the time rate of sales growth ($\dot{x}(t)$) can be expressed as a dynamic function of advertising expenditure

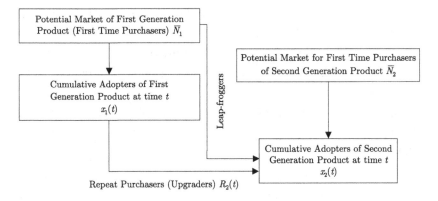

Figure: Components of the Sales Model

$(A(t))$ and cumulative sales volume $(x(t))$:

$$g(t) = \dot{x}(t) = \dot{x}(A(t), x(t)) \tag{8.1}$$

For, single generation product, we can assume that the demand growth function is twice differentiable and increases with advertising expenditure and can be written as $\frac{\delta g}{\delta A} > 0$. Notations used in this section are as follows: Let 't' denote the time such that $0 \leq t \leq T$. The length of the planning period, T, is fixed.

$x_i\ (i = 1, 2) = x_i(t)$: Cumulative sales volume (demand) of the ith generation product by time 't'.

$R_2(t)$: Cumulative repeat purchasers (upgraders) of second-generation product.

$\bar{N}_i$: Potential market size of the ith generation product for the first time purchasers.

$A_i\ (i = 1, 2) = A_i(t)$: The firm's rate of advertising expenditure on ith generation product by time 't'.

$\dot{x}_i (i = 1, 2) = \dot{x}_i(t)$: Sales (demand) rate of the ith generation product at time 't'.

$c_i(t)\ (i = 1, 2)$: Marginal cost of production of the ith generation product by time 't', which is a function of x_i over time.

$p_i\ (i = 1, 2)$: Constant price of the ith generation product over the planning period.

r: The discount rate.

$\lambda = \lambda(t)$ and $\mu = \mu(t)$: The current-value costate variable of the first and second generation products respectively.

t_1: The introduction time of first generation product.

$t_2 = t - \tau, \tau$ is the introduction time of second generation product.

[$\dot{x}(t)$ denotes the derivative of $x(t)$ with respect to 't' in this chapter]

The simultaneous equation diffusion models for two generation case can be given as (the modelling frame-work for multi-generation diffusion is discussed in detail in section 8.4):

$$\dot{x}_1 = \dot{x}_1(t) = g_1(x_1(t_1), x_2(t_2), A_1(t_1), A_2(t_2));$$
$$x_1(0) = x_{10} \geq 0 = \text{a fixed value} \qquad (8.2)$$
$$\dot{x}_2 = \dot{x}_2(t) = g_2(x_1(t_2), x_2(t_2), A_1(t_2), A_2(t_2));$$
$$x_2(0) = x_{20} \geq 0 = \text{a fixed value} \qquad (8.3)$$

Functions (8.2) and (8.3) are twice differentiable, and

$$g_{1A_1}, g_{2A_2} > 0;\ g_{1A_1A_1}, g_{2A_2A_2} < 0 \text{ and}$$
$$g_{1A_1A_2} = g_{1A_2A_1};\ g_{2A_1A_2} = g_{2A_2A_1} \qquad (8.4)$$

It is assumed that $(p_i - c_i(t)) > 0;\ \forall\ t$ over the planning period. Ideally, the optimization problem should include prices and other marketing-mix variables (Kalish 1983, 1985; Thompson and Teng 1996), but as the objective of this chapter is determining the influence of advertising expenditures affect on two successive generations of durables, the cost pricing effects are deliberately avoided to keep the terms simple.

The instantaneous profit stream to the firm is given by

$$P(x, A) = (p_1 - c_1(x_1))g_1(x_1, x_2, A_1, A_2) + (p_2 - c_2(x_2))$$
$$\times g_2(x_1, x_2, A_1, A_2) - A_1 - A_2 \qquad (8.5)$$

The functional for total discounted profit is

$$J(x, A) = \int_0^T e^{-rt}[(p_1 - c_1)g_1 + (p_2 - c_2)g_2 - A_1 - A_2]dt \qquad (8.6)$$

The objective is to determine an optimal policy for Advertisement expenditure over the fixed time interval from $t = 0$ to $t = T$ such that (8.6) is maximized, subject to the constraints (8.2) and (8.3).

The control variable $A_i(i = 1, 2) = A_i(t)$ are twice differentiable in 't' and satisfy $A_i(t) \geq 0\ \forall\ t$.

To solve the problem, Pontryagin Maximum principle is applied. Dropping the time notation, the current value Hamiltonian can be written as:

$$H = H(x_1, x_2, A_1, A_2, \lambda, \mu)$$
$$= (p_1 - c_1 + \lambda)g_1 + (p_2 - c_2 + \mu)g_2 - A_1 - A_2 \quad (8.7)$$

where, $\lambda = \lambda(t)$ and $\mu = \mu(t)$ represent the shadow prices of x_1 and x_2 respectively that satisfy the differential equations:

$$\dot{\lambda} = r\lambda + c_{1x_1}g_1 - g_{1x_1}(p_1 - c_1 + \lambda) - g_{2x_1}(p_2 - c_2 + \mu) \quad (8.8)$$
$$\dot{\mu} = r\mu - g_{1x_2}(p_1 - c_1 + \lambda) + c_{2x_2}g_2 - g_{2x_2}(p_2 - c_2 + \mu) \quad (8.9)$$

with the transversality condition at $t = T$

$$\lambda(T) = 0 \text{ and } \mu(T) = 0 \quad (8.10)$$

The first order necessary conditions for deriving optimal values of A_i ($i = 1, 2$) are $H_{A_i} = 0$. i.e.,

$$H_{A_1} = (p_1 - c_1 + \lambda)g_{1A_1} + (p_2 - c_2 + \mu)g_{2A_1} - 1 = 0 \quad (8.11)$$
$$H_{A_2} = (p_1 - c_1 + \lambda)g_{1A_2} + (p_2 - c_2 + \mu)g_{2A_2} - 1 = 0 \quad (8.12)$$

The second order conditions are:

$$H_{A_1 A_1} < 0 \Rightarrow (p_1 - c_1 + \lambda)g_{1A_1 A_1} + (p_2 - c_2 + \mu)g_{2A_1 A_1} < 0 \quad (8.13)$$
$$H_{A_2 A_2} < 0 \Rightarrow (p_1 - c_1 + \lambda)g_{1A_2 A_2} + (p_2 - c_2 + \mu)g_{2A_2 A_2} < 0 \quad (8.14)$$

and,

$$\Delta = \begin{vmatrix} H_{A_1 A_1} & H_{A_1 A_2} \\ H_{A_2 A_1} & H_{A_2 A_2} \end{vmatrix} > 0 \quad (8.15)$$

also,

$$H_{A_1 A_2} = (p_1 - c_1 + \lambda)g_{1A_1 A_2} + (p_2 - c_2 + \mu)g_{2A_1 A_2} \quad (8.16)$$
$$H_{A_2 A_1} = (p_1 - c_1 + \lambda)g_{1A_2 A_1} + (p_2 - c_2 + \mu)g_{2A_2 A_1} \quad (8.17)$$

Now integrating (8.8) and (8.9) with condition (8.10), the future benefits of producing one more unit from each generation are obtained as follows:

$$\lambda = \int_t^T e^{-rs}[g_{1x_1}(p_1 - c_1 + \lambda) + g_{2x_1}(p_2 - c_2 + \mu) - c_{1x_1}g_1]ds \quad (8.18)$$

and,

$$\mu = \int_t^T e^{-rs}[g_{1x_2}(p_1 - c_1 + \lambda) + g_{2x_2}(p_2 - c_2 + \mu) - c_{2x_2}g_2]ds \quad (8.19)$$

Defining $\beta_i (i = 1, 2)$ as elasticity of demand with respect to the advertisement expenditure of the ith generation product $\left[= \left(\dfrac{A_i}{g_i} \right) g_{iA_i} \right]$, the cross-elasticities as $\beta_{ij}(i, j = 1, 2) = \left[-\left(\dfrac{A_j}{g_i} \right) g_{iA_j} \right]$ and substituting (8.11) and (8.12), the expression for optimal advertising expenditure are obtained as follows:

$$A_1^* = (p_1 - c_1 + \lambda)g_1 \left[\beta_1 - \left(\frac{(p_2 - c_2 + \mu)g_2}{(p_1 - c_1 + \lambda)g_1} \right) \beta_{21} \right] \quad (8.20)$$

$$A_2^* = (p_2 - c_2 + \mu)g_2 \left[\beta_2 - \left(\frac{(p_1 - c_1 + \lambda)g_1}{(p_2 - c_2 + \mu)g_2} \right) \beta_{12} \right] \quad (8.21)$$

It is observed (8.20) and (8.21), advertising elasticities directly influence the optimal advertising spending.

8.3.1 A Subclass of the General Formulation

The sales model used in the optimization problem defined above is more general and includes the social communications. For a more concentrated study on effect of advertising on sales, a scaled down model is proposed below that excludes the impact of word-of-mouth on sales. In this case equation (8.18) and (8.19) become

$$\lambda = -\int_t^T e^{-rs} c_{1x_1} g_1 ds \quad (8.22)$$

and,

$$\mu = -\int_t^T e^{-rs} c_{2x_2} g_2 ds \quad (8.23)$$

These equations represent the future benefit of selling one more unit each from the two competing generations. Taking time-derivatives of (8.11) and (8.12), we have (for details, refer to Appendix A)

$$\dot{A}_1 = -\frac{r\alpha}{\Delta}[H_{A_2 A_2} - \beta H_{A_1 A_2}], \quad (8.24)$$

$$\dot{A}_2 = \frac{r\alpha}{\Delta}[H_{A_2 A_1} - \beta H_{A_1 A_1}] \quad (8.25)$$

where, $\alpha = \lambda g_{1A_1} + \mu g_{2A_1}$; $\beta = \dfrac{\lambda g_{1A_2} + \mu g_{2A_2}}{\lambda g_{1A_1} + \mu g_{2A_1}}$ and $\Delta > 0$

α can be interpreted as an indicator towards the usefulness of promoting the first generation product when both the generations are in the market. β is the ratio that compares the future benefits due to second generation advertising to that of first generation.

In (8.24) and (8.25) it is observed that for the case $r = 0$ optimal policies reduce to that of a static problem and the advertising expenditure of both the generations remain constant for the entire planning horizon. Now, if the planning horizon is long enough and also the discount rate is positive, then depending on the signs of α and β we have the following general advertising expenditure strategies. The objective of this chapter is to find the time dependent optimal policies. A more detailed derivation of which are provided in the appendix.

Theorem 8.1. *If $\beta < 0$; $g_1 = g_1(A_1, A_2)$ and $g_2 = g_2(A_1, A_2)$*

	Conditions	Results	
		$a > 0$	$a < 0$
Case 1	$H_{A_2 A_2} - \beta H_{A_1 A_2} > 0$	$\dot{A}_1 < 0$ & $\dot{A}_2 < 0$	$\dot{A}_1 > 0$ & $\dot{A}_2 > 0$
Case 2	$H_{A_2 A_1} - \beta H_{A_1 A_1} > 0$	$\dot{A}_1 > 0$ & $\dot{A}_2 > 0$	$\dot{A}_1 < 0$ & $\dot{A}_2 < 0$
Case 3	$H_{A_2 A_1} = H_{A_1 A_2} < 0$	$\dot{A}_1 > 0$ & $\dot{A}_2 < 0$	$\dot{A}_1 < 0$ & $\dot{A}_2 > 0$

Proof. See Appendix A. □

Case 1 suggest that when $H_{A_2 A_2} - \beta H_{A_1 A_2} > 0 \Rightarrow H_{A_1 A_2} > 0$ and from case 2, when $H_{A_2 A_1} - \beta H_{A_1 A_1} > 0 \Rightarrow H_{A_2 A_1} > 0$. These results suggest that the total profit H at any time 't' can be optimized by simultaneous increasing (decreasing) advertisement expenditure for both generations. From case 3, when $H_{A_2 A_1} = H_{A_1 A_2} < 0$, optimal advertising expenditure of first and second generations go up or down in opposite directions.

The next theorem summarizes optimization results when $\beta > 0$.

Theorem 8.2. *If $\beta > 0$; $g_1 = g_1(A_1, A_2)$ and $g_2 = g_2(A_1, A_2)$*

	Conditions	Results	
		$a > 0$	$a < 0$
Case 1	$H_{A_2 A_2} - \beta H_{A_1 A_2} > 0$	$\dot{A}_1 < 0$ & $\dot{A}_2 > 0$	$\dot{A}_1 > 0$ & $\dot{A}_2 < 0$
Case 2	$H_{A_2 A_1} - \beta H_{A_1 A_1} < 0$	$\dot{A}_1 > 0$ & $\dot{A}_2 < 0$	$\dot{A}_1 < 0$ & $\dot{A}_2 > 0$
Case 3	$H_{A_2 A_1} = H_{A_1 A_2} > 0$	$\dot{A}_1 > 0$ & $\dot{A}_2 > 0$	$\dot{A}_1 < 0$ & $\dot{A}_2 < 0$

Proof. See Appendix A. □

Here case 1 suggest that when $H_{A_2A_2} - \beta H_{A_1A_2} > 0 \Rightarrow H_{A_1A_2} < 0$ and from case 2, $H_{A_2A_1} - \beta H_{A_1A_1} < 0 \Rightarrow H_{A_2A_1} < 0$. Thus under the conditions, to optimize the total profit H at any time 't', the firm should adopt a policy that the optimal advertising expenditure path of first and second generation should move in opposite directions. Case 3 suggest that when $H_{A_2A_1} = H_{A_1A_2} > 0$, total profit H at any time 't' can be optimized by increasing (decreasing) advertisement expenditure of both the generational products simultaneously. To illustrate the applications of results from dynamic optimization carried out above, they would be applied on an explicit demand model. A diffusion model is therefore proposed in the following section to measure the sales growth of two competing technological generations.

8.4 Special Functional Form

A diffusion model in Marketing is a mathematical structure (often an equation or a sequence of equations) that relates the adoption of a new product with explanatory variables like time, marketing-mix variables etc. The model described in this section is based on the following basic assumptions:

- Once an adopter adopts a new technology, he/she doesn't again revert to earlier generations in future.
- New adopters (first time buyers) will purchase only that particular generation product for which he/she will get the maximum utility. Utility can be expressed as a function of advertising expenditure and the goodwill in the market about all the available generations.
- Sales of a second-generation durable come from two sources:
  (1) *New Purchasers* (*First time Buyers*): Those who have for the first time adopted the product (from any generation).
  (2) *Repeat Purchasers*: Those adopters who have bought the earlier generation and now upgrade to latest technology.
- Each adopter can purchase maximum one unit from each generation. They can skip a generation and wait for more advanced one.

The incorporation of a new technology into a social system is a process, which involves the diffusion of knowledge about the characteristic of the technology. High technology have generations, with each generation having its own characteristic features. The first generation is a disruptive innovation (Norton and Bass 1987), which in an adopter's perception, provides features, utilities and functions not offered by an existing product. Second generation with some improvements and additional features is introduced into the market before its predecessor is withdrawn. Both the competing generations attract potential adopters. A potential adopter makes a mental evaluation of goodwill, price and benefits in offer before making the purchase decision. There would also be purchasers of first generation product who can upgrade to second generation. Assuming there are only two generations in the market, the two components of the model are adoption due to the new purchasers and repeat purchasers. Thus,

$$\text{Cumulative Adopters}_j(t)$$
$$= \text{New Purchasers}_j(t) + \text{Repeat Purchasers}_j(t)$$

where 'j' is the index representing the generation of a particular technology and x_j is the cumulative number of adopters in the jth generation. Thus,

$$x_j(t) = x'_j(t) + R_j(t) \qquad (8.26)$$

$x'_j(t)$ is the cumulative number of first time purchasers and $R_j(t)$ is the cumulative number of repeat purchasers. A monopoly market situation is assumed where each adopter can adopt a one unit of product from each generation. In general we do expect the potential market size to increase monotonically after every τ_j (where, τ_j is the introduction time of jth generation product) but it may not hold for all generation. The model for different market situations can be built as follows:

Case 1. When a single generation product is in the market place

When there exist only a single generation product in the market, the cumulative sales pattern of that generation can be described by the following model, which is an extension of basic mixed innovation diffusion model proposed by Bass (1969):

$$\dot{x}_1(t) = \frac{dx_1(t)}{dt} = (\overline{N}_1 - x_1(t))f_1(A_1(t), x_1(t)) \qquad (8.27)$$

where,

$\bar{N}_1$: Potential of the first generation product.

$f_i(A_i(t), x_i(t))$: Diffusion effect on ith generation demand at time t ($i = 1, 2$).

$f(A, x)$ can be specified as in Horsky-Simon (1983) model. The functional form of diffusion equation then becomes:

$$f_1(A_1, x_1) = a_1 + a_1' \ln A_1 + b_1 \left(\frac{x_1}{\overline{N}_1} \right)$$

a_1: innovation coefficient, a_1': reaction coefficient for advertising, b_1: imitation coefficient.

Case 2. When two generation products are in the market place

When there are two generations of the product in the market, the potential purchasers of first generation technology product come to the influence of both the advertising/promotion and word-of-mouth influence of the second generation product. As a result a fraction of the adopters (say, $\gamma(t)$) who would have otherwise adopted the first generation product instead adopt the latest technology and the remaining $[1 - \gamma(t)]$ will adopt the first generation product. Let us define the parameter $\gamma(t)$ as the switching parameter. Using it the two generational demand functions are:

$$\dot{x}_1(t) = \frac{dx_1(t)}{dt} = (\overline{N}_1 - x_1(t)) f_1(A_1(t), x_1(t))(1 - \gamma(t)) \qquad (8.28)$$

$$\begin{aligned}\dot{x}_2(t) = \frac{dx_2(t)}{dt} &= (\overline{N}_2 - x_2'(t_2)) f_2(A_2(t_2), x_2(t_2)) \\ &+ (\overline{R}_2(t_2) - R_2(t_2)) f_2'(A_2(t_2), x_2(t_2)) \\ &+ \gamma(t)(\overline{N}_1 - x_1(t_2)) f_1(A_1(t_2), x_1(t_2)) \end{aligned} \qquad (8.29)$$

Define the skipping parameter ($\gamma(t)$) as:

$$\gamma(t) = \omega(t) \left[\frac{f_2(A_2(t_2), x_2(t_2))}{f_1(A_1(t_2), x_1(t_2)) + f_2(A_2(t_2), x_2(t_2))} \right];$$

$$\omega(t) = \begin{cases} 0, & \text{if } t < \tau \\ 1, & \text{if } t \geqslant \tau \end{cases} \qquad (8.30)$$

$f_2'(A_2(t), x_2(t))$: Diffusion effect on repeat purchasers of 2nd generation product at time t.

The potential repeat purchasers of the second generation product is a function of the adopters of the first generation.

$$\text{Potential repeat purchasers} = \overline{R}_2(t_2) = \left[\sum_{i=1}^{t_2-1} n_1(i)\right] \quad (8.31)$$

$n_1(i)$: Sales of first generation product due to first time purchasers at time t.

Now, if we assume that $f_2(A_2(t), x_2(t)) = f_2'(A_2(t), x_2(t))$, then equation (8.29) can be written as:

$$\begin{aligned}\dot{x}_2(t) &= \frac{dx_2(t_2)}{dt} \\ &= (\bar{x}_2(t_2) - x_2(t_2))f_2(A_2(t_2), x_2(t_2)) \\ &\quad + \gamma(t)(\bar{N}_1 - x_1(t_2))f_1(A_1(t_2), x_1(t_2))\end{aligned} \quad (8.32)$$

where, $\bar{x}_2(t_2) = \bar{N}_2 + \bar{R}_2(t_2)$; $x_2(t) = x_2'(t) + R_2(t_2)$ and

$$f_2(A_2(t_2), x_2(t_2)) = a_2 + a_2' \ln A_2(t_2) + b_2\left(\frac{x_2(t_2)}{\bar{x}_2(t_2)}\right)$$

where,

a_2: innovation coefficient, a_2': reaction coefficient for advertising,
b_2: imitation coefficient.

Therefore,

$$\begin{aligned}\dot{x}_1(t) &= g_1(x_1(t), x_2(t), A_1(t), A_2(t)) \\ &= f_1(A_1(t), x_1(t))(\overline{N}_1 - x_1(t))[1 - \gamma(t)] \quad (8.33)\\ \dot{x}_2(t) &= g_2(x_1(t_2), x_2(t_2), A_1(t_2), A_2(t_2)) \\ &= f_2(A_2(t_2), x_2(t_2))[(\overline{x}_2(t_2)) - x_2(t_2)] \\ &\quad + \gamma(t)f_1(A_1(t_2), x_1(t_2))(\overline{N}_1 - x_1(t_2)) \quad (8.34)\end{aligned}$$

where,

$$\begin{aligned} g_{iA_i} &> 0; g_{iA_j} < 0; g_{iA_iA_i} < 0 \text{ and} \\ g_{iA_iA_j} &= g_{iA_jA_i}, \text{ where } (i, j = 1, 2) \end{aligned} \quad (8.35)$$

also

$$\gamma_{A_1} < 0; \gamma_{A_2} < 0; \gamma_{A_1A_1} > 0; \gamma_{A_2A_2} > 0 \text{ and } \gamma_{A_1A_2} = \gamma_{A_2A_1} \quad (8.36)$$

Thus from (8.13)-(8.17), we have

$$H_{A_1A_1} = (p_1 - c_1 + \lambda)g_{1A_1A_1} + (p_2 - c_2 + \mu)g_{2A_1A_1} < 0$$
$$H_{A_2A_2} = (p_1 - c_1 + \lambda)g_{1A_2A_2} + (p_2 - c_2 + \mu)g_{2A_2A_2} < 0$$
$$H_{A_1A_2} = H_{A_2A_1}$$

8.4.1 *Optimal Advertisement Expenditure Strategies*

In this sub-section, the optimal advertisement expenditure policies for the two product generations are analyzed for a monopolist subject to the equations (8.33) and (8.34). Applying Theorems 8.1 and 8.2 to demand equations (8.33) and (8.34), the following results were obtained as summarized in a theorem.

Theorem 8.3. *For constant price $p_i (i = 1, 2)$ of the ith generation with $g_1 = g_1(A_1, A_2)$ and $g_2 = g_2(A_1, A_2)$, the following holds:*

If the future benefit on selling first generation product is less than or equal to that of second generation product than the optimal advertisement expenditure paths of both the generational products depend on the following conditions.

Conditions		Results
$H_{A_2A_2} - \beta H_{A_1A_2} > 0$	$\dot{A}_1 < 0$ & $\dot{A}_2 > 0$	$A_2(t)$ increasing, $A_1(t)$ decreasing
$H_{A_2A_1} - \beta H_{A_1A_1} < 0$	$\dot{A}_1 > 0$ & $\dot{A}_2 < 0$	$A_2(t)$ decreasing, $A_1(t)$ increasing

Conditions	Results
$H_{A_2A_1} = H_{A_1A_2} > 0$ $\quad$ $\dot{A}_1 > 0$ & $\dot{A}_2 > 0$	$A_2(t)$ increasing, $A_1(t)$ increasing

Proof. See Appendix B. $\square$

From the first case if $H_{A_2A_2} - \beta H_{A_1A_2} > 0$, the firm should maintain a policy that advertisement expenditures of first and second generations move in opposite directions. In this situation the optimal strategy is to increase the advertising of second generation product gradually over the planning horizon and decrease the advertising expenditure on first generation product. On the other hand if $H_{A_2A_1} - \beta H_{A_1A_1} < 0$ the suggested optimal strategy is to increase the advertising of first generation product gradually over the planning horizon and decrease the advertising expenditure of second generation product, so that first generation sales attain maturity stage early before the second generation product cannibalizes its market. Finally for $H_{A_2A_1} = H_{A_1A_2} > 0$ the suggested optimal policy in such situation is to increase the advertising of both the product generations gradually over the planning horizon to counter balance the decrease in current sales caused by penetration of latest generation.

Theorem 8.4. *For $\lambda \gg \mu$ the optimal advertisement expenditures path of both the generational product depend on the following conditions.*

Conditions	Results
$H_{A_2A_2} - \beta H_{A_1A_2} > 0$ $\quad$ $\dot{A}_1 < 0$ & $\dot{A}_2 < 0$	$A_2(t)$ decreasing, $A_1(t)$ decreasing

Conditions		Results
$H_{A_2A_1} - \beta H_{A_1A_1} > 0$	$\dot{A}_1 > 0$ & $\dot{A}_2 > 0$	$A_2(t)$ increasing above, $A_1(t)$ increasing below
$H_{A_2A_1} = H_{A_1A_2} < 0$	$\dot{A}_1 > 0$ & $\dot{A}_2 < 0$	$A_2(t)$ decreasing above, $A_1(t)$ increasing below

Proof. See Appendix B. □

The results of Theorem 8.4 can be interpreted as: so long as $H_{A_2A_2} - \beta H_{A_1A_2} > 0$ varying advertising expenditure of second generation product is more beneficial than first generation in increasing the overall profitability. The optimal policy in such situation is to advertise heavily in the beginning of the planning-horizon in order to stimulate innovators or early adopters and then as the market settles down the advertisement expenditure goes down steadily. In this situation the optimal price path of both the generations will follow the same trend as $H_{A_1A_2} > 0$. This pattern can be seen when the diffusion effects are positive for both the product generations throughout their planning horizons or when the initial heavy advertising/promotion targeted towards a particular market segment has done well. On the other hand if $H_{A_2A_1} - \beta H_{A_1A_1} > 0$ investing more on advertising expenditure for first generation product is more beneficial than second generation in increasing the overall profitability. The suggested optimal policy in this situation is to increase advertising expenditure gradually over the planning horizon to counter balance the decrease in current sales caused by current penetration levels. The optimal price paths of both the generations will follow the same trend as $H_{A_2A_1} > 0$. Policies for increasing advertising expenditures occur when the diffusion effects are negative for both the product generations throughout their planning horizons. This can be the situation when higher levels of promotional efforts are required to dispel the negative perceptions (may be due to word-of-mouth) about the product. Finally $H_{A_2A_1} = H_{A_1A_2} < 0$ indicates that the firm should maintain a policy that advertising expenditures of first and second generations move in opposite directions. In this situation the optimal strategy is to intensify the advertising of first generation and reduce the advertising of second generation product, so that the first generation product can attain saturation point as early as possible before the sales of second generation can pick up. Otherwise, the first generation product maybe forced to go out of the market rather early due to high consumer expectations from the latest technology.

The above results are consistent with previous research on single generation products. In this chapter, it was shown that the investment pattern on advertising or promotions of earlier generation product can also depend on the diffusion rate of the latest generation and vice-versa. Moreover for both the generations the optimal advertisement expenditure increases or decreases monotonically over the entire planning horizon. Thus, the optimal investment on advertising/promotion either be initially low and pick up later or it is very high initially and slows down subsequently. In Theorem 8.3 and 8.4, the results have been summarized diagrammatically, to show the structure of the time path of optimal advertising expenditures for two generation diffusion models (8.33) and (8.34).

8.5 Conclusions

This chapter discusses the optimal advertising expenditure strategy for products with technological generations under dynamic situation. A diffusion model has been proposed based on the basic Horsky and Simon (1983) model. The results discussed here are consistent with the work of Dockner and Jorgensen (1988) for single generation product. The proposed model incorporates the replacement behavior of first generation purchasers when an advanced generation is introduced in the market. Conditions are found in which optimal advertising expenditures can decrease and increase. In all the cases it was observed that optimal policies behave monotonically. The theoretical results obtained here confirm and extend the prior results in the literature. Finally, the model can be extended in several ways, e.g. by extending the monopolistic model for a duopolistic or oligopolistic market. The model can also be extended for n-generation product situations.

Acknowledgements. The authors are thankful to the referee for useful comments on the original draft. The chapter was revised based on these comments.

Appendix

A. Proof of conditions and theorem of General Formulation

Taking time-derivative on (8.11), we have

$$(\dot{c}_1 \dot{\lambda}) g_{1A_1} + (p_1 - c_1 + \lambda)[g_{1A_1A_1}\dot{A}_1 + g_{1A_1A_2}\dot{A}_2 + g_{1A_1x_1}\dot{x}_1 + g_{1A_1x_2}\dot{x}_2]$$
$$+ (\dot{c}_2 + \dot{\mu}) g_{2A_1} + (p_2 - c_2 + \mu)[g_{2A_1A_1}\dot{A}_1 + g_{2A_1A_2}\dot{A}_2$$
$$+ g_{2A_1x_1}\dot{x}_1 + g_{2A_1x_2}\dot{x}_2] = 0$$
$$\Rightarrow (-c_{1x_1}\dot{x}_1 + r\lambda + c_{1x_1}g_1 - (p_1 - c_1 + \lambda)g_{1x_1} - (p_2 - c_2 + \mu)g_{2x_1})g_{1A_1}$$
$$+ \dot{A}_1[(p_1 - c_1 + \lambda)g_{1A_1A_1} + (p_2 - c_2 + \mu)g_{2A_1A_1}]$$
$$+ \dot{A}_2[(p_1 - c_1 + \lambda)g_{1A_1A_2} + (p_2 - c_2 + \mu)g_{2A_1A_2}]$$
$$+ (-c_{2x_2}\dot{x}_2 + r\mu + c_{2x_2}g_2 - (p_1 - c_1 + \lambda)g_{1x_2}$$
$$- (p_2 - c_2 + \mu)g_{2x_2})g_{2A_1} + (p_1 - c_1 + \lambda)[g_{1A_1x_1}\dot{x}_1 + g_{1A_1x_2}\dot{x}_2]$$
$$+ (p_2 - c_2 + \mu)[g_{2A_1x_1}\dot{x}_1 + g_{2A_1x_2}\dot{x}_2] = 0$$
$$\Rightarrow H_{A_1A_1}\dot{A}_1 + H_{A_1A_2}\dot{A}_2 = d_1 \tag{A.1}$$

where,

$$d_1 = -r\lambda g_{1A_1} - r\mu g_{2A_1} + g_{1A_1}[(p_1 - c_1 + \lambda)g_{1x_1} + (p_2 - c_2 + \mu)g_{2x_1}]$$
$$+ g_{2A_1}[(p_1 - c_1 + \lambda)g_{1x_2} + (p_2 - c_2 + \mu)g_{2x_2}] + (p_1 - c_1 + \lambda)$$
$$[g_{1A_1x_1}\dot{x}_1 + g_{1A_1x_2}\dot{x}_2] + (p_2 - c_2 + \mu)[g_{2A_1x_1}\dot{x}_1 + g_{2A_1x_2}\dot{x}_2]$$

Taking time-derivative on (8.12), we have

$$(\dot{c}_1 + \dot{\lambda}) g_{1A_2} + (p_1 - c_1 + \lambda)[g_{1A_2A_1}\dot{A}_1 + g_{1A_2A_2}\dot{A}_2 + g_{1A_2x_1}\dot{x}_1 + g_{1A_2x_2}\dot{x}_2]$$
$$+ (\dot{c}_2 + \dot{\mu}) g_{2A_2} + (p_2 - c_2 + \mu)[g_{2A_2A_1}\dot{A}_1 + g_{2A_2A_2}\dot{A}_2 + g_{2A_2x_1}\dot{x}_1 + g_{2A_2x_2}\dot{x}_2] = 0$$
$$\Rightarrow (-c_{1x_1}\dot{x}_1 + r\lambda + c_{1x_1}g_1 - (p_1 - c_1 + \lambda)g_{1x_1} - (p_2 - c_2 + \mu)g_{2x_1})g_{1A_2}$$
$$+ \dot{A}_1[(p_1 - c_1 + \lambda)g_{1A_2A_1} + (p_2 - c_2 + \mu)g_{2A_2A_1}] + \dot{A}_2[(p_1 - c_1 + \lambda)g_{1A_2A_2}$$
$$+ (p_2 - c_2 + \mu)g_{2A_2A_2}] + (-c_{2x_2}\dot{x}_2 + r\mu + c_{2x_2}g_2 - (p_1 - c_1 + \lambda)g_{1x_2}$$
$$- (p_2 - c_2 + \mu)g_{2x_2})g_{2A_2} + (p_1 - c_1 + \lambda)[g_{1A_2x_1}\dot{x}_1 + g_{1A_2x_2}\dot{x}_2]$$
$$+ (p_2 - c_2 + \mu)[g_{2A_2x_1}\dot{x}_1 + g_{2A_2x_2}\dot{x}_2] = 0$$
$$\Rightarrow H_{A_2A_1}\dot{A}_1 + H_{A_2A_2}\dot{A}_2 = d_2 \tag{A.2}$$

where,

$$d_2 = -r\lambda g_{1A_2} - r\mu g_{2A_2} + g_{1A_2}[(p_1 - c_1 + \lambda)g_{1x_1} + (p_2 - c_2 + \mu)g_{2x_1}]$$
$$+ g_{2A_2}[(p_1 - c_1 + \lambda)g_{1x_2} + (p_2 - c_2 + \mu)g_{2x_2}] + (p_1 - c_1 + \lambda)$$
$$[g_{1A_2x_1}\dot{x}_1 + g_{1A_2x_2}\dot{x}_2] + (p_2 - c_2 + \mu)[g_{2A_2x_1}\dot{x}_1 + g_{2A_2x_2}\dot{x}_2]$$

Solving (A.1) and (A.2), we have

$$\dot{A}_1 = \frac{1}{\Delta}[d_1 H_{A_2 A_2} - d_2 H_{A_1 A_2}] \qquad (A.3)$$

$$\dot{A}_2 = -\frac{1}{\Delta}[d_1 H_{A_2 A_1} - d_2 H_{A_1 A_1}] \qquad (A.4)$$

Assuming that the cumulative sales of both the generations are function of advertising only and word-of-mouth influence are not important, then

$$d_1 = -r\lambda g_{1A_1} - r\mu g_{2A_1} \quad \text{and} \quad d_2 = -r\lambda g_{1A_2} - r\mu g_{2A_2}$$

Solving (A.3) and (A.4), we have

$$\dot{A}_1 = \frac{1}{\Delta}[(-r\lambda g_{1A_1} - r\mu g_{2A_1})H_{A_2 A_2} - (-r\lambda g_{1A_2} - r\mu g_{2A_2})H_{A_1 A_2}]$$
$$= -\frac{r(\lambda g_{1A_1} + \mu g_{2A_1})}{\Delta}\left[H_{A_2 A_2} - \left(\frac{\lambda g_{1A_2} + \mu g_{2A_2}}{\lambda g_{1A_1} + \mu g_{2A_1}}\right)H_{A_1 A_2}\right]$$

$$\dot{A}_2 = -\frac{1}{\Delta}[(-r\lambda g_{1A_1} - r\mu g_{2A_1})H_{A_2 A_1} - (-r\lambda g_{1A_2} - r\mu g_{2A_2})H_{A_1 A_1}]$$
$$= \frac{r(\lambda g_{1A_1} + \mu g_{2A_1})}{\Delta}\left[H_{A_2 A_1} - \left(\frac{\lambda g_{1A_2} + \mu g_{2A_2}}{\lambda g_{1A_1} + \mu g_{2A_1}}\right)H_{A_1 A_1}\right]$$

Let, $\alpha = \lambda g_{1A_1} + \mu g_{2A_1}$ and $\beta = \dfrac{\lambda g_{1A_2} + \mu g_{2A_2}}{\lambda g_{1A_1} + \mu g_{2A_1}}$; then

$$\dot{A}_1 = -\frac{r\alpha}{\Delta}[H_{A_2 A_2} - \beta H_{A_1 A_2}] \qquad (A.5)$$

$$\dot{A}_2 = \frac{r\alpha}{\Delta}[H_{A_2 A_1} - \beta H_{A_1 A_1}] \qquad (A.6)$$

Using the general-model properties we have the following results:

Lemma 8.1. When $\beta < 0$

8.1.1. If $H_{A_2 A_2} - \beta H_{A_1 A_2} > 0$ then $H_{A_1 A_2} > 0$ and $H_{A_2 A_1} - \beta H_{A_1 A_1} < 0$

Proof. We have $H_{A_2A_2} < 0$ and $H_{A_2A_2} - \beta H_{A_1A_2} > 0 \Rightarrow H_{A_1A_2} > 0$. Also,

$$\begin{vmatrix} H_{A_1A_1} & H_{A_1A_2} \\ H_{A_2A_1} - \beta H_{A_1A_1} & H_{A_2A_2} - \beta H_{A_1A_2} \end{vmatrix} = \begin{vmatrix} H_{A_1A_1} & H_{A_1A_2} \\ H_{A_2A_1} & H_{A_2A_2} \end{vmatrix} = \Delta > 0$$

$$\Rightarrow H_{A_1A_1}(H_{A_2A_2} - \beta H_{A_1A_2}) - H_{A_1A_2}(H_{A_2A_1} - \beta H_{A_1A_1}) > 0 \quad (A.7)$$

Now, as $H_{A_1A_1} < 0$, (A.7) will be true iff $H_{A_2A_1} - \beta H_{A_1A_1} < 0$. □

8.1.2. If $H_{A_2A_1} - \beta H_{A_1A_1} > 0$ then $H_{A_2A_1} > 0$ and $H_{A_2A_2} - \beta H_{A_1A_2} < 0$

Proof. Similar to Lemma 8.1.1. □

8.1.3. If $H_{A_2A_1} = H_{A_1A_2} < 0$ then $H_{A_2A_1} - \beta H_{A_1A_1} < 0$ and $H_{A_2A_2} - \beta H_{A_1A_2} < 0$

Proof. It can be proved using Lemma 8.1.1 and 8.1.2. □

Lemma 8.2. *When $\beta > 0$*

8.2.1. If $H_{A_2A_2} - \beta H_{A_1A_2} > 0$ then $H_{A_1A_2} < 0$ and $H_{A_2A_1} - \beta H_{A_1A_1} > 0$
8.2.2. If $H_{A_2A_1} - \beta H_{A_1A_1} < 0$ then $H_{A_2A_1} < 0$ and $H_{A_2A_2} - \beta H_{A_1A_2} < 0$
8.2.3. If $H_{A_2A_1} = H_{A_1A_2} > 0$ then $H_{A_2A_1} - \beta H_{A_1A_1} > 0$ and $H_{A_2A_2} - \beta H_{A_1A_2} < 0$

Proof. Similar to Lemma 8.1. □

Proof of Theorem 8.1 and 8.2. From Lemma 8.1 and Lemma 8.2, we can easily prove these theorems.

B. Proof of theorem of Special Functional Form

$$\alpha = \lambda g_{1A_1} + \mu g_{2A_1}$$
$$= \lambda \frac{a'_1}{A_1}(\overline{N}_1 - x_1)(1 - \gamma^2) + \mu \frac{a'_1}{A_1}(\overline{N}_1 - x_1)\gamma^2 \quad (B.1)$$

$$= \frac{a'_1}{A_1}(\overline{N}_1 - x_1)[(1 - \gamma^2)\lambda + \gamma^2 \mu] > 0 \quad (B.2)$$

and,

$$\beta = \frac{\lambda g_{1A_2} + \mu g_{2A_2}}{\lambda g_{1A_1} + \mu g_{2A_1}} \quad (B.3)$$

Now,

$$\lambda g_{1A_2} + \mu g_{2A_2}$$
$$= \lambda[-f_1(\overline{N}_1 - x_1)\gamma_{A_2}] + \mu[f_{2A_2}\{\overline{x_2} - x_2\}] + \mu[f_1(\overline{N}_1 - x_1)\gamma_{A_2}]$$
$$= f_1(\overline{N}_1 - x_1)\frac{a_2'}{A_2 f_2}\gamma(1-\gamma)(\mu - \lambda) + \mu\frac{a_2'}{A_2}[\overline{x}_2 - x_2]$$
$$= \mu\frac{a_2'}{A_2}\left[\{\overline{x}_2 - x_2\} + \frac{f_1}{f_2}(\overline{N}_1 - x_1)\gamma(1-\gamma)\left(1 - \frac{\lambda}{\mu}\right)\right] \qquad (B.4)$$

therefore

$$\beta = \begin{cases} \text{positive when} & \lambda \leq \mu \\ \text{negative when} & \lambda \gg \mu \end{cases} \qquad (B.5)$$

Proof of Theorem 8.3 and 8.4. Theorem 8.3 and 8.4 can be proved by using condition (B.3) along with Lemma 8.1 and Lemma 8.2.

Bibliography

Bardhan, A. K. and Chanda, U. (2008). *Dynamic Optimal Control Policy in Price and Quality for High Technology Product*, in *Mathematical Programming and Game Theory for Decision Making*, S.K. Neogy, R.B. Bapat, A.K. Das, T. Parthasarthy (eds.), Delhi, India: World Scientific. Chapter 13, pp. 201–219.

Bass, F. M. (1969). A New-Product Growth Model for Consumer Durables, *Management Science* **15**, pp. 215–227.

Bass, F. M. and Bultez, A. V. (1982). A Note on Optimal Strategic Pricing of Technological Innovations, *Marketing Science* **1**, pp. 371–378.

Bass, F. M., Trichy, V. K. and Jain, D. C. (1994). Why the Bass Model Fits Without the Decision Variables, *Marketing Science* **13**, pp. 203–223.

Bayus, L. B. (1992). The Dynamic Pricing of Next Generation Consumer Durables, *Marketing Science* **11**, pp. 251–265.

Chanda, U. and Bardhan, A. K. (2008). Modelling Innovation and Imitation Sales of Products with Multiple Technological Generations, *Journal of High Technology Management Research* **18**, pp. 173–190.

Danaher, P. J., Hardie, B. G. S. and William (Jr.), P. P. (2001). Marketing-Mix Variables and the Diffusion of Successive Generations of a Technology Innovation, *Journal of Marketing Research* **XXXVIII**, pp. 501–514.

Dockner, E. and Jørgensen, S. (1988). Optimal Advertising Policies for Diffusion Models of New Product Innovation in Monopolistic Situations, *Management Science* **34**, pp. 119–131.

Feichtinger, G., Hartl, R. F. and Sethi, S. P. (1994). Dynamic Optimal Control Models in Advertising: Recent Developments, *Management Science* **40**, pp. 195–226.

Horsky, D. and Simon, L. S. (1983). Advertising and the Diffusion of a New Products, *Marketing Science* **2**, pp. 1–17.

Horsky, D. (1990). A Diffusion Model Incorporating Product Benefits, Price, Income and Information, *Marketing Science* **9**, pp. 342–365.

Islam, T. and Meade, N. (1997). The Diffusion of Successive Generation of a Technology; a More General Model, *Technological Forecasting and Social Change* **56**, pp. 49–60.

Kalish, S. (1983). Monopolist Pricing with Dynamic Demand and Product Cost, *Marketing Science* **2**, pp. 135–159.

Kalish, S. (1985). A New Product Adoption Model with Price Advertising, and Uncertainty. *Management Science*, **31**, pp. 1569–1585.

Kamakura, W. and Balasubramanium, S. K (1988). Long-Term View of the Diffusion of Durables: a Study of the Role of Price and Adoption Influence Processes via Tests of Nested Models, *International Journal of Research in Marketing*, **5**, pp. 1–13.

Lilly, B. and Walters, R. (1997). Toward a Model of New Product Preannouncement Timing. *Journal of Product Innovation Management* **14**, pp. 4–20.

Mahajan, V. and Muller, E. (1996). Timing, Diffusion, and Substitution of Successive Generations of Technological Innovations: The IBM Mainframe Case. *Technological Forecasting and Social Change*, **51**, pp. 109–132.

Mahajan, V., Muller, E. and Wind, Y. (2000). *New Product Diffusion Models*, (eds.).pp. 1–355, (Kluwer Academic Publishers, Boston).

Norton, J. A. and Bass, F. M. (1987). A Diffusion Theory Model of Adoption and Substitution for Successive Generation of High-Technology Products, *Management Science* **33**, pp. 1069–1086.

Robinson, B. and Lakhani, C. (1975). Dynamic Price Models for New Product Planning, *Management Science*, **21**, pp. 1113-1122.

Rogers, E. M. (1962). *Diffusion of Innovations*, (The Free Press, New York).

Sethi, S. P. (1977). Dynamic Optimal Control Models in Advertising: A Survey, *Siam Review*, **19**, pp. 685–725.

Sethi, S.P. and Bass, F.M. (2003). Optimal Pricing in a Hazard Rate Model of Demand. *Optimal Control Applications and Methods* **24**, pp. 183–196.

Simon, H. and Sebastian, K. (1987). Diffusion and Advertising: German Telephone Company, *Management Science*, **33**, pp. 451–466.

Thompson, G. L. and Teng, J. T.(1984). Optimal Pricing and Advertising Policies for New Product Oligopoly Market, *Marketing Science*, **3**, pp. 148–168.

Thompson, G. L. and Teng, J. T.(1996). Optimal Strategies for General Price-Quality Decision Models of New Products with Learning Production Costs, *European Journal of Operational Research*, **93**, pp. 476–489.

Chapter 9

Nonconvex Vector Minimization with Set Inclusion Constraint

Anjana Gupta
*Department of Operational Research, University of Delhi,
Delhi-110007, India
e-mail: guptaanjana2003@yahoo.co.in*
Aparna Mehra
*Department of Mathematics, Indian Institute of Technology Delhi
Hauz Khas, New Delhi-110016, India
e-mail: apmehra@maths.iitd.ac.in*
Davinder Bhatia
*Department of Operational Research, University of Delhi,
Delhi-110007, India*

Abstract

In this chapter we employ a nonconvex separation theorem to scalarize the vector minimization problem subject to the constraint given in the form of set inclusion. A new Lagrange function is formulated for the scalarized problem. Saddle point criteria is developed which ensure the existence of the Lagrange multipliers. We also discuss the Lagrange duality.

Key Words: Lagrange duality, vector minimization problem, nonlinear scalarization scheme, saddle point criteria

9.1 Introduction

One of the most fundamental approach to study vector optimization problem is to convert it into an equivalent scalar optimization problem by using an appropriate scalarization technique. Plenty of literature is available on linear scalarization schemes that are designed for convex or generalized convex vector optimization problems. However moving away from convexity eventually lead us to use the idea of nonlinear scalarization. In this direction a pioneer work has been done by [Gerth and Weidner (1990)]. They developed a nonlinear scalarization scheme by developing nonconvex separation theorems in linear spaces. Their ideas were carried forward by [Dutta and Tammer (2006)] who derived the Lagrange multiplier rules for vector optimization problems in Banach spaces. Later, [Hernández and RodrÍGuez-Marin (2007)] employed the nonlinear scalarization scheme to study set-valued optimization problems. Despite some successful attempts, we found that there have been scarcely few research articles on nonlinear scalarization in contrast to the linear scalarization.

Our principal aim in this article is to employ the nonlinear scalarization scheme to study the saddle point criteria and the Lagrange duality results for the following vector minimization problem involving set inclusion constraint

$$(VPSIC) \quad V - \min_B f(x)$$
$$\text{subject to } g(x) \in P$$

$f : X \to R^p$, $g : X \to R^m$, X and B are nonempty subsets of R^n and R^p, respectively, and P is a nonempty closed convex subset of R^m.

Let $S = \{x \in X \mid g(x) \in P\}$ denotes the feasible set, and minimization is taken with respect to B in the sense described below.

Definition 9.1.1 (see [Gerth and Weidner (1990)], Definition 3.1). $x^\star \in S$ is called an efficient solution of $(VPSIC)$ with respect to B if there exists no $x \in S$ such that

$$f(x^\star) \in f(x) + (B \setminus \{0\}).$$

The set of all such efficient solutions is denoted as $E((VPSIC), B)$.

If $B = R^p_+$ then the Definition 9.1.1 reduces to the classic notion of efficiency while if $B = int\, R^p_+$ then it reduces to the notion of weak efficiency.

Problems of the type $(VPSIC)$ cater to a large class of optimization problems. In particular, it includes the classes of vector optimization problems with inequality constraints, vector optimization problems with cone constraints and vector optimization problems with abstract constraints. In all the mentioned cases, the saddle point optimality conditions have been thoroughly investigated in numerous research articles in the past, see for example [Adán and Novo (2005); Duc, Hoang and Nguyen (2006); Ehrgott and Wiecek (2005); Li (1988); Sposito and David (1971); Tadeusz (2005); Zlobec (2004)] and references cited therein.

We believe that the present study unifies the existing saddle point theories in multiobjective optimization and simultaneously defines a new notion of nonlinear Lagrange saddle point.

We now sketch an outline of the chapter. The section to follow uses the nonlinear separating functional [Gerth and Weidner (1990); Tammer (1992)] to scalarize $(VPSIC)$. Section 9.3 presents a new Lagrange function in which the multiplier associated with the objective function is nonlinear convex functional whereas the multiplier associated with the constraint is in the normal cone to the set P. We obtained the sufficient saddle point optimality condition and illustrate it by an example. Section 9.4 is devoted to establish the necessary saddle point criteria while section 9.5 derives the Lagrange duality results using the Fuchssteiner and König's nonconvex minmax theorem. Some concluding remarks and further suggestions are presented in section 9.6.

9.2 Nonlinear Scalarization

The primary aim of this section is to construct an equivalent scalar optimization problem of $(VPSIC)$. We present below a result due to [Gerth and Weidner (1990)] that has been recognized as a separation theorem for arbitrary sets which are not necessarily convex. Let us first recall that a function $\psi : R^p \to R$ is called D-monotone on a set $E \subset R^p$ if $y^1, y^2 \in E$, $y^1 \in y^2 + D$ implies $\psi(y^1) \geq \psi(y^2)$, and is called strictly D-monotone on E if $y^1, y^2 \in X$, $y^1 \in y^2 + D \setminus \{0\}$ implies $\psi(y^1) > \psi(y^2)$. Here D is a nonempty subset of R^p. Moreover, for a nonempty subset $\mathbb{A}$ of R^p, the interior, closure and boundary of $\mathbb{A}$ are denoted respectively by $int\,\mathbb{A}$, $cl\,\mathbb{A}$ and $\delta \mathbb{A}$.

Theorem 9.2.1 (see [Gerth and Weidner (1990)], Theorem 2.2).
Let the following conditions hold:

(H-1) C *be a proper open convex subset of* R^p;
(H-2) $\exists\, k^0 \in R^p$ *such that* $cl\,C - \alpha k^0 \subset cl\,C$, $\forall\, \alpha \in R_+$; *where* R_+ *is the set of all nonnegative reals;*
(H-3) $R^p = \bigcup \{cl\,C + \alpha k^0 \,|\, \alpha \in R\}$;
(H-4) A *is a nonempty subset of* R^p.

Then we have

(a) $A \cap C = \emptyset$ *if and only if* $\exists$ *a continuous convex functional* $z : R^p \to R$ *with the range* $(-\infty, +\infty)$ *and* $z(A) \geq 0$, $z(int\, A) > 0$, $z(C) < 0$, $z(\delta C) = 0$.

(b) *Let D be a nonempty subset of R^p. If, $\delta C - D \subset cl\, C$, then z in* (a) *can be chosen such that it is D-monotone. Further, if $\delta C - (D \setminus \{0\}) \subset C$, then z in* (a) *is strictly D-monotone.*

(c) *If $\delta C + \delta C \subset cl\, C$ then z in* (a) *can be chosen such that it is subadditive.*

The separating functional z in the theorem is described by $z(\xi) = \inf\{\lambda \,|\, \xi \in S_\lambda\}$, $S_\lambda = cl\, C + \lambda k^0, \lambda \in R$, and is called Gerstewitz's function. This function is known to exhibit certain nice geometrical properties which have been exploited by many researchers. We refer the readers to [Gerth and Weidner (1990); Hernández and RodrÍGuez-Marin (2007); Luc (1989); Tammer (1992)] and references therein for a complete analysis of properties of Gerstewitz's function as well as its applications in nonlinear optimization.

We illustrate the definition of z by means of two examples. In the first example, the set C is taken to be a convex cone while in the next case it is only a convex set and not a cone.

Example 9.2.2. Let $C = int\, R_-^2$, $k^0 = (1,1)^t$.

Here, $z(\xi_1, \xi_2) = \inf\{\lambda \,|\, (\xi_1, \xi_2) \in S_\lambda\}$ is given by

$$z(\xi_1, \xi_2) = \begin{cases} \xi_1, & \xi_1 \geq \xi_2 \\ \xi_2, & \xi_1 < \xi_2. \end{cases}$$

Clearly z is a continuous convex subadditive strictly $(int\, R_+^2)$-monotone.

Example 9.2.3. Let $C = \{(x, y) \in R^2 \,|\, y > x^2\}$ and $k^0 = (0, -1)^t$.

Let $\xi = (\xi_1, \xi_2) \in S_\lambda$. Then, $\xi_2 > \xi_1^2 - \lambda$, and hence, $z(\xi_1, \xi_2) = \xi_1^2 - \xi_2$.

Observe that z is a continuous convex functional which is neither C-monotone nor subadditive.

In the discussion to follow we assume the set B satisfies (H-1), (H-3) of Theorem 9.2.1 along with the following two conditions:

(H-5) $0 \in cl\, B \setminus B$;
(H-6) $\exists\, k^0 \in R^p$ such that $cl\, B + \alpha k^0 \subset cl\, B,\ \forall\, \alpha \in R_+$.

With this ground work we are ready to present an equivalent scalar version of $(VPSIC)$ in the form of two theorems. The equivalence has been explained in section 3.1.2 of the monograph of [Göpfert, Riahi, Tammer, Zăliomesci (2003)]. However for the sake of convenience we provide below the proofs of the theorems.

Theorem 9.2.4. *Let $x_0 \in E((VPSIC), B)$. Then there exists a continuous convex functional $\bar{z} : R^p \to R$, $\bar{z} \not\equiv 0$, such that x_0 is an optimal solution of the following scalarized problem*

$$(SP)_{\bar{z}}\quad \text{Min } (\bar{z} \circ f)(x)$$
$$\text{subject to } x \in S.$$

Further, if $B + \delta B \subset B$ then $\bar{z}$ is a strictly B-monotone functional.

Proof. It follows from the given hypothesis that

$$f(S) \cap (f(x_0) - B) = \emptyset.$$

Define two sets, $A = f(S)$ and $C = f(x_0) - B$, and invoking Theorem 9.2.1 to get the existence of a continuous convex functional $\bar{z} : R^p \to R$ such that

$$\bar{z}(f(S)) \geq 0 \text{ and } \bar{z}(f(x_0) - B) < 0.$$

Clearly, $\bar{z} \not\equiv 0$. Also continuity of $\bar{z}$ at $0 \in cl\, B \setminus B$ yields

$$(\bar{z} \circ f)(x_0) \leq (\bar{z} \circ f)(x),\ \forall\, x \in S,$$

implying that x_0 is an optimal solution of $(SP)_{\bar{z}}$. The latter part of the theorem follows from Theorem 9.2.1(b). $\square$

Theorem 9.2.5. *Let $z : R^p \to R$ be a strictly B-monotone continuous functional. If $x_0 \in S$ is an optimal solution for $(SP)_z$ then $x_0 \in E((VPSIC), B)$.*

Proof. Let $y_0 = f(x_0)$. Define a functional $\bar{z} : R^p \to R$ as

$$\bar{z}(y) = z(y + y_0),\ \forall\, y \in R^p.$$

Then, for $x \in S$,

$$\bar{z}(f(x) - y_0) = z(f(x)) \geq z(f(x_0)) = \bar{z}(0). \tag{9.1}$$

Let $b \in B$. Then, $y_0 \in -b + y_0 + B \setminus \{0\}$. As z is strictly B-monotone therefore we have

$$z(-b+y_0) < z(y_0) \Rightarrow \bar{z}(-b) < \bar{z}(0). \tag{9.2}$$

The desired relation

$$f(S) \cap (f(x_0) - B) = \emptyset$$

follows on account of (9.1) and (9.2). Hence $x_0 \in E((VPSIC), B)$. □

We use the notation $\mathcal{T}$ to denote the set $\{z \,|\, z : R^p \to R,\ z \text{ is a continuous convex strictly } B\text{-monotone functional}\}$.

From now onward we assume that $z \in \mathcal{T}$.

9.3 Sufficient Optimality Criteria

The above results inspire us to construct a new Lagrange function of $(VPSIC)$ and try to work out the sufficient saddle point criteria. We first recall the notion of support function (refer, Definition 2.3.10 of [Bector, Chandra and Dutta (2005)]) of a set. The support function $\sigma_P : R^m \to R \cup \{+\infty\}$ of P is defined as

$$\sigma_P(x^*) = \sigma(x^*, P) = \sup_{x \in P} x^{*t} x.$$

It is well known that the support function is a lower semicontinuous convex function.

The extended Lagrange function $\mathcal{L} : X \times \mathcal{T} \times R^m \to R \cup \{-\infty\}$ associated with $(SP)_z$ is defined as

$$\mathcal{L}(x, z, \mu) = (z \circ f)(x) + \mu^t g(x) - \sigma(\mu, P).$$

Although saddle point for the scalar Lagrange function is well defined we restate the definition for the sake of completeness.

Definition 9.3.1. A triplet $(x_0, z_0, \mu_0) \in X \times \mathcal{T} \times R^m$ is called a saddle point of $\mathcal{L}$ if

$$\mathcal{L}(x_0, z_0, \mu) \leq \mathcal{L}(x_0, z_0, \mu_0) \leq \mathcal{L}(x, z_0, \mu_0),\ \forall\, x \in X, \forall\, \mu \in R^m.$$

Remark 9.3.2. (i) Letting $\mu = \mu_0 + \nu$, $\nu \in R^m$ in the saddle point inequality, we get
$$\nu^t g(x_0) - \sigma(\mu_0 + \nu, P) \leq -\sigma(\mu_0, P).$$
On account of subadditivity of $\sigma(\cdot, P)$, we obtain
$$\nu^t g(x_0) \leq \sigma(\nu, P), \quad \forall \, \nu \in R^m. \tag{9.3}$$

Now suppose $g(x_0) \notin P$. Then by the strong separation theorem of convex sets there exist $\alpha \in R^m$ and $\beta \in R$ such that
$$\alpha^t p \leq \beta, \; \forall \, p \in P \quad \text{and} \quad \alpha^t g(x_0) > \beta,$$
implying, $\sigma(\alpha, P) < \alpha^t g(x_0)$, a contradiction to (9.3). Consequently, $g(x_0) \in P$.

(ii) Taking $\mu = 0$ in the first inequality of the saddle point yields
$$\mu_0^t g(x_0) \geq \mu_0^t p, \quad \forall \, p \in P.$$
Hence, $\mu_0 \in N_P(g(x_0))$, $N_P(g(x_0))$ is the normal cone to P at $g(x_0)$ described by $\{\xi \in R^m \mid \xi^t(p - g(x_0)) \leq 0, \, \forall \, p \in P\}$.

Remark 9.3.3. Observe that when P is a closed convex cone, $N_P(g(x_0))$ is same as the polar cone $P^* = \{\xi \in R^m \mid \xi^t p \leq 0, \, \forall \, p \in P\}$. In this scenario, $\mu_0^t g(x_0) = 0$ and $\sigma(\mu, P) = 0$, $\forall \, \mu \in P^*$. As a consequence, the saddle point inequalities become
$$(z_0 \circ f)(x_0) + \mu^t g(x_0) \leq (z_0 \circ f)(x_0)$$
$$\leq (z_0 \circ f)(x) + \mu_0^t g(x), \; \forall \, x \in X, \; \forall \, \mu \in P^*.$$
Thus a particular case of cone constraint follows on the expected lines.

It is clear from the above discussion that the support function and the normal cone to P will be instrumental in the subsequent results. We provide below two examples. The first example demonstrates the role of support function in the saddle point criteria while the second example illustrates the importance of the normal cone in case of set inclusion constraint.

Example 9.3.4. Consider the optimization problem
$$\text{Min}_{R_+^2} \; (-x_1, -x_2 + 1)$$
$$\text{subject to } (x_1 + x_2, x_1 - x_2) \in P$$
$$P = \{\xi \in R^2 \mid \xi_1 \leq 0, \, \xi_1^2 + 2\xi_2 \leq 0\}.$$

The feasible set S is described by
$$S = \{x \in R^2 \,|\, x_1 \leq -x_2,\ (x_1 + x_2 + 1)^2 - 4(x_2 + \tfrac{1}{4}) \leq 0\}.$$
It may be noted that $x_0 = (-1/2, 1/2)$ is an efficient solution of the problem. The support function of P is given by
$$\sigma(\mu, P) = \sup\{\xi_1\mu_1 + \xi_2\mu_2 \,|\, (\xi_1, \xi_2) \in P\}$$
$$= \begin{cases} \dfrac{\mu_1^2}{2\mu_2}, & \mu_1 \leq 0, \mu_2 > 0 \\ 0, & \mu_1 \geq 0, \mu_2 \geq 0 \\ +\infty, & \text{otherwise.} \end{cases}$$
For $\mu_0 = (1,0) \in N_P(g(x_0)) = \{\xi \in R^2 \,|\, \xi_1 \geq 0,\ \xi_2 = 0\}$,
$$(\mu - \mu_0)g(x_0) - \sigma(\mu, P) + \sigma(\mu_0, P) \leq 0$$
$$\Rightarrow \quad -\mu_2 - \sigma(\mu, P) \leq 0.$$
Observe that the last inequality holds for all $\mu \in R^2$.

Example 9.3.5. Consider the optimization problem
$$\operatorname{Min}_{R_+^2}\ (x_1, -x_2)$$
$$\text{subject to}\ g(x) = (x_1 + x_2,\ x_1 - x_2) \in P$$
where $P = \{\xi \in R^2 \,|\, \xi_1 + \xi_2 \geq 0,\ \xi_1 - \xi_2 \leq 0,\ 0 \leq \xi_2 \leq 1\}$.

The feasible set S is given as
$$S = \{x \in R^2 \,|\, 0 \leq x_1 \leq 1 + x_2,\ x_2 \leq 0\}.$$
It may be noted that $x_0 = (0,0)$ is an efficient solution of the problem. The standard Lagrange function of the problem is given by
$$L(x, \lambda, \mu) = \lambda^t f(x) + \mu^t g(x)$$
$$= (\lambda_1 - \mu_2 + \mu_3)x_1 + (-\lambda_2 + \mu_1 - \mu_3)x_2 - \mu_3,$$
$$\lambda_1, \lambda_2, \mu_i \geq 0,\ \forall\ i = 1, 2, 3.$$
For $\lambda_1 = \mu_2,\ \lambda_2 = \mu_1,\ \mu_3 = 0$, the inequality
$$L(x, \lambda, \mu) \geq L(x_0, \lambda, \mu),\ \forall\ x \in R^2,$$
holds. So, L possesses a saddle point of KKT type. However, the set P was completely ignored while constructing the Lagrange function. It is therefore not possible to dig out a relationship between μ and $N_P(g(x_0))$.

It may be noted that $N_P(g(x_0)) = \{\xi \in R^2 \mid \xi_2 \leq -|\xi_1|\}$, and

$$\sigma(\mu, P) = \begin{cases} \mu_1 + \mu_2, & \mu_1, \mu_2 \geq 0 \text{ or } \mu_1 \geq -\mu_2 \geq 0 \\ \mu_2 - \mu_1, & \mu_1 \leq 0, \mu_2 \geq 0 \text{ or } \mu_1 \leq \mu_2 \leq 0 \\ 0, & \mu_2 \leq \mu_1 \leq 0 \text{ or } 0 \leq \mu_1 \leq -\mu_2. \end{cases}$$

Recalling Example 9.2.2,

$$z_0(\xi_1, \xi_2) = \begin{cases} \xi_1, & \xi_1 \geq \xi_2 \\ \xi_2, & \xi_1 < \xi_2. \end{cases}$$

For saddle point, we must have

$$\mathcal{L}(x, z_0, \mu_0) \geq \mathcal{L}(x_0, z_0, \mu_0), \ \forall \ x \in R^2. \tag{9.4}$$

Choosing $\mu_0 = (1/2, -1/2)$, the inequality (9.4) holds. The second inequality of the saddle point can easily be verified. Hence, (x_0, z_0, μ_0) is a saddle point of $\mathcal{L}$. Observe that $\mu_0 \in N_P(g(x_0))$.

The following example is included to show that there does not exist any $\lambda_0 \in R_+^p$ and $\mu_0 \in R_+^m$ such that (x_0, λ_0, μ_0) is a saddle point of the standard Lagrange function L of the vector problem under consideration, whereas there exist $z_0 \in \mathcal{T}$ and $\mu_0 \in N_P(g(x_0))$ such that (x_0, z_0, μ_0) is a saddle point of the Lagrange function $\mathcal{L}$. In this way, the example signify the role of nonlinear Lagrange multipliers in nonlinear optimization problems. It is worth to remark here that the vector problem in the example is nonconvex.

Example 9.3.6. Consider the following vector optimization problem

$$\text{Min } \{x_1 x_2, \max(x_1, x_2)\}$$
$$\text{subject to } x_1^2 - x_2^2 \leq 0.$$

Clearly, $x_0 = (0, 0)$ is an efficient solution of the problem. The standard Lagrange function is

$$L(x, \lambda, \mu) = \lambda_1(x_1 x_2) + \lambda_2(\max(x_1, x_2)) + \mu(x_1^2 - x_2^2),$$

$\lambda_1, \lambda_2, \mu \geq 0$. It can be verified that for any $\lambda_0 \in R_+^2 \setminus \{0\}, \mu_0 \in R_+$, the saddle point inequality, $L(x, \lambda_0, \mu_0) \geq L(x_0, \lambda_0, \mu_0), \forall x \in R^2$, fails to hold. Consequently, the KKT saddle point of L does not exist.

Using Example 9.2.2,

$$(SP)_{z_0} \quad \text{Min}(z_0 \circ f)(x) = \begin{cases} f_1(x), & f_1(x) \geq f_2(x) \\ f_2(x), & f_1(x) < f_2(x) \end{cases}$$
$$\text{subject to } x_1^2 - x_2^2 \in R_-.$$

Note that $x_0 = (0, 0)$ is an optimal solution of $(SP)_{z_0}$.

The Lagrange function of $(SP)_{z_0}$ is given by

$$\mathcal{L}(x, z_0, \mu) = (z_0 \circ f)(x) + \mu(x_1^2 - x_2^2) - \sigma(\mu, P),$$

where $\mu \geq 0$. Observe that $\sigma(\mu, P) = 0$. It is clear that $(x_0, z_0, 0)$ is a saddle point of $\mathcal{L}$.

The following theorem provides the sufficient condition for the saddle point criteria.

Theorem 9.3.7. *Let $(x_0, z_0, \mu_0) \in X \times \mathcal{T} \times R^m$ be the saddle point of $\mathcal{L}$. Then x_0 is an optimal solution of $(SP)_{z_0}$, and hence $x_0 \in E((VPSIC), B)$.*

Proof. Using Remark 9.3.2, we have, $g(x_0) \in P$ and $\mu_0 \in N_p(g(x_0))$. Furthermore, the inequality $\mathcal{L}(x_0, z_0, \mu_0) \leq \mathcal{L}(x, z_0, \mu_0)$ of the saddle point gives

$$(z_0 \circ f)(x_0) + \mu_0^t g(x_0) \leq (z_0 \circ f)(x) + \mu_0^t g(x), \ \forall \ x \in X.$$

Since $\mu_0 \in N_P(g(x_0))$, we obtain

$$(z_0 \circ f)(x_0) \leq (z_0 \circ f)(x), \ \forall \ x \in S.$$

Therefore, x_0 is the optimal solution for $(SP)_{z_0}$. The requisite result follows from Theorem 9.2.5. $\square$

We next present an example to illustrate Theorem 9.3.7.

Example 9.3.8. Consider the following vector optimization problem

$$\text{Min}_B \ f(x) = (-x_1, -x_2)$$
$$\text{subject to } g(x) = (x_1 - x_2, x_1 + x_2) \in P$$

where $P = \{\xi \in R^2 \,|\, \xi_1^2 + 2\xi_2 \leq 0\}$ and $B = \{\xi \in R^2 \,|\, \xi_1 + \xi_2 > 0\}$.

Observe that the feasible set is given by $\{x \in R^2 \mid (x_1 - x_2 + 1)^2 + 4(x_2 - \frac{1}{4}) \leq 0\}$. It can easily be verified that $x_0 = (0,0)$ is an efficient solution of the problem with respect to B.

For $C = -B$, $k^0 = (1/2, 1/2)^t$, the separating function z_0 is given by
$$z_0(\xi) = \xi_1 + \xi_2.$$

The support function of P is as follows
$$\sigma(\mu, P) = \begin{cases} +\infty, & \mu_2 \leq 0, (\mu_1, \mu_2) \neq (0,0) \\ 0, & \mu_1 = 0, \mu_2 \geq 0 \\ \dfrac{\mu_1^2}{2\mu_2}, & \mu_1 > 0, \mu_2 > 0 \text{ or } \mu_1 < 0, \mu_2 > 0. \end{cases}$$

Let us choose $\mu_0 = (0,1)^t \in N_P(g(x_0))$. Then,
$$(z_0 \circ f)(x) + \mu_0^t g(x) \geq 0, \forall\, x \in R^2;$$
$$\sigma(\mu_0, P) = 0,\ \sigma(\mu, P) \geq 0, \forall\, \mu \in R^2,$$
and hence we obtain that (x_0, z_0, μ_0) is a saddle point of $\mathcal{L}$.

Obviously $x_0 = (0,0)$ is an optimal solution of the scalarized problem.
$$\text{Min } (-x_1 - x_2)$$
$$\text{subject to } g(x) \in P.$$

9.4 Necessary Optimality Criteria

Our next aim is to derive the necessary saddle point characterization to ensure the existence of efficient solution of $(VPSIC)$. Generally, necessary saddle point criteria require convexity or some kind of generalized convexity assumptions. As we plan to work in a general setting, without involving any kind of differentiability, and at the same time avoiding the convexity conditions on the domain set of the function, we looked for a suitable generalized convexity conditions relevant to the present context. For this reason, we work under convexlike-concavelike assumptions. These notions are well known and widely studied in literature. For instance, one can refer to [Adán and Novo (2005); Craven, Gwinner and Jeyakumar (1987); Jeyakumar (1985, 1986); Li (1988)] and references therein. Below we present the concept of vector convexlike function for ready reference.

Definition 9.4.1 (see [Craven, Gwinner and Jeyakumar (1987)]).
Let X, K_1 and K_2 be nonempty subsets of R^n, R^q and R^l respectively. Let $\psi_1 : X \to R^q$ and $\psi_2 : X \to R^l$. Then the vector function $(\psi_1, \psi_2) : X \to R^q \times R^l$ is called $K_1 \times K_2$-convexlike on X if

$$(\exists\, \lambda \in (0,1))(\forall\, x_1, x_2 \in X)(\exists\, x_3 \in X)$$
$$\lambda \psi_1(x_1) + (1-\lambda)\psi_1(x_2) - \psi_1(x_3) \in K_1,$$
$$\lambda \psi_2(x_1) + (1-\lambda)\psi_2(x_2) - \psi_2(x_3) \in K_2.$$

Observe that the condition $(\exists\, \alpha \in (0,1))$ has replaced the condition $(\forall\, \alpha \in (0,1))$ in the earlier definition of convexlike as given by [Jeyakumar (1986)]. In this sense, the definition of [Craven, Gwinner and Jeyakumar (1987)] is a weaker version of convexlike condition of [Jeyakumar (1986)].

It is worth mentioning here that in order to have a vector function (ψ_1, ψ_2) to be convexlike on X we need to have the existence of a common point $x_3 \in X$, depending on x_1, x_2, α, satisfying the above two inequalities. Thus, it is very much possible that the two individual functions are convexlike on X but if taken together to form a vector function, the resultant vector function fails to be convexlike on X. For instance, let $X = \{x \in R_+^2 \,|\, x_1 + x_2 > 2\} \bigcup \{(2,0),(0,2)\}$, $\psi_1(x) = x_1$, $\psi_2(x) = x_2$. Then, $\psi_1(\cdot)$ and $\psi_2(\cdot)$ are R_+-convexlike functions on X but $\psi(\cdot) = (\psi_1(\cdot), \psi_2(\cdot))$ fails to be R_+^2-convexlike vector function on X by considering the points $x_1 = (2,0)$, $x_2 = (0,2)$.

In a process to prove the main result of this section, we will be needing the following lemma and a theorem of alternative.

Define sets $U = \{\varsigma \in R^m \,|\, ||\varsigma|| = 1\}$ and $W = \{w = (\alpha, \xi) \,|\, 0 \leq \alpha \leq 1, \xi \in R^m, ||\xi|| \leq 1\}$. For any $w = (\alpha, \xi) \in W$, we can always find some $\beta \in [0,1]$ and $\varsigma \in U$ such that $\xi = \beta \varsigma$, and conversely.

For $z \in \mathcal{T}$ and $x_0 \in X$, define a function $H : X \times W \to R$ as
$$H(x, w) = \alpha(z \circ f(x) - z \circ f(x_0)) + \xi^t g(x) - \sigma(\xi, P)\,. \qquad (9.5)$$

Lemma 9.4.2. *If, for every $\varsigma \in U$, $(f, \varsigma^t g)$ is $B \times R_+$-convexlike on X then $H(\cdot, w)$ is R_+-convexlike on X.*

Proof. Let $w = (\alpha, \xi) \in W$ be any arbitrary point. Then there exist $\beta \in [0,1]$ and $\varsigma \in U$ such that $\xi = \beta \varsigma$. For this $\varsigma \in U$, using the given hypothesis, we obtain

$$(\exists\ \lambda \in (0,1))(\forall\ x_1, x_2 \in X)(\exists\ x_3 \in X)$$
$$\lambda f(x_1) + (1-\lambda)f(x_2) - f(x_3) \in B,$$
$$\lambda \varsigma^t g(x_1) + (1-\lambda)\varsigma^t g(x_2) - \varsigma^t g(x_3) \geq 0.$$

Using B-monotonocity and convexity of z along with $\alpha,\ \beta \geq 0$, we get

$$\alpha\lambda(z(f(x_1)) - z(f(x_0))) + \alpha(1-\lambda)(z(f(x_2)) - z(f(x_0)))$$
$$\geq \alpha(z(f(x_3)) - z(f(x_0))), \quad (9.6)$$
$$\beta\lambda(\varsigma^t g(x_1) - \sigma(\varsigma, P)) + \beta(1-\lambda)(\varsigma^t g(x_2) - \sigma(\varsigma, P))$$
$$\geq \beta(\varsigma^t g(x_3) - \sigma(\varsigma, P)). \quad (9.7)$$

Adding (9.6) and (9.7) and using $\xi = \beta\varsigma$ we obtain

$$\lambda H(x_1, w) + (1-\lambda)H(x_2, w) \geq H(x_3, w).$$

Thus, $H(\cdot, w)$ is convexlike on X. □

Remark 9.4.3. For every $x \in X$, $H(x, \cdot)$ is linear function with respect to $w \in W$.

Theorem 9.4.4. *Let ϑ be the optimal value of the problem $(SP)_{\bar{z}}$, for some $\bar{z} \in \mathcal{T}$. Suppose, for every $\varsigma \in U$, the vector function $(f, \varsigma^t g)$ is $B \times R_+$-convexlike on X. Then there exist $\lambda_0 \in R_+$ and $\mu_0 \in R^m$, $(\lambda_0, \mu_0) \neq 0$, such that*

$$\lambda_0 \vartheta = \inf\{\lambda_0(\bar{z} \circ f)(x) + \mu_0^t g(x) - \sigma(\mu_0, P) \mid x \in X\}.$$

Proof. In view of Lemma 9.4.2 and Remark 9.4.3, the function $H(\cdot, \cdot)$ satisfies the following symmetrical F and K conditions (SFK3) and (SFK4) of [Craven, Gwinner and Jeyakumar (1987)]

$\forall\ \epsilon > 0$, we have

$$(\exists\ \lambda \in (0,1))(\forall\ w_1, w_2 \in W)(\exists\ w_3 \in W)$$
$$\lambda H(x, w_1) + (1-\lambda)H(x, w_2) \leq H(x, w_3) + \epsilon,\ \forall\ x \in X \quad \text{(SFK3)}$$

$$(\exists\ \lambda \in (0,1))(\forall\ x_1, x_2 \in X)(\exists\ x_3 \in X)$$
$$H(x_3, w) \leq \lambda H(x_1, w) + (1-\lambda)H(x_2, w) + \epsilon,\ \forall\ w \in W. \quad \text{(SFK4)}$$

Furthermore, for each $x \in X$, $\varsigma^t g(x) - \sigma(\varsigma, P)$ is a linear (so positively homogenous and continuous) function on U. Obviously $R^m = R_+ U$ and

$0 \notin U$. Thus, we observe that all the conditions required in Theorem 3.1 of [Craven, Gwinner and Jeyakumar (1987)] are met. Hence the proof of the present theorem can now be immediately worked out on the lines of Theorem 3.1 of [Craven, Gwinner and Jeyakumar (1987)]. □

We are now in a position to establish the necessary optimality criteria in terms of the saddle point.

Theorem 9.4.5. *Let x_0 be an optimal solution for $(SP)_{\bar{z}}$, for some $\bar{z} \in \mathcal{T}$. Further, for any $\varsigma \in U$, let $(f, \varsigma^t g)$ be $B \times R_+$-convexlike function on X and the following constraint qualification holds*

$$(CQ) \quad \exists\, \hat{x} \in S \text{ such that } N_P(g(\hat{x})) = \{0\}.$$

Then there exists $\mu_0 \in N_P(g(x_0))$ such that $(x_0, \bar{z}, \mu_0)$ is a saddle point of $\mathcal{L}$.

Proof. Invoking Lemma 9.4.2 and Remark 9.4.3, the function $H : X \times W \to R$ given by (9.5) satisfies the symmetrical Fuchssteiner and König's conditions $(SFK3) - (SFK4)$. It now follows from Theorem 9.4.4 that $\exists\, \lambda_0 \in R_+,\, \mu_0 \in R^m,\, (\lambda_0, \mu_0) \neq 0$ such that

$$\lambda_0(\bar{z} \circ f)(x_0) = \inf\{\lambda_0(\bar{z} \circ f)(x) + \mu_0^t g(x) - \sigma(\mu_0, P) | x \in X\}.$$

Using the definition of the support function and the imposed (CQ), it is a routine to show that $\lambda_0 \neq 0$, and hence it can be taken as unity. Thereby yielding

$$(\bar{z} \circ f)(x_0) \leq (\bar{z} \circ f)(x) + \mu_0^t g(x) - \sigma(\mu_0, P), \ \forall\, x \in X. \tag{9.8}$$

As $g(x_0) \in P$, we have

$$\mu_0^t g(x_0) - \sigma(\mu_0, P) \leq 0. \tag{9.9}$$

Adding inequalities (9.8) and (9.9), we obtain

$$\mathcal{L}(x_0, \bar{z}, \mu_0) \leq \mathcal{L}(x, \bar{z}, \mu_0), \quad \forall\, x \in X.$$

By taking $x = x_0$ in (9.8) and using (9.9), we obtain, $\sigma(\mu_0, P) - \mu_0^t g(x_0) = 0$.

Furthermore, for any $\mu \in R^m$, $\sigma(\mu, P) \geq \mu^t p, \forall\, p \in P$. Thus, $\mu^t g(x_0) - \sigma(\mu, P) \leq 0, \forall\, \mu \in R^m$. Consequently, $\mathcal{L}(x_0, \bar{z}, \mu) \leq \mathcal{L}(x_0, \bar{z}, \mu_0), \forall\, \mu \in R^m$. Thus the desired saddle point inequalities are achieved, completing the proof of the theorem. □

9.5 Duality

The optimality criteria obtained in the previous two sections can easily be applied to derive the Lagrange duality results for $(SP)_{\bar{z}}$, $\bar{z} \in \mathcal{T}$.

$$(SP)_{\bar{z}} \quad \text{Min}(\bar{z} \circ f)(x)$$
$$\text{subject to } g(x) \in P.$$

The dual function is given by

$$\underset{\mu \in R^m}{\text{Max}} \mathcal{L}(x, \bar{z}, \mu) = \underset{\mu \in R^m}{\text{Max}} \left((\bar{z} \circ f)(x) + \mu^t g(x) - \sigma(\mu, P) \right).$$

It may be noted that

$$\underset{\mu \in R^m}{\text{Max}} \left(\mu^t g(x) - \sigma(\mu, P) \right) = 0, \ \forall \ x \in S.$$

If $g(x) \notin P$ then by the strong separation theorem $\exists \ \mu \in R^m$ and $\delta \in R$ such that

$$\mu^t g(x) > \delta \ \text{ and } \ \mu^t p \leq \delta, \ \forall \ p \in P.$$

Consequently,

$$\underset{\mu \in R^m}{\text{Max}} \left(\mu^t g(x) - \sigma(\mu, P) \right) \to +\infty.$$

Thus, $(SP)_{\bar{z}}$ can be reformulated as

$$\underset{x \in X}{\text{Min}} \underset{\mu \in R^m}{\text{Max}} \mathcal{L}(x, \bar{z}, \mu).$$

We propose the dual of $(SP)_{\bar{z}}$ as

$$(SD)_{\bar{z}} \qquad \underset{\mu \in R^m}{\text{Max}} \underset{x \in X}{\text{Min}} \mathcal{L}(x, \bar{z}, \mu).$$

It is always true that

$$\underset{\mu \in R^m}{\text{Max}} \underset{x \in X}{\text{Min}} \mathcal{L}(x, \bar{z}, \mu) \leq \underset{x \in X}{\text{Min}} \underset{\mu \in R^m}{\text{Max}} \mathcal{L}(x, \bar{z}, \mu). \tag{9.10}$$

The next theorem provides conditions which ensure the above inequality to hold as an equation thereby establishing the strong duality between the primal-dual pair of problems.

Theorem 9.5.1. *Let $x_0 \in E((VPSIC), B)$. Assume that for every $\varsigma \in U$, $(f, \varsigma^t g)$ is $B \times R_+$-convexlike on X and let the constraint qualification (CQ) holds. Then $\exists \ \mu_0 \in N_P(g(x_0))$ such that*

$$\underset{x \in X}{\text{Min}} \underset{\mu \in R^m}{\text{Max}} \mathcal{L}(x, \bar{z}, \mu) = \underset{\mu \in R^m}{\text{Max}} \underset{x \in X}{\text{Min}} \mathcal{L}(x, \bar{z}, \mu) = \mathcal{L}(x_0, \bar{z}, \mu_0).$$

Proof. It follows from Theorem 9.2.4 that $\exists \, \bar{z} \in \mathcal{T}$ such that x_0 is an optimal solution of $(SP)_{\bar{z}}$. Invoking Theorem 9.4.5, there exists $\mu_0 \in N_P(g(x_0))$ such that $(x_0, \bar{z}, \mu_0)$ is a saddle point of $\mathcal{L}$. Therefore

$$\underset{\mu \in R^m}{\text{Max}} \left(\underset{x \in X}{\text{Min}} \, \mathcal{L}(x, \bar{z}, \mu) \right) \geq \underset{x \in X}{\text{Min}} \, \mathcal{L}(x, \bar{z}, \mu_0) \geq \mathcal{L}(x_0, \bar{z}, \mu_0),$$

$$\underset{x \in X}{\text{Min}} \left(\underset{\mu \in R^m}{\text{Max}} \, \mathcal{L}(x, \bar{z}, \mu) \right) \leq \underset{\mu \in R^m}{\text{Max}} \, \mathcal{L}(x_0, \bar{z}, \mu) \leq \mathcal{L}(x_0, \bar{z}, \mu_0).$$

Thus,

$$\underset{\mu \in R^m}{\text{Max}} \underset{x \in X}{\text{Min}} \, \mathcal{L}(x, \bar{z}, \mu) \geq \underset{x \in X}{\text{Min}} \underset{\mu \in R^m}{\text{Max}} \, \mathcal{L}(x, \bar{z}, \mu).$$

The above inequality along with (9.10) yields the requisite result. □

9.6 Concluding Remarks

Many research papers have appeared in literature where the saddle point principles are established for vector optimization problems. The main line of investigation in these papers has been to first derive a separation theorem by assuming convexity or some form of generalized convexity conditions on the functions and later as an application of this theorem obtain the optimality criteria in terms of the saddle point. Instead, in the present chapter, we apply the nonconvex separation theorem to study the saddle point criteria for vector optimization problem with set inclusion constraint $(VPSIC)$. The problem $(VPSIC)$ is scalarized using the nonconvex separation theorem. Once the scalarized problem is obtained, it facilitates us to formulate a Lagrange function and introduce the notion of nonlinear saddle point. One important aspect is the interpretation of the Lagrange multipliers. The multiplier associated with the objective function is a convex functional while the multiplier corresponding to the constraint function lies in the normal cone to the set involved in the constraint of $(VPSIC)$. Although in order to work out the complete characterization of the saddle point we lately took help of the symmetrical Fuchssteiner and König's conditions and a generalized Motzkin's theorem of alternative of [Craven, Gwinner and Jeyakumar (1987)]. The Lagrange duality follows as an immediate consequence of the saddle point characterization.

We would like to remark here that without restoring to scalarization, the concept of vector Lagrange function and thereby the concept of vector saddle point for $(VPSIC)$ can be discussed. It would be interesting to develop vector Lagrange duality results for $(VPSIC)$. Moreover one can also attempt to extend the nonsmooth necessary optimality conditions of [Dutta and Tammer (2006)] from the locally Lipschitz's scalar optimization problems with set constraint to the vector optimization problems. It would be worth to explore the nature and physical interpretation of the Lagrange multipliers for the vector Lagrangian of $(VPSIC)$.

Acknowledgements. The authors are thankful to the referees for their useful suggestions.

Bibliography

Adán, M. and Novo, V. (2005). Duality and saddle-points for convex-like vector optimization problems on real linear spaces, *TOP* **13**, pp. 343–357.

Bector, C. R., Chandra, S. and Dutta, J. (2005). *Principles of Optimization Theory*, (Narosa Publishing House, India).

Craven, B. D., Gwinner, J. and Jeyakumar, V. (1987). Nonconvex theorems of the alternative and minimization, *Optimization* **18**, pp. 151–163.

Duc, D. M., Hoang, N. D. and Nguyen, L. H. (2006). Lagrange multipliers theorem and saddle point optimality criteria in mathematical programming, *J. Math. Anal. Appl.* **323**, pp. 441–455.

Dutta, J. and Tammer, C. (2006). Lagrangian conditions for vector optimization in Banach spaces, *Math. Meth. Oper. Res.* **64**, pp. 521–540.

Ehrgott, M. and Wiecek, M. M. (2005). Saddle points and Pareto points in multiple objective programming, *J. Global Optim.* **32**, pp. 11–33.

Gerth, C. and Weidner, P. (1990). Nonconvex separation theorems and some applications in vector optimization, *J. Optim. Theory Appl.* **67**, pp. 297–320.

Göpfert, A., Riahi, H., Tammer, C., Zăliomesci, C. (2003). *Variational Methods in Partially Ordered Spaces*, (Springer, Berlin).

Hernández, E. and RodrÍGuez-Marin, L. (2007). Nonconvex scalarization in set optimization with set-valued maps, *J. Math. Anal. Appl.* **323**, pp. 1–18.

Jeyakumar, V. (1985). Convexlike alternative theorems and mathematical programming, *Optimization* **16**, pp. 643–652.

Jeyakumar, V. (1986). A generalization of a minmax theorem of Fan via a theorem of alternative, *J. Optim. Theory Appl.* **48**, pp. 525–533.

Li, S. (1988). Saddle points for convex programming in order vector spaces, *Optimization* **21**, pp. 343–349.

Luc, D. T. (1989). *Theory of Vector Optimization*, Lecture Notes in Economics and Mathematical Systems, 319, (Springer-Verlag, Berlin).

Sposito, V. A. and David, H. T. (1971). Saddle point optimality criteria of nonlinear programming problems over cones without differentiability, *SIAM J. Appl. Math.* **20**, pp. 698–702.

Tadeusz, A. (2005). Saddle point criteria and duality in multiobjective programming via an η-approximation method, *ANZIAM J.* **47**, pp. 155–172.

Tammer, Chr. (1992). A generalization of Ekeland's variational principle, *Optimization* **25**, pp. 129–141.

Wang, S. and Li, Z. F. (1992). Scalarization and Lagrange duality in multiobjective programming, *Optimization* **26**, pp. 315–324.

Zlobec, S. (2004). Saddle-point optimality: a look beyond convexity, *J. Global Optim* **29**, pp. 97–112.

Chapter 10

Approximate Optimality in Semi Infinite Programming

Deepali Gupta* and Aparna Mehra
Department of Mathematics, Indian Institute of Technology Delhi
Hauz Khas, New Delhi-110016, India
e-mail:iit_deepali@yahoo.com*

Abstract

In this chapter we define the concept of approximate optimal solutions for semi infinite programming problems. The KKT type necessary optimality conditions, characterizing approximate optimal solutions, are derived using the exact penalty function approach. Finally we provide the bound on the penalty parameter in terms of dual variables so as to obtain an almost approximate solution for the primal semi infinite programming problem.

Key Words: Semi infinite programming, approximate optimal solution, approximate optimality conditions, exact penalty function

10.1 Introduction

Numerous models in physical and social sciences require consideration of the constraints on the state or the control of the system during the whole period of time or at every point in the geometric region. Mathematical formulation of these models leads to the functional inequality constraints which are characteristic of the "Semi Infinite Programming (SIP)". The term SIP refers to the class of problems involving infinite number of constraints while the number of variables remain finite. The specific semi infinite programming problem considered herein is of the following form:

$$\text{(SIP)} \quad \min f(x)$$
$$\text{subject to } g(x,y) \leq 0, \ y \in Y,$$
$$x \in X$$

where $f : \mathbb{R}^n \to \mathbb{R}$, $g : \mathbb{R}^n \times Y \to \mathbb{R}$, $X \subset \mathbb{R}^n$ is a convex set, $Y \subset \mathbb{R}^m$ is an arbitrary index set.

The aforementioned (SIP) problem can be seen as an extension of the finite non-linear programming problem with corresponding $Y = \{1,\ldots,m\}$. Extensive research work has been done for (SIP). The rich bibliographies in the review articles by [Hettich and Kortanek (1993)] and very recent one by [Lopez and Still (2007)], reflect the interest (SIP) and its variants have generated over the years. The theoretical research articles [Goberna and Lopez (2001); Price and Coope (1990); Conn and Gould (1987)] are also worth mentioning. On the computational side, various globally convergent algorithms [Price and Coope (1996); Watson, G.A. (1981)] have been developed for (SIP).

It is often noticed that the iterative algorithms designed to solve optimization problems converge to the exact optimal solutions in the limiting scenario only. Though this is theoretically acceptable, but in practice, it sounds more reasonable for the algorithms to produce approximate solutions of the problems in finite number of iterations. Moreover, it is very common that an optimization problem fails to possess an optimal solution. One such simple problem of linear (SIP) was constructed way back in 1962 by [Charnes, Cooper and Kortanek (1963)]. These observations are strong enough for us to talk about the approximate solutions in semi infinite programming. We found that no serious attempt has been made in this direction and hence in this chapter we venture into the zone of approximate solutions for (SIP).

The concept of ϵ-solutions can be considered as a satisfactory compromise with a given prescribed error $\epsilon > 0$. Due to the importance of ϵ-solutions in applications [Kazmi (2001); Loridan (1984)], a great deal of attention has been given on developing conditions identifying such solutions in nonlinear programming problems (see, for instance, [Hiriart-Urruty and Lemarechal (1993)], [Loridan (1984)], [Loridan and Morgan (1993)]).

The purpose of this chapter is to develop the necessary optimality conditions for a feasible point to be an approximate solution for (SIP). The L_∞ exact penalty function is used as a main tool for achieving this aim.

The organization of the chapter is described as follows. Section 10.2 provides some examples to demonstrate precisely what constitutes an approximate solution of (SIP), while Section 10.3 presents the prerequisites necessary to understand the subsequent discussion and derives the necessary optimality conditions for approximate optimality of (SIP) using the L_∞ exact penalty function. Section 10.4 develops the bound on the penalty parameter in terms of the Lagrange multipliers. The penultimate section concludes with some remarks.

10.2 Approximate Solution

The feasible set of (SIP) is denoted by $C = \{x \in X : g(x,y) \leq 0,\ \forall\, y \in Y\}$. We assume C is nonempty. We denote the optimal value of (SIP) by

$$v := \inf_{x \in C} f(x).$$

We also assume that $v > -\infty$. For $\epsilon \geq 0$, we define the set of ϵ-minimizers of (SIP) as

$$\mathcal{O}_\epsilon(SIP) = \left\{x \in C : f(x) \leq \inf_{x \in C} f(x) + \epsilon\right\}.$$

Let $f : \mathbb{R}^n \to \mathbb{R} \cup \{+\infty\}$ be a convex function, finite at $\bar{x}$. The ϵ-subdifferential of f at $\bar{x}$ is the set $\partial_\epsilon f(\bar{x})$ defined by

$$\partial_\epsilon f(\bar{x}) = \{\xi \in \mathbb{R}^n\,|\, f(x) \geq f(\bar{x}) + <\xi, x - \bar{x}> -\epsilon, \forall\, x \in \mathbb{R}^n\},$$

and the ϵ-normal set to the set X at $\bar{x} \in X$ is given as

$$\mathcal{N}_{X,\epsilon}(\bar{x}) = \{\xi \in \mathbb{R}^n\,|\, <\xi, x - \bar{x}> \leq \epsilon, \forall\, x \in X\}.$$

Before beginning with our main discussion, it is important to understand the need and meaning of approximate solution for (SIP).

We present below the linear (SIP) formulated by [Charnes, Cooper and Kortanek (1963)], and also mentioned in the introduction, which does not possess an optimal solution though the ϵ-optimal solution can easily be computed.

Example 10.1. Consider

$$-\min -t$$
$$\text{subject to} \quad x(1+y^2)^{-1} - t \geq -\tan^{-1} y + y(1+y^2)^{-1},\ y \in [0, \infty)$$
$$0 \leq t \leq \pi/2.$$

The points of the form $(y, \tan^{-1} y)$, $y \in [0, \infty)$, are feasible points. The optimal value of the problem is $\pi/2$ but is not attained, as when $y \to \infty$, $t = \tan^{-1} y \to \pi/2$. But the problem possess an ϵ-optimal solution for any ϵ, $0 \leq \epsilon \leq \pi/2$. In fact, it can be shown that $(\hat{x}, \hat{t}) = (\cot \epsilon, \pi/2 - \epsilon)$ is an ϵ-optimal solution.

The above phenomenon is not only restricted to the linear case but in fact is more prevalent in the nonlinear setting.

Example 10.2. Consider

$$\begin{aligned} \min \quad & exp\,(x_1)/2x_2 \\ \text{subject to} \quad & x_2 - x_1 \geq y \\ & x_1 + 2x_2 \geq 3y \\ & x_1 \geq 0,\, x_2 \geq 0,\, y \in Y = [0, 1]. \end{aligned}$$

The problem does not possess an exact optimal solution. But for $\epsilon = 1/3$, $\hat{x} = (0, 3/2)$ is an ϵ-optimal solution.

In lieu of above simple examples we must appreciate the concept of approximate solutions and recognize that it may be more appropriate to talk about them in context of (SIP).

10.3 Approximate Optimality Conditions

We begin this section by presenting some basic calculus results concerning ϵ-subdifferential and ϵ-normal cone.

Lemma 10.1. *[see, Example 4.1 in [Hiriart-Urruty and Lemarechal (1993)]] Let $h : \mathbb{R}^n \to \mathbb{R}$ be a convex function and $\bar{x} \in \mathbb{R}^n$. Then the exact expression for $\partial_\epsilon (max(h(.), 0))(\bar{x})$ is given as*

$$\partial_\epsilon (max(h(.), 0))(\bar{x}) = \bigcup_{0 \leq \alpha \leq 1} \partial_{\eta(\alpha)} (\alpha h)(\bar{x})$$

where $\eta(\alpha) = \epsilon - max(h(\bar{x}), 0) + \alpha h(\bar{x})$.

Lemma 10.2. *[see, Theorem 4.2 in [Strodiot, Nguyen Hien and Heukemes (1983)]] Let Y be a compact subset of $\mathbb{R}^m$, and $g : \mathbb{R}^n \times Y \to \mathbb{R}$ be convex in $x \in X$ and upper semicontinuous in $y \in Y$. Let $\bar{x} \in \mathbb{R}^n$. Then $\xi \in$*

$\partial_\epsilon(\max_{y\in Y} g(.,y))(\bar{x})$ if and only if there exists $n+1$ vectors $y_1,\ldots,y_{n+1} \in Y$, scalars $\lambda_i \geq 0$ for $i = 1,\ldots,n+1$, with $\sum_{i=1}^{n+1}\lambda_i = 1$ and $\epsilon_i \geq 0$ for $i = 1,\ldots,n+1$ such that

(i) $\xi \in \sum_{i=1}^{n+1} \partial_{\epsilon_i}(\lambda_i g(\cdot,y_i))(\bar{x}).$

(ii) $0 \leq \max_{y\in Y} g(\bar{x},y) - \sum_{i=1}^{n+1}\lambda_i g(\bar{x},y_i) \leq \epsilon - \sum_{i=1}^{n+1}\epsilon_i.$

We note the following important facts concerning the approximate subdifferential of sum of functions.

Lemma 10.3. *(Sum-Rule)[see, Theorem 3.1.1 in [Hiriart-Urruty and Lemarechal (1993)]] If $h_1, h_2 : \mathbb{R}^n \to \mathbb{R}$ are lower semicontinuous convex functions then we have*

$$\partial_\epsilon (h_1+h_2)(x) \supset \bigcup_{\epsilon_1,\epsilon_2\geq 0,\ \epsilon_1+\epsilon_2=\epsilon}\{\partial_{\epsilon_1}h_1(x) + \partial_{\epsilon_2}h_2(x)\}.$$

If in addition one of them is continuous, then

$$\partial_\epsilon (h_1+h_2)(x) = \bigcup_{\epsilon_1,\epsilon_2\geq 0,\ \epsilon_1+\epsilon_2=\epsilon}\{\partial_{\epsilon_1}h_1(x) + \partial_{\epsilon_2}h_2(x)\}.$$

Lemma 10.4. *[see, Proposition 4.1.1 in [Hiriart-Urruty and Lemarechal (1993)]] Let $h : \mathbb{R}^n \to \mathbb{R}$. Suppose (ϵ_k,x_k,ξ_k) is a sequence in $\mathbb{R}_+ \times \mathbb{R}^n \times \mathbb{R}^n$ converging to some (ϵ,x,ξ) with $\xi_k \in \partial_{\epsilon_k}h(x_k)$, $\forall\ k$. Then $\xi \in \partial_\epsilon h(x)$.*

Remark 10.1. If $h(x) = \delta_X(x)$, where $\delta_X(\cdot)$ is the indicator function of set X, then the consequence of the above Lemma is, $\xi \in \mathcal{N}_{X,\epsilon}(x)$.

For finite nonlinear programs, the optimal solutions are characterized as points at which the Karush-Kuhn Tucker (KKT) conditions hold. A common approach to solve the constrained nonlinear programming problem is to employ the L_1 exact penalty function and convert the problem into the unconstrained programming problem. Thereafter iteratively solve the problem by sequentially adjusting the penalty parameter. The penalty function method works exceptionally well and proves to be very handy for obtaining the optimal solutions of nonlinear programs.

In contrast, in (SIP), the L_1 exact penalty function is found to possess discontinuities in the infeasible region even if the objective and the constraint functions are continuous everywhere, whereas, the L_∞ exact penalty function prevents such discontinuities and potentially a better choice for SIP problems.

In this section, we derive the first order KKT type necessary optimality conditions for approximate solutions for (SIP) using the L_∞ exact penalty function approach under some assumptions which make (SIP) tractable.

Assumption 10.1 (A1). *Let $f : \mathbb{R}^n \to \mathbb{R}$ be lower semicontinuous and level bounded (hence bounded below). Also, let f be convex on X.*

Assumption 10.2 (A2). *Let $g : \mathbb{R}^n \times Y \to \mathbb{R}$ be convex and continuous in $x \in X$ and upper semicontinuous in $y \in Y$.*

In studying KKT type of optimality conditions, constraint qualification plays an important role. In the classical sense, that is when Y is finite and each g_y is differentiable, the Mangasarin Fromovitz constraint qualification (MFCQ), and the basic constraint qualification (BCQ) are well known. The MFCQ have two equivalent forms: one is expressed in terms of the functions g_y, while the other is expressed in terms of the gradient of g_y. It is the latter form which led to the origination of the BCQ. The interrelation among them have been studied both in smooth and nonsmooth cases. Furthermore, MFCQ and BCQ have also been studied for general case: Y is a compact set and g_y are smooth and/or non smooth convex functions [see [Jongen, Twilt and Weber (1992)] and [Li and Ng (2005)]]. The following constraint qualification ensures that each finite subsystem of (SIP) satisfies the BCQ. We call this constraint qualification as extended BCQ.

Assumption 10.3 (EBCQ). *If, for $\delta_i \geq 0$, $y_i \in Y$, $i = 1, \ldots, n+1$,*

$$0 \in \sum_{i=1}^{n+1} \partial \left(\delta_i g(\cdot, y_i) \right)(x^*) + \mathcal{N}_X(x^*),$$

then $\delta_i = 0$, $\forall\, i = 1, \ldots, n+1$.

Theorem 10.1. *Consider (SIP) with X a closed set and Y a compact set. Let (A1), (A2) and (EBCQ) be satisfied and $x_0 \in \mathcal{O}_\epsilon(SIP)$. Then there* exist $x^* \in \mathcal{O}_\epsilon(SIP)$ and scalars $\tilde{\epsilon}^*$, ϵ'_i, β_i, $\bar{\epsilon}$, all non-negative and vectors $y_i \in Y$, $i = 1, \ldots, n+1$ such that

$$(3(a)) \quad 0 \in \partial_{\tilde{\epsilon}^*} f(x^*) + \sum_{i=1}^{n+1} \partial_{\epsilon'_i} \left(\beta_i g(., y_i)\right)(x^*) + \mathcal{N}_{X,\bar{\epsilon}}(x^*),$$

$$(3(b)) \quad \sum_{i=1}^{n+1} \epsilon'_i + \tilde{\epsilon}^* + \bar{\epsilon} - \epsilon \leq \sum_{i=1}^{n+1} \beta_i g(x^*, y_i) \leq 0.$$

Proof. Consider the L_∞-penalty problem

$$\min F^k(x) = f(x) + k \max(\max_{y \in Y} g(x,y), 0), \text{ as } k \to \infty, \; x \in X,$$

where $k > 0$ is the penalty parameter.
Since F^k is bounded below on X, for any $\epsilon_k > 0$ there exists $x_k \in X$ such that

$$F^k(x_k) \leq F^k(x) + \epsilon_k, \; \forall \, k, \; \forall \, x \in X. \tag{10.1}$$

We choose the sequence $\{\epsilon_k\}$ such that $\epsilon_k \to \epsilon$. Rewriting (10.1), we have

$$f(x_k) + k \max(\max_{y \in Y} g(x_k, y), 0) \leq f(x) + k \max(\max_{y \in Y} g(x, y), 0) + \epsilon_k.$$

In particular for $x = x_0$, we get

$$f(x_k) + k \max(\max_{y \in Y} g(x_k, y), 0) \leq f(x_0) + \epsilon_k, \; \forall \, k, \tag{10.2}$$

and thus $\max(\max_{y \in Y} g(x_k, y), 0) \to 0$.

Also as f is level bounded on a closed set X, x_k converges to some $x^* \in X$. Using continuity of $g(., y)$ on X, we get, as $k \to \infty$, $\max(\max_{y \in Y} g(x_k, y), 0)$ converges to $\max(\max_{y \in Y} g(x^*, y), 0)$. By uniqueness of limit, $\max(\max_{y \in Y} g(x^*, y), 0) = 0$, implying $x^* \in C$.

Consequently, (10.2) and $x_0 \in \mathcal{O}_\epsilon(SIP)$ yields

$$f(x_0) - \epsilon \leq f(x^*) \leq f(x_0) + \epsilon.$$

Also, from (10.1) we may note that $\lim_{k \to \infty} k \max(\max_{y \in Y} g(x_k, y), 0)$ is finite say, l. Then, $l \in [0, \epsilon]$. Furthermore, from (10.1) we have that x_k is an ϵ_k-optimal solution of (SIP), as

$$f(x_k) \leq F^k(x_k) \leq f(x) + \epsilon_k, \; \forall \, k, \; \forall \, x \in C. \tag{10.3}$$

Taking limit as $k \to \infty$ and using the lower semicontinuity of f we get that $x^* \in \mathcal{O}_\epsilon(SIP)$. Moreover, x_k is an ϵ_k-optimal solution of the problem: $\min_{x \in X} F^k(x)$, thus

$$0 \in \partial_{\epsilon_k}(F^k(\cdot) + \delta_X(\cdot))(x_k).$$

Using the sum-rule (Lemma 10.3),

$$0 \in \partial_{\epsilon_k^*} F^k(x_k) + \mathcal{N}_{X,\bar{\epsilon}_k}(x_k),$$

with $\epsilon_k = \epsilon_k^* + \bar{\epsilon}_k$, $\forall\, k$.

Again applying Lemma 10.3, we get

$$0 \in (\partial_{\tilde{\epsilon}_k^*} f)(x_k) + \partial_{\hat{\epsilon}_k^*}(k\, \max(\max_{y \in Y} g(\cdot, y), 0))(x_k) + \mathcal{N}_{X,\bar{\epsilon}_k}(x_k),$$

where $\epsilon_k = \epsilon_k^* + \bar{\epsilon}_k = \tilde{\epsilon}_k^* + \hat{\epsilon}_k^* + \bar{\epsilon}_k$.

Now, from Lemma 10.1,

$$\partial_{\hat{\epsilon}_k^*}(k\, \max(\max_{y \in Y} g(\cdot, y), 0))(x_k) = \bigcup_{0 \leq \alpha_k \leq 1} \partial_{\eta(\alpha_k)}(k\, \alpha_k \max_{y \in Y} g(\cdot, y))(x_k), \text{ where}$$

$$\eta(\alpha_k) = \hat{\epsilon}_k^* - k\, \max(\max_{y \in Y} g(x_k, y), 0) + k\, \alpha_k \max_{y \in Y} g(x_k, y). \quad (10.4)$$

So, there exists $\alpha_k \in [0,1]$ such that (10.4) holds and

$$0 \in (\partial_{\tilde{\epsilon}_k^*} f)(x_k) + \partial_{\epsilon_k'}(k\, \alpha_k \max_{y \in Y} g(\cdot, y))(x_k) + \mathcal{N}_{X,\bar{\epsilon}_k}(x_k), \text{ where } \epsilon_k' = \eta(\alpha_k).$$

Now using Lemma 10.2, there exist $y_i^k \in Y$, $i = 1, \ldots, n+1$, and scalars $\lambda_i^k \geq 0$, $i = 1, \ldots, n+1$, $\sum_{i=1}^{n+1} \lambda_i^k = 1$, such that

$$0 \in \partial_{\tilde{\epsilon}_k^*} f(x_k) + \sum_{i=1}^{n+1} \partial_{\epsilon_{k_i}'} \left(k\, \alpha_k \lambda_i^k g(\cdot, y_i^k)\right)(x_k) + \mathcal{N}_{X,\bar{\epsilon}_k}(x_k), \quad (10.5)$$

where

$$\left.\begin{array}{c} 0 \leq k\, \alpha_k \max_{y \in Y} g(x_k, y) - \sum_{i=1}^{n+1} k\, \alpha_k \lambda_i^k g(x_k, y_i^k) \leq \epsilon_k' - \sum_{i=1}^{n+1} \epsilon_{k_i}', \\ \epsilon_{k_i}' \geq 0,\ \forall\, i = 1, \ldots, n+1. \end{array}\right\} \quad (10.6)$$

Substituting (10.4) in (10.6) and simplifying, we get

$$0 \leq k \max(\max_{y \in Y} g(x_k, y), 0) - \sum_{i=1}^{n+1} k\, \alpha_k \lambda_i^k g(x_k, y_i^k) \leq \hat{\epsilon}_k^* - \sum_{i=1}^{n+1} \epsilon'_{k_i}. \quad (10.7)$$

The sequence on RHS of the above inequality is bounded, therefore the sequence on LHS must also be bounded, and being a non-negative sequence we get that the sequence $\left\{ k \max(\max_{y \in Y} g(x_k, y), 0) - \sum_{i=1}^{n+1} k\, \alpha_k \lambda_i^k g(x_k, y_i^k) \right\}$ has a convergent subsequence. We retain the index k to denote the subsequence.

Note that the sequences $\{\hat{\epsilon}_k^*\}$, $\{\epsilon'_{k_i}\}$ are bounded thus without loss of generality we may assume that $\hat{\epsilon}_k^* \to \epsilon^*$ and $\epsilon'_{k_i} \to \epsilon'_i$. Now, taking limit as $k \to \infty$ in (10.7) and using the definition of l, we get

$$0 \leq l - \lim_{k \to \infty} \sum_{i=1}^{n+1} k\, \alpha_k \lambda_i^k g(x_k, y_i^k) \leq \hat{\epsilon}^* - \sum_{i=1}^{n+1} \epsilon'_i.$$

Let $\beta_i^k = k\, \alpha_k \lambda_i^k$, $\forall\, i = 1, \ldots, n+1$. Then,

$$\sum_{i=1}^{n+1} \epsilon'_i - \hat{\epsilon}^* \leq \lim_{k \to \infty} \sum_{i=1}^{n+1} \beta_i^k g(x_k, y_i^k) - l \leq 0. \quad (10.8)$$

We take a pause here and go back to recall (10.5), i.e.,

$$0 \in \partial_{\bar{\epsilon}_k^*} f(x_k) + \sum_{i=1}^{n+1} \partial_{\epsilon'_{k_i}} \left(\beta_i^k g(., y_i^k)\right)(x_k) + \mathcal{N}_{X, \bar{\epsilon}_k}(x_k). \quad (10.9)$$

Moreover, $y_i^k \in Y$, Y being compact, yields $y_i^k \to y_i$, for some $y_i \in Y$, $i = 1, \ldots, n+1$. Also, let $\bar{\epsilon}_k \to \bar{\epsilon}$.

If, for any $i = 1, \ldots, n+1$, the sequence $\{\beta_i^k\}_k$ is unbounded then it leads to the violation of (EBCQ) on account of Lemma 10.4 and Remark 10.1. So, $\{\beta_i^k\}$ are bounded sequences for all $i = 1, \ldots, n+1$. We may assume, $\beta_i^k \to \beta_i \geq 0$, $\forall\, i = 1, \ldots, n+1$.

It is important to observe that for each k, $\partial_{\epsilon'_{k_i}} \left(\beta_i^k g(., y_i^k)\right)(x_k)$, $i = 1, \ldots, n+1$ are closed convex bounded sets. Hence invoking Lemma 10.4 and in view of Theorem 3.5 (page 81) and 4.29(d) (page 127) of [Rockafellar and Wets (1998)], we get

$$\lim_{k \to \infty} \sum_{i=1}^{n+1} \partial_{\epsilon'_{k_i}} \left(\beta_i^k g(., y_i^k)\right)(x_k) = \sum_{i=1}^{n+1} \partial_{\epsilon'_i} \left(\beta_i g(., y_i)\right)(x^*),$$

in the sense of set convergence.

Thus, condition (10.9) on taking the limit and using Remark 10.1 reduces to

$$0 \in \partial_{\bar{\epsilon}^*} f(x^*) + \sum_{i=1}^{n+1} \partial_{\epsilon'_i} \left(\beta_i g(., y_i) \right)(x^*) + \mathcal{N}_{X,\bar{\epsilon}}(x^*),$$

thereby yielding (3(a)).

Moreover, (10.8) along with the facts that $\beta_i \geq 0$, $x^* \in C$, $l \in [0, \epsilon]$, implies

$$\sum_{i=1}^{n+1} \epsilon'_i + \tilde{\epsilon}^* + \bar{\epsilon} - \epsilon \leq \sum_{i=1}^{n+1} \beta_i g(x^*, y_i) \leq 0.$$

Thus (3(b)) holds. □

The following example illustrates the above result.

Example 10.3. Consider

$$\begin{aligned}
\min \quad & f(x_1, x_2) = x_1^2 + x_2 - 1 \\
\text{subject to} \quad & \\
& g_1(x_1, x_2, y) = x_1^2 y - x_2 \leq 0 \\
& g_2(x_1, x_2, y) = y - (x_1 + x_2) \leq 0 \\
& x_1 \geq 0, x_2 \geq 0, y \in Y = [0, 1].
\end{aligned}$$

Note that (A1), (A2) are all satisfied. Let $\epsilon = 1/3$. Then, $x_0 = (0, 1)$ is an ϵ-optimal solution to the problem.

Also, there exists $x^* = (0.25, 0.75)$, feasible for the problem, such that x^* is an ϵ-optimal solution, $|f(x^*) - f(x_0)| = 0.1875 < \epsilon$ and (EBCQ) is satisfied at x^*.

Also, $g_1(x^*, y) = 0.0625y - 0.75$, $g_2(x^*, y) = y - 1$. We will be using the double subscripts to denote the scalars corresponding to two semi infinite constraints $g_1(x_1, x_2, .)$ and $g_2(x_1, x_2, .)$.

Let $\tilde{\epsilon} = 0.25$, $\hat{\epsilon} = 0.04$, $\bar{\epsilon} = 0.04$,
$\epsilon_{11} = 0.01$, $\epsilon_{12} = 0.005$, $\epsilon_{13} = 0$,
$\epsilon_{21} = 0$, $\epsilon_{22} = 0.01$, $\epsilon_{23} = 0.01$,
$\beta_{11} = 0.002$, $\beta_{12} = 0$, $\beta_{13} = 0.001$,
$\beta_{21} = 0.98$, $\beta_{22} = 0.004$, $\beta_{23} = 0$,
$y_{11} = 0.2$, $y_{12} = 0.5$, $y_{13} = 0.333$,
$y_{21} = 1.0$, $y_{22} = 0.75$, $y_{23} = 0.111$.

From ϵ-subdifferential calculus, we have that
$\partial_{\bar{\epsilon}} f(x^*) = ([-0.5, 1.5], 1)$
$\partial_{\epsilon_{11}} (\beta_{11}\ g_1)(x^*, y_{11}) = ([-0.0038, 0.0042], -0.002)$
$\partial_{\epsilon_{12}} (\beta_{12}\ g_1)(x^*, y_{12}) = (0, 0)$
$\partial_{\epsilon_{13}} (\beta_{13}\ g_1)(x^*, y_{13}) = (0.0001667, -0.001)$
$\partial_{\epsilon_{21}} (\beta_{21}\ g_2)(x^*, y_{21}) = (-0.98, -0.98)$
$\partial_{\epsilon_{22}} (\beta_{22}\ g_2)(x^*, y_{22}) = (-0.004, -0.004)$
$\partial_{\epsilon_{23}} (\beta_{23}\ g_2)(x^*, y_{23}) = (0, 0)$
$\mathcal{N}_{X,\bar{\epsilon}}(x^*) = (\xi_1, [-0.04, 0]), \xi_1 \leq 0$, we choose $\xi_1 = -0.1$.
Then,
$(0,0) \in ([-1.58763, 0.42037], [-0.027, 0.013]) = \partial_{\bar{\epsilon}} f(x^*) +$
$\partial_{\epsilon_{11}}(\beta_{11}\ g_1)(x^*, y_{11}) + \partial_{\epsilon_{12}}(\beta_{12}\ g_1)(x^*, y_{12}) + \partial_{\epsilon_{13}}(\beta_{13}\ g_1)(x^*, y_{13}) +$
$\partial_{\epsilon_{21}}(\beta_{21}\ g_2)(x^*, y_{21}) + \partial_{\epsilon_{22}}(\beta_{22}\ g_2)(x^*, y_{22}) + \partial_{\epsilon_{23}}(\beta_{23}\ g_2)(x^*, y_{23}) +$
$\mathcal{N}_{X,\bar{\epsilon}}(x^*)$

and,
$\epsilon_{11} + \epsilon_{12} + \epsilon_{13} + \epsilon_{21} + \epsilon_{22} + \epsilon_{23} - \hat{\epsilon} = -0.005 < -0.003204 = \beta_{11}\ g_1(x^*, y_{11}) + \beta_{12}\ g_1(x^*, y_{12}) + \beta_{13}\ g_1(x^*, y_{13}) + \beta_{21}\ g_2(x^*, y_{21}) + \beta_{22}\ g_2(x^*, y_{22}) + \beta_{23}\ g_2(x^*, y_{23})$.

Hence (3(a)) and (3(b)) holds.

10.4 Bounded KKT Multipliers

In the section we investigate a natural question that what is the the size of the penalty parameter k such that a single unconstrained minimization of the penalty problem, $\min F^k(x)$, yields an approximate solution to (SIP). This threshold is desirable as it makes the performance of the penalty method less dependent on the strategy for updating the penalty parameter. For utility of this concept, one can refer to [Han and Mangasarian (1979)].

Let $\mathbb{R}^Y$ be the space of all functions $\lambda : Y \to \mathbb{R}$. Consider the space of functions $\lambda \in \mathbb{R}^Y$ such that only a finite number of $\lambda(y)$ are non-zero. We denote this space by $\mathbb{R}^{(Y)}$. For $\lambda \in \mathbb{R}^Y$, $\lambda^* \in \mathbb{R}^{(Y)}$, define the scalar product $<\lambda^*, \lambda> := \sum_{y \in Y} \lambda^*(y)\lambda(y)$, where the summation is such $y \in Y$ that $\lambda(y) \neq 0$.

Define the set $\mathbb{R}^Y_+$ as follows:
$$\mathbb{R}^Y_+ := \{\lambda : Y \to \mathbb{R} : \lambda(y) \geq 0,\ \forall\ y \in Y\}.$$

Then (SIP) can be written as
$$\min_{x \in X} f(x) \text{ subject to } -g(x,.) \in \mathbb{R}_+^Y.$$

Consider the dual of the cone $\mathbb{R}_+^Y$,
$$(\mathbb{R}_+^Y)^* := \left\{ \lambda^* \in \mathbb{R}^{(Y)} : \langle \lambda^*, \lambda \rangle \geq 0, \, \forall \lambda \in \mathbb{R}_+^Y \right\},$$

and the Lagrangian function $L : X \times (\mathbb{R}_+^Y)^* \to \mathbb{R}$, where $L(x, \lambda^*) := f(x) + \langle \lambda^*, g(x,.) \rangle$, hence problem (SIP) can be written in the min-max form
$$\min_{x \in X} \max_{\lambda^* \in (\mathbb{R}_+^Y)^*} L(x, \lambda^*).$$

Then the (Lagrangian) dual of (SIP) is the problem
$$\text{(DSIP)} \quad \sup_{\lambda^* \in (\mathbb{R}_+^Y)^*} \left\{ \psi(\lambda^*) := \inf_{x \in X} L(x, \lambda^*) \right\}.$$

The set of ϵ-maximizers of (DSIP) is defined as follows:
$$\mathcal{O}_\epsilon(DSIP) = \left\{ \lambda^* \in (\mathbb{R}_+^Y)^* : \psi(\lambda^*) \geq \sup_{\lambda^* \in (\mathbb{R}_+^Y)^*} \psi(\lambda^*) - \epsilon \right\}.$$

A point $\bar{x}$ is said to be an almost ϵ-optimal solution to (SIP) if $\bar{x} \in C_\epsilon = \{x \in X : g(\bar{x}, y) \leq \epsilon, \, \forall y \in Y\}$ and
$$f(\bar{x}) \leq \inf_{x \in C} f(x) + \epsilon.$$

Remark 10.2. From the assumptions $(A1)$, $(A2)$ and (EBCQ) and closedness of X it follows that f is bounded below on C and attains its bound. Thus, the optimal solution set of (SIP) is non-empty and bounded. Consequently from Proposition 3.4 in [Shapiro (2005)], the duality gap between (SIP) and (DSIP) is zero.

Theorem 10.2. *Let $\bar{x}$ be an ϵ-optimal solution to the penalty problem*
$$\min F^k(x) = f(x) + k \, \max(\max_{y \in Y} g(x, y), 0) \text{ subject to } x \in X,$$
and $\bar{\lambda}^* \in \mathcal{O}_\epsilon(DSIP)$.

We set the penalty parameter k as
$$k \geq 3 + \|\bar{\lambda}^*\|_\infty,$$
where $\|\bar{\lambda}^*\|_\infty = \max_{\bar{\lambda}^* \in (\mathbb{R}_+^Y)^*} |\bar{\lambda}^*|$. Then, $\bar{x}$ is an almost ϵ-solution to (SIP).

Proof. By given hypothesis,

$$f(\bar{x}) + k \max(\max_{y \in Y} g(\bar{x}, y), 0) \leq f(x) + \epsilon, \; \forall \, x \in C.$$

Furthermore, there exists $\hat{x} \in X$ such that

$$f(\hat{x}) - \epsilon \leq \inf_{x \in C} f(x).$$

Thus,

$$f(\bar{x}) + k \max(\max_{y \in Y} g(\bar{x}, y), 0) \leq f(\hat{x}) + \epsilon \leq \inf_{x \in C} f(x) + 2\epsilon. \quad (10.10)$$

Moreover, from Remark 10.2 and $\bar{\lambda} \in \mathcal{O}_\epsilon(DSIP)$, we get

$$\inf_{x \in C} f(x) = \sup_{\lambda \in (\mathbb{R}_+^Y)^*} \psi(\lambda) \leq \psi(\bar{\lambda}) + \epsilon.$$

So that

$$\inf_{x \in C} f(x) \leq \psi(\bar{\lambda}) + \epsilon \leq f(\bar{x}) + \bar{\lambda} g(\bar{x}, y) + \epsilon.$$

Now, using (10.10) we get

$$f(\bar{x}) + k \max(\max_{y \in Y} g(\bar{x}, y), 0) \leq f(\bar{x}) + \bar{\lambda} g(\bar{x}, y) + 3\epsilon$$

This implies that

$$k \max(\max_{y \in Y} g(\bar{x}, y), 0) \leq \bar{\lambda} g(\bar{x}, y) + 3\epsilon$$
$$\leq \bar{\lambda} \max(\max_{y \in Y} g(\bar{x}, y), 0) + 3\epsilon,$$

giving

$$(k - \bar{\lambda}) \max(\max_{y \in Y} g(\bar{x}, y), 0) \leq 3\epsilon.$$

So for all $k \geq 3 + \|\bar{\lambda}\|_\infty$, we get

$$\max(\max_{y \in Y} g(\bar{x}, y), 0) \leq \epsilon.$$

Hence, $\bar{x} \in C_\epsilon$. $\square$

10.5 Concluding Remarks

As published papers in literature confirm, approximate optimality is significant and have played important role in nonlinear programming. In this chapter, we have shown the importance of approximate optimal solutions in semi infinite programming. The approximate optimal solution for (SIP) is characterized in terms of the necessary KKT optimality conditions using the L_∞ exact penalty function approach.

We believe that the exact penalty function approach can also be employed to (SIP) with locally Lipschitz objective and constraint functions. The reason being that the Clarke subdifferential enjoys the similar properties as the ϵ-subdifferential. So, by taking the aid of variational principles, optimality conditions for Lipschitz (SIP) can be worked out as well.

Acknowledgements. The authors are thankful to the referees for their useful suggestions.

Bibliography

Charnes, A., Cooper, W. W. and Kortanek, K. (1963). Duality in semi-infinite programming and some works of Haar and Carathéodory, *Management Science* **9**, pp. 209–228.

Conn, A. R. and Gould, N. I. M. (1987). An exact penalty function for semi-infinite programming, *Mathematical Programming* **37**, pp. 19–40.

Goberna, M.A. and Lopez, M.A. (Eds) (2001). *Semi Infinite Programming. Recent Advances*, (Kluwer Verlag).

Han, S. P. and Mangasarian, O. L. (1979). Exact penalty functions in nonlinear programming, *Mathematical Programming* **17**, pp. 251–269.

Hettich, R. and Kortanek, K. O. (1993). Semi infinite programming: Theory, methods and applications, *SIAM Review* **35**, pp. 380–429.

Hiriart-Urruty, J. B. and Lemarechal, C. (1993). *Convex analysis and minimization algorithms II*, A Series of Comprehensive Studies in Mathematics, (Springer-Verlag, Berlin).

Jongen, H. T., Twilt, F. and Weber, G. W. (1992). Semi infinite optimization : structure and stability of the feasible set, *Journal of Optimization Theory and Applications* **72** , pp. 529–552.

Kazmi, K. R. (2001). Existence of ϵ-minima for vector optimization problem, *Journal of Optimization Theory and Applications* **109**, pp. 667–674.

Li, C and Ng, K. F. (2005). On constraint qualifications for an infinite system of convex inequalities in a Banach space, *SIAM Journal of Optimization* **15**, pp. 488–512.

Lopez, M. and Still, G. (2007). Semi-infinite programming, *European Journal of Operations Research* **180**, pp. 491–578.

Loridan, P. (1984). ϵ-solutions in vector minimization problems, *Journal of Optimization Theory and Applications* **43**, pp. 265–276.

Loridan, P. and Morgan, J. (1993). Penalty functions in ϵ-programming and ϵ-minimax problems, *Mathematical Programming* **26**, pp. 213–231.

Price, C. J. and Coope, I. D. (1990). An exact penalty function algorithm for semi-infinite programmes, *BIT* **30**, pp. 723–734.

Price, C. J. and Coope, I. D. (1996). Numerical experiments in semi-infinite programming, *Computational Optimization and Application* **6**, pp. 169–189.

Rockafellar, R. T. and Wets, R. J. B. (1998). Variational Analysis, (Springer, Heidelberg).

Shapiro, A. (2005). On duality theory of convex semi-infinite programming, *Optimization* **54**, 535–543.

Strodiot, J-J., Nguyen Hien, V. and Heukemes, N. (1983). ϵ-optimal solutions in non-differentiable convex programming and some related questions, *Mathematical Programming* **25**, pp. 307–328.

Watson, G.A. (1981). Globally convergent methods for semi-infinite programming, *BIT* **21**, pp. 362–373.

Chapter 11

A Relaxation Based Solution Approach for the Inventory Control and Vehicle Routing Problem in Vendor Managed Systems

Oğuz Solyalı and Haldun Süral
Department of Industrial Engineering
Middle East Technical University, Ankara 06531 Turkey,
e-mail: solyali@metu.edu.tr, sural@ie.metu.edu.tr

Abstract

We consider a one supplier - multiple retailers system over a finite planning horizon. Retailers have external demands for a single product and their inventories are controlled by the supplier based on order-up-to level inventory policy. The problem is to determine the time and the quantity of product to order for the supplier, the retailers to be visited in any period, the quantity of product to be delivered in these visits and the vehicle routes for deliveries so as to minimize system-wide inventory and routing costs. We present a Lagrangian relaxation based solution procedure and implement the procedure on test instances. Computational study shows that fairly good solutions are found in reasonable time.

Key Words: Inventory routing problem; order-up-to level inventory policy; Lagrangian relaxation

11.1 Introduction

Vendor managed inventory (VMI) systems have been implemented in several industries and their benefits are well acknowledged [Cetinkaya and Lee (2000); Campbell and Savelsbergh (2004)]. VMI systems where the vendor decides on when to replenish customers entail coordination among different companies or different echelons of the same company. Numerous researchers have investigated coordination of inventory, production and distribution related decisions on a single model. A review of such models can be found in [Thomas and Griffin (1996); Sarmiento and Nagi (1999); Baita, Ukovich, Pesenti and Favaretto (1998)]. Some of them have empirically shown the savings that can be obtained by a coordinated approach over a decoupled (sequential) approach [Chandra (1993); Chandra and Fisher (1994); Fumero and Vercellis (1999)].

Integration of inventory management and vehicle routing problem, referred to as Inventory Routing Problem (IRP), is an important problem in a VMI setting. IRP has been studied both for deterministic demand [Fumero and Vercellis (1999); Bertazzi, Paletta and Speranza (2002, 2005); Campbell, Clarke and Savelsbergh (2002); Campbell and Savelsbergh (2004); Pınar and Süral (2006); Archetti, Bertazzi, Laporte and Speranza (2007)] and stochastic demand [Kleywegt, Nori and Savelsbergh (2002, 2004); Berman and Larson (2001)] under different inventory control policies. For a recent review on inventory routing problems, the reader may refer to [Moin and Salhi (2007)].

Bertazzi et al. [Bertazzi, Paletta and Speranza (2002)] introduce a deterministic order-up-to level inventory control policy for a multiple period IRP and present a heuristic procedure to solve the problem with different objective functions. Pınar and Süral [Pınar and Süral (2006)] develop a mathematical programming model for the problem and suggest a solution approach based on Lagrangian relaxation. Bertazzi et al. [Bertazzi, Paletta and Speranza (2005)], similar to the present paper, consider an extension of the problem in which the supplier has a replenishment (production) problem and propose heuristic procedures to solve the problem. Solyalı and Süral [Solyalı and Süral (2008)] propose a polynomial algorithm for the single supplier-single retailer version of the problem considered therein. Lastly, Archetti et al. [Archetti, Bertazzi, Laporte and Speranza (2007)] develop an exact branch-and-cut algorithm for the problem considered in [Bertazzi, Paletta and Speranza (2002); Pınar and Süral (2006)].

In this study, we consider a VMI system in which a single supplier delivers a product to multiple retailers over a finite planning horizon. Retailers that have external demands are replenished by the supplier via a fleet of vehicles departing from the supplier's depot and visiting retailers on the route to make deliveries. Each retailer has given minimum and maximum levels of inventory, it has to be replenished before its inventory level drops below the minimum level, and its inventory is brought up to the maximum level in each replenishment, called deterministic order-up-to level inventory policy. In addition, the supplier has to order from a higher echelon (or manufacture) in order to be able to service retailers. This problem arises frequently in distribution of industrial gases and soft drinks and in replenishment of vending machines as well as shelves in grocery stores, where tanks (or shelf spaces or vending machines) are filled up to their capacity whenever replenished. It makes inventory control and distribution routing decisions concurrently so as to exploit benefits of coordination such as reduced costs and enhanced service levels. As demand is mostly stochastic in real-life, the minimum inventory level that must be maintained at retailers due to the deterministic policy provides a hedging against the variability of demand. Accurate demand forecasts justify usage of such a policy. Another aspect of the present study is the supplier's order policy. Instead of assuming a given amount made available in each period at the supplier as in [Bertazzi, Paletta and Speranza (2002); Pınar and Süral (2006); Archetti, Bertazzi, Laporte and Speranza (2007)], we consider that the supplier does replenish itself and determines when and how much to order, as in [Bertazzi, Paletta and Speranza (2005)]. Ours is the first study that attempts to develop a mathematical programming based approach to solve such a problem. In short, the problem addressed in this study, referred to as the inventory routing problem with order-up-to level policy (IRO), is to simultaneously determine (i) retailers to be visited, (ii) quantity of product to be delivered to retailers, (iii) routes of vehicles, (iv) quantity of product to be ordered to supplier in each period so that the total of fixed order and inventory holding costs at the supplier and fixed vehicle dispatching, routing and inventory holding costs at the retailers is minimized. Note that even if the decisions concerning (i), (ii) and (iv) are already given, determining only routes of vehicles is NP-hard. Thus, the problem we consider is extremely difficult to solve.

Motivation of the study is to develop a mathematical programming based approach to the IRO whose lower bounding procedure has not been proposed yet. We introduce a general model that can handle different cost structures as well as replenishment policies and propose Lagrangian relaxation based bounding procedures for the model. Our main contribution is that the suggested algorithm decomposes the overall problem into disaggregated replenishment problems for the retailers and the supplier, which enable us to address various replenishment policies at the retailers and the supplier with slight modifications of the model. The chapter is organized as follows. In Section 11.2, we give a mathematical programming formulation of the problem. The solution approach is described next in Section 11.3. Section 11.4 is devoted to the computational study on test instances and to the discussion of results of those test instances. Finally, we conclude the chapter and discuss future research issues in Section 11.5.

11.2 Problem Definition and Formulation

We consider a VMI system, in which a single product is delivered from supplier, denoted by 0, to a set of $N = \{1, \ldots, R\}$ retailers over a set $\tau = \{1, \ldots, T\}$ of discrete time periods. Retailer $i \in N$ faces deterministic dynamic demand d_{it} in period $t \in \tau$ and has a minimum (maximum) inventory level $s_i(S'_i)$ of the product. A fixed cost $f_{0t}(f_t)$ is incurred every time $t \in \tau$ supplier (or any retailer) places an order. An inventory holding cost h_{it} is charged for every unit of product at the end of period $t \in \tau$ at location $i \in \overline{N}$ where $\overline{N} = \{0\} \cup N$. Shipments from the supplier to the retailers are made by a set of $\vartheta = \{1, \ldots, V\}$ vehicles, each with capacity K. We assume that V is a large number. There are transportation costs c_{ij} from location $i \in \overline{N}$ to $j \in \overline{N}$. The problem is to find the optimal replenishment policy both at supplier and retailers as well as the routes of vehicles that minimize the total of fixed, inventory holding and transportation costs.

Let $y_t = 1$ if an order of size P_t for supplier is given in period $t \in \tau$ and 0 otherwise. Let I'_{it} and I_{0t} be the inventory levels of retailer $i \in N$ and supplier at the end of $t \in \tau$, respectively. The retailer has to be replenished before its inventory level I'_{it} drops below the minimum level s_i. The beginning inventory levels both at the supplier I_{00} and at the retailers I'_{i0} are assumed to be known. Also, let X_{ijvt} be the amount of product transported from location $i \in \overline{N}$ to $j \in \overline{N}$ by vehicle $v \in \vartheta$ in period $t \in \tau$, Q_{ivt} the amount of product delivered to retailer $i \in N$ by vehicle $v \in \vartheta$ in period $t \in \tau$, $w_{ijvt} = 1$ if location $j \in \overline{N}$ is visited immediately after location $i \in \overline{N}$ by vehicle $v \in \vartheta$ in period $t \in \tau$, 0 otherwise, and $z_{it} = 1$ if a quantity is delivered to retailer $i \in N$ in period $t \in \tau$ up to the maximum level, 0 otherwise. Then, if retailer $i \in N$ receives a shipment $\sum_{v \in V} Q_{ivt}$ from supplier in period $t \in \tau$, its inventory level I'_{it} is brought up to S'_i, i.e. $\sum_{v \in V} Q_{ivt} = S'_i - I'_{it}$. Without loss of generality, we do a simplification as follows. Let $M_i = S'_i - s_i$, $S_i = S'_i - s_i$, $I'_{it} - s_i = I_{it} \geq 0$, and $I'_{i0} - s_i = I_{i0}$ for $i \in N, t \in \tau$. We thus transform the problem into an equivalent problem without s_i. Although we consider the same problem as [Bertazzi, Paletta and Speranza (2005)], there exist differences in the cost structure and the receiving quantities of retailers. They assume a unit production cost at the supplier and a fixed transportation cost that is incurred if a vehicle is used at least once throughout the planning horizon. We do not consider unit production cost and incur a fixed vehicle dispatching cost whenever a vehicle is used in a period. Bertazzi et al. [Bertazzi, Paletta and Speranza (2005)] assume that if a retailer is replenished in period $t \in \tau$ then it receives the quantity $S_i - I_{i,t-1} + d_{it}$ whereas this paper assume that the quantity received is equal to $S_i - I_{i,t-1}$. The formulation is as follows.

$$IRO: Min \sum_{t \in \tau} f_{0t} y_t + \sum_{t \in \tau} h_{0t} I_{0t} + \sum_{i \in N} \sum_{t \in \tau} h_{it}(I_{it} + s_i)$$

$$+ \sum_{j \in N} \sum_{v \in \vartheta} \sum_{t \in \tau} f_t w_{0jvt} + \sum_{i \in \overline{N}, i \neq j} \sum_{j \in \overline{N}} \sum_{v \in \vartheta} \sum_{t \in \tau} c_{ij} w_{ijvt} \quad (11.1)$$

$$s.t. \quad I_{0t} = I_{0,t-1} + P_t - \sum_{i \in N} \sum_{v \in \vartheta} Q_{ivt} \quad \forall t \in \tau \quad (11.2)$$

$$I_{it} = I_{i,t-1} + \sum_{v \in \vartheta} Q_{ivt} - d_{it} \quad \forall i \in N, t \in \tau \quad (11.3)$$

$$P_t \leq M_0 y_t \quad \forall t \in \tau \quad (11.4)$$

$$\sum_{j \in \overline{N}, j \neq i} X_{jivt} - \sum_{j \in \overline{N}, i \neq j} X_{ijvt} = Q_{ivt} \quad \forall i \in N, v \in \vartheta, t \in \tau \quad (11.5)$$

$$\sum_{j \in N} X_{0jvt} = \sum_{i \in N} Q_{ivt} \quad \forall v \in \vartheta, t \in \tau \quad (11.6)$$

$$\sum_{j \in N} X_{j0vt} = 0 \quad \forall v \in \vartheta, t \in \tau \quad (11.7)$$

$$\sum_{j \in N} w_{0jvt} \leq 1 \quad \forall v \in \vartheta, t \in \tau \quad (11.8)$$

$$\sum_{j \in \overline{N}, i \neq j} w_{ijvt} = \sum_{j \in \overline{N}, j \neq i} w_{jivt} \quad \forall i \in \overline{N}, v \in \vartheta, t \in \tau \quad (11.9)$$

$$X_{ijvt} \leq K w_{ijvt} \quad \forall i, j (i \neq j) \in \overline{N}, v \in \vartheta, t \in \tau \quad (11.10)$$

$$\sum_{v \in \vartheta} Q_{ivt} + I_{i,t-1} \leq S_i \quad \forall i \in N, t \in \tau \quad (11.11)$$

$$\sum_{v \in \vartheta} Q_{ivt} \leq S_i z_{it} \quad \forall i \in N, t \in \tau \quad (11.12)$$

$$S_i z_{it} - I_{i,t-1} \leq \sum_{v \in \vartheta} Q_{ivt} \quad \forall i \in N, t \in \tau \quad (11.13)$$

$I_{it} \geq 0, \forall i \in \overline{N}, t \in \tau; Q_{ivt} \geq 0, \forall i \in N, v \in \vartheta, t \in \tau;$
$P_t \geq 0, \forall t \in \tau; X_{ijvt} \geq 0, \forall i, j(i \neq j) \in \overline{N}, v \in \vartheta, t \in \tau;$
$w_{ijvt} \in \{0,1\} \forall i, j(i \neq j) \in \overline{N}, v \in \vartheta, t \in \tau;$
$y_t \in \{0,1\}, \forall t \in \tau; z_{it} \in \{0,1\}, \forall i \in N, t \in \tau \quad (11.14)$

The objective function (11.1) of the model consists of fixed order and inventory holding costs at the supplier, and inventory holding, fixed vehicle dispatching and transportation costs at the retailers. Constraints (11.2) and (11.3) are the inventory balance equations for the supplier and retailers, respectively. Constraint (11.4) ensures that a fixed order cost is incurred if supplier places an order in a period. Constraints (11.5), (11.6) and (11.7) are the flow conservation equations at the retailers and supplier, respectively. Constraint (11.6) stipulates that the total amount of product sent from the supplier should be equal to the total amount left to the visited retailers while constraint (11.7) ensures that no amount of product are transported back to the supplier. Constraint (11.8) states that each vehicle can perform at most one trip in a period. Constraint (11.9) and (11.10) together with (11.5), (11.6) and (11.7) assure feasibility of tours. Constraint (11.11) stipulates that the total amount delivered to retailer plus inventory carried from previous period cannot exceed the maximum level. Constraint (11.11) together with (11.12) and (11.13) are the order-up-to level constraints. They ensure that the inventory level at the retailer is brought up to the maximum level if a delivery is made and no amount is shipped unless a delivery is made. Constraints (11.14) are for nonnegativity and integrality of variables.

The model is flexible so that one can insert a cost term composed of X variables instead of w's into the objective function to represent transportation costs proportional to the amount shipped. In addition, if the fixed order and vehicle dispatching costs are removed, P_t is set as a constant, and a constraint accounting for delivery lead times is added, one would come up with a model for the problem in [Archetti, Bertazzi, Laporte and Speranza (2007); Bertazzi, Paletta and Speranza (2002); Pınar and Süral (2006)]. Moreover, it can be easily extended to account for multiple products by appending a subscript denoting product type to the related variables.

IRO is a mixed integer programming (MIP) model involving $2N^2VT + 3NVT + 2NT + 3T$ variables, $N^2VT + NVT + NT + T$ of which are integer (binary). The formulation has $N^2VT + 2NVT + 4NT + 4VT + 2T$ constraints. In a problem instance with 50 retailers, 30 periods and a single vehicle, which is the instance size we experiment with (see Section 11.4), the model would have over 150,000 variables made up of about 75,000 binary ones, in addition to approximately 84,000 constraints. Therefore, it is difficult to solve such a large scale MIP formulation to optimality.

11.3 Lagrangian Relaxation Based Solution Approach

There are various successful applications of Lagrangian relaxation to the difficult MIP problems in the literature (e.g. see [Fisher (1981, 1985)]). This study relaxes a set of constraints in IRO in a Lagrangian manner such that the replenishment and the distribution planning problems are separated from each other. It thus disaggregates the problem of multiple retailers into N single retailer problems. It is achieved by relaxing constraints (11.2), (11.5) and (11.6) in the model, using the multipliers μ_t, α_{ivt} and β_{vt}, respectively. The resulting relaxed problem, referred to as PR, decomposes into four subproblems: (i) supplier's order problem, (ii) supplier's inventory problem, (iii) distribution problem, (iv) retailers' replenishment problem. The solution of four subproblems yields a lower bound to the problem (its solution value is denoted by LB). Also, using the information obtained from the solution of subproblems, a feasible solution (its solution value results in an upper bound on the objective function of the problem, denoted by UB) is constructed for IRO.

11.3.1 *Lower Bounding Procedure*

Supplier's Order Problem ($ORDER$)

It involves only lot-sizing and setup variables of the supplier as given below.

$$ORDER:$$
$$Min \quad \sum_{t \in \tau}(f_{0t}y_t - \mu_t P_t) \quad (11.15)$$
$$s.t. \quad (11.4), \ y \in \{0,1\}; P \geq 0 \quad (11.16)$$

The $ORDER$ problem decomposes with respect to time periods, each of T problems can be solved using a simple decision rule. If $f_{0t}y_t - \mu_t P_t$ is less than zero for any $t \in \tau$, then let $P_t = M_0$ and $y_t = 1$, otherwise let both P_t and y_t be zero. To make $ORDER$ tighter, the following valid inequalities are proposed.

$$\sum_{t \in \tau} P_t \leq \sum_{i \in N}(S_i - I_{i0} + \sum_{r=1}^{T-1} d_{ir}) - I_{00} \quad (11.17)$$

$$\sum_{r=1}^{t} P_r \geq \sum_{i \in N}(\sum_{r=1}^{t} d_{ir} - I_{i0}|\sum_{r=1}^{t} d_{ir} > I_{i0}) - I_{00} \quad \forall t \in \tau \quad (11.18)$$

Inequality (11.17) ensures that the total amount of product that can be ordered to the supplier cannot exceed the total of the maximum requirements of customers over the horizon less the beginning inventory level at the supplier. $S_i - I_{i0} + \sum_{r=1}^{T-1} d_{ir}$ is the maximum amount of product that can be delivered to retailer i over the horizon. Assuming that all retailers are replenished in every period, $S_i - I_{i0}$ equals the replenishment amount in $t = 1$ while $d_{i,t-1}$ equals the replenishment amount in $t \geq 2$ for retailer $i \in N$. Inequality (11.18) stipulates that the total amount of product to be ordered to supplier up to period t should be at least the total of minimal requirements of retailers less the initial amount at the supplier. Here, the minimal requirement of retailers up to t is the amount demanded from i up to t less the initial inventory available at i. If the total of minimal requirements over all retailers up to t exceeds the initial inventory level of the supplier, then an amount at least equal to this difference must be ordered to the supplier up to t. These inequalities make problem tighter at the expense of disrupting the decomposition feature with respect to time periods.

Supplier's Inventory Problem ($SINV$)

It consists of only inventory variables of supplier with nonnegativity constraints.

$SINV$:

$$Min \sum_{t=1}^{T-1} (h_{0t} + \mu_t - \mu_{t+1})I_{0t} + (h_{0T} + \mu_T)I_{0T} - \mu_1 I_{00} \quad (11.19)$$

$$s.t. \quad I_0 \geq 0 \quad (11.20)$$

The problem further decomposes into T independent single period problems whose solutions are trivial. Given that $\mu_t \geq 0$ and if $(h_{0t} + \mu_t - \mu_{t+1}) \geq 0$ from $t = 1$ to $T - 1$, then $I_{0t} = 0$ else $I_{0t} = \infty$. For $t = T$, $I_{0T} = 0$ and the last term is constant. Therefore, one can provide that $h_{0t} + \mu_t \geq \mu_{t+1}$, where $I_{0t} = 0$ for every t in the optimal solution.

Distribution Problem (*DIST*)

It involves variables regarding vehicle routes.

DIST :

$$\text{Min} \sum_{j \in N} \sum_{v \in \vartheta} \sum_{t \in \tau} (f_t + c_{0j}) w_{0jvt} + \sum_{i \in \overline{N}, i \neq j} \sum_{j \in \overline{N}} \sum_{v \in \vartheta} \sum_{t \in \tau} c_{ij} w_{ijvt}$$

$$- \sum_{i \in N} \sum_{v \in \vartheta} \sum_{t \in \tau} (\alpha_{ivt} + \beta_{vt}) X_{0ivt}$$

$$+ \sum_{i \in N, i \neq j} \sum_{j \in N} \sum_{v \in \vartheta} \sum_{t \in \tau} (\alpha_{ivt} - \alpha_{jvt}) X_{ijvt} \qquad (11.21)$$

$$\text{s.t.} \quad (11.7), (11.8), (11.9), (11.10), X \geq 0; w \in \{0,1\} \qquad (11.22)$$

The *DIST* problem decomposes with respect to vehicles and time periods, but not easy to solve. In order to make the model tighter, we add the following valid inequalities.

$$\sum_{j \in \overline{N}} w_{ijvt} \leq 1 \quad \forall i \in N, v \in \vartheta, t \in \tau \qquad (11.23)$$

$$\sum_{j \in \overline{N}} w_{ijvt} \leq \sum_{j \in N} w_{0jvt} \quad \forall i \in N, v \in \vartheta, t \in \tau \qquad (11.24)$$

$$X_{0jvt} \leq K w_{0jvt} \quad \forall j \in N, v \in \vartheta, t \in \tau \qquad (11.25)$$

$$X_{ijvt} \leq (K - S_i + I_{i0}) w_{ijvt} \quad \forall i, j (i \neq j) \in N, v \in \vartheta, t = 1 \qquad (11.26)$$

$$X_{ijvt} \leq (K - d_{i,t-1}) w_{ijvt} \; \forall i, j (i \neq j) \in N, v \in \vartheta, t \in \tau - \{1\} \qquad (11.27)$$

$$X_{ijvt} - S_j \leq \sum_{k \in N, k \neq i \neq j} X_{jkvt} \; \forall i \in \overline{N}, j \in N, i \neq j, v \in \vartheta, t \in \tau \qquad (11.28)$$

Inequality (11.23) ensures that only one retailer j (or supplier) can be visited from retailer i via vehicle v in period t. Inequality (11.24) states that visiting from retailer i can only be possible if the vehicle v departs from the supplier in period t. Inequalities (11.25), (11.26) and (11.27) replaces (11.10). Inequality (11.25) guarantees that if a visit from the supplier to retailer j occurs, its transportation cost is charged. Inequalities (11.26) and (11.27) have the same purpose as (11.25) and upgrade the variable bound by the minimum amount that can be delivered to retailer i in period t, which is $S_i - I_{i0}$ for $t = 1$ and $d_{i,t-1}$ for $t \in \{2, \ldots, T\}$. Note that if vehicle v visits i before j, it should carry an amount at most capacity of the vehicle less the minimum amount delivered to i. Last inequality (11.28) ensures that for each vehicle-period pair (v, t), the amount transported from retailer j to k is at least the amount transported from i to j less the maximum amount that can be delivered to $j(S_j)$.

Retailers' Replenishment Problem (*RET*)

It comprises variables concerning delivered amounts to retailers as well as inventory levels of retailers.

RET :

$$Min \sum_{i \in N} \sum_{v \in \vartheta} \sum_{t \in \tau} (\mu_t + \alpha_{\text{ivt}} + \beta_{\text{vt}}) Q_{\text{ivt}} + \sum_{i \in N} \sum_{t \in \tau} h_{\text{it}} I_{\text{it}} \qquad (11.29)$$

$$s.t. \quad (11.3), (11.11), (11.12), (11.13), \quad I \geq 0; Q \geq 0; z \in \{0,1\} \quad (11.30)$$

The *RET* problem decomposes into N separate single retailer problems. We transform the single retailer problem into a shortest path problem that can be solved by an $O(T^2)$ algorithm based on dynamic programming (see [Solyalı and Süral (2008)]).

11.3.2 Upper Bounding Procedure

To construct a feasible solution to the *IRO*, we basically rely on the solution to the *RET* subproblem in *PR* because its solution provides the set of retailers to be visited as well as the amounts to be shipped for each vehicle-period pair (i.e. Q_{ivt} values). Using Q_{ivt} values, one can construct a feasible tour for a vehicle-period pair and determine the amounts ordered to supplier in a period. The latter issue is resolved by defining a Wagner-Whitin type inventory problem (see [Wagner and Whitin (1958)]) for the supplier. The procedure using Q_{ivt} values to obtain a feasible solution to *IRO* is described below in detail.

The total of Q_{ivt} values over i may be infeasible due to the capacity of vehicles. To cope with such an infeasibility, a linear programming model (*LP*) is developed to reallocate the already computed Q_{it} to vehicles (ignoring the vehicle to which the amount assigned) by considering the vehicle capacities. Given $Q_{\text{it}} = \sum_{v \in \vartheta} Q_{\text{ivt}}$ for each i, t, the proposed *LP* is as follows:

$$Min \sum_{i \in N} \sum_{v \in \vartheta} \sum_{t \in \tau} (\mu_t + \alpha_{\text{ivt}} + \beta_{\text{vt}}) Q_{\text{it}} B_{\text{ivt}} \qquad (11.31)$$

$$s.t. \quad \sum_{i \in N} Q_{\text{it}} B_{\text{ivt}} \leq K \quad \forall v \in \vartheta, t \in \tau \qquad (11.32)$$

$$\sum_{v \in \vartheta} B_{\text{ivt}} = 1 \quad \forall i \in N, t \in \tau \ni Q_{\text{it}} > 0 \qquad (11.33)$$

$$B_{\text{ivt}} \geq 0 \quad \forall i \in N, v \in \vartheta, t \in \tau \qquad (11.34)$$

where B_{ivt} is the fraction of replenishment size shipped to retailer i via vehicle v in period t. The above LP model can be cast into a transportation model.

Since the set of retailers to be visited for each vehicle-period pair with feasible B_{ivt} values are known, the problem of finding feasible tours transforms into solving the VT independent traveling salesman problems (TSPs). TSP is a well-studied NP-hard combinatorial optimization problem[E. L. Lawler, Lenstra, Rinnooy Kan and Shmoys (1985)] and can be stated as: Given n nodes and costs of traveling between these nodes, it seeks the least cost tour subject to visiting all of nodes and returning to the starting base. If inventory balance at the supplier may not hold due to a possible imbalance between the ordered and distributed amounts, then given that for any t the sum $\sum_{i \in N} \sum_{v \in \vartheta} Q_{ivt}$ corresponds a demand at the supplier in t, a lot-sizing problem is defined. It is solved to decide on when and how much to order for the supplier. After performing these steps, we obtain an upper bound to the problem IRO.

11.3.3 Solution of the Lagrangian Dual Problem

The Lagrangian dual problem is $Max_{\mu,\alpha,\beta} PR$, referred to as LDP. The solution of LDP is obtained using standard subgradient optimization algorithm. In each iteration of the subgradient optimization for solving LDP, step length ρ^k together with gradients and current values of multipliers (μ^k, α^k, β^k) update the values of multipliers as follows.

Let $\rho^k = a^k(UB^* - LB^k)/(g_1 + g_2 + g_3)$, where a^k is a scalar in k^{th} iteration ($0 \leq a^k \leq 2$), UB^* is the best upper bound, LB^k refers to the lower bound computed in k^{th} iteration, $g_1 = \sum_{t \in \tau}(I_{0t}^k - I_{0,t-1}^k - P_t^k + \sum_{i \in N}\sum_{v \in \vartheta} Q_{ivt}^k)^2$, $g_2 = \sum_{i \in N}\sum_{v \in \vartheta}\sum_{t \in \tau}(-\sum_{j \in \overline{N}, j \neq i} X_{jivt}^k + \sum_{j \in \overline{N}, i \neq j} X_{ijvt}^k + Q_{ivt}^k)^2$ and $g_3 = \sum_{v \in \vartheta}\sum_{t \in \tau}(-\sum_{j \in N} X_{0jvt}^k + \sum_{i \in N} Q_{ivt}^k)^2$. The new values of multipliers are computed as follows. Below we satisfy the condition $h_{0t} + \mu_t \geq \mu_{t+1}$ for newly computed μ_t values.

$$\mu_t^{k+1} = \mu_t^k + \rho^k(I_{0t}^k - I_{0,t-1}^k - P_t^k + \sum_{i \in N}\sum_{v \in \vartheta} Q_{ivt}^k) \quad \forall t \in \tau \quad (11.35)$$

$$\alpha_{ivt}^{k+1} = \alpha_{ivt}^k + \rho^k(-\sum_{j \in \overline{N}, j \neq i} X_{jivt}^k + \sum_{j \in \overline{N}, i \neq j} X_{ijvt}^k + Q_{ivt}^k)$$

$$\forall i \in N, v \in \vartheta, t \in \tau \quad (11.36)$$

$$\beta_{vt}^{k+1} = \beta_{vt}^{k} + \rho^{k}(-\sum_{j \in N} X_{0jvt}^{k} + \sum_{i \in N} Q_{ivt}^{k}) \quad \forall v \in \vartheta, t \in \tau \quad (11.37)$$

In Figure 11.1, we present our entire Lagrangian relaxed based solution approach LR. Implementation details of the procedures will be discussed in Section 11.4.

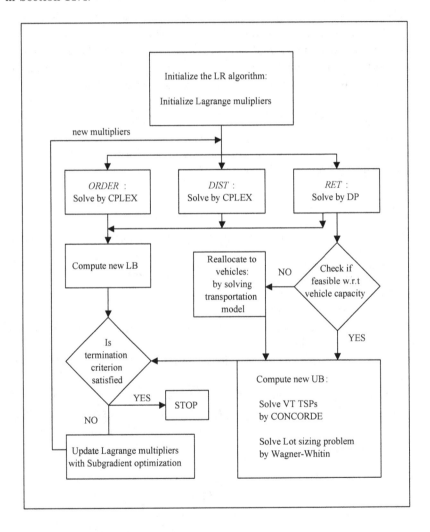

Fig. 11.1 An overview of the solution approach

11.4 Computational Experiments

All computational experiments have been implemented on instances generated by [Bertazzi, Paletta and Speranza (2002)]. They have randomly generated test instances for each combination of inventory carrying cost of supplier (h_0), inventory carrying cost of retailers (h_i), and transportation cost (c_{ij}). Instances involve 50 retailers, 30 periods, and constant demands over time at retailers (i.e. $d_{it} = d_i, \forall t$). A single vehicle is assumed. We have modified these instances to make them appropriate for use in our context. We assume a sufficiently large vehicle capacity ($K = \sum_{i \in N} S_i$), and generate fixed order cost for supplier's replenishment and fixed vehicle dispatching costs for deliveries to retailers.

The procedures are coded in C. For solving LP and MIP problems, Callable Library of CPLEX 8.1 is embedded into the code. CONCORDE (http:// www.tsp.gatech.edu/concorde) is called from the code for solving $TSPs$. Experiments are conducted on Pentium IV 1.6 GHz CPU PCs with 256 MB RAM.

11.4.1 *Small Instances*

For the first part of our experiments, we have derived 18 test problems with eight retailers and five periods from basic test instances. They were largest manageable sizes to solve the problems to optimality. Retailers selected among 50 are the first eight retailers that require at least one shipment during the 5-period horizon. For all instances, we added fixed order cost $f_{0t} = \sum_{i \in N} h_0 d_i$ and fixed vehicle dispatching costs $f_t = m * \sum_{i \in N} h_i d_i$ where $m=1$, 0.8 and 0.2 for the 1st, 2nd and 3rd set of six instances respectively.

We first solve the MIP formulation of IRO and its LP relaxation to optimality. Then LR is run to compare its lower bounding quality with the optimal solution value and the LP relaxation solution value. We also compare the performance of the upper bounding by LR with the optimal solution value and the two greedy heuristics solutions, called *Every* and *Latest*, which are adapted from [Bertazzi, Paletta and Speranza (2002)] to solve IRO. LR is run 100 iterations under three different settings for the scalar of the subgradient optimization algorithm: initially scalar $a = 2$ and it is divided (i) by 1.005 per iteration, (ii) by 2 per 30 consecutive non-improving iterations, and (iii) by 2 per 10 consecutive non-improving iterations. In *Every*, retailers are replenished every period and only one

TSP solution is needed to find the optimal sequence of retailers. *Latest* replenishes a retailer at the last possible period just before it stock outs and T many TSPs are solved to find optimal sequence of retailers to be visited over the horizon. For the supplier's replenishment problem for both heuristics, we solve an uncapacitated lot-sizing problem by Wagner-Whitin algorithm to find replenishment amounts. Results are given in Table 11.1.

Table 11.1 shows that the deviation of the LP relaxation solution value from the optimal solution value is quite large (71% on average). The average solution time for MIP is about 112 minutes. The best implementation of LR is halving scalar after 30 consecutive non-improving iterations and results in 33.2% (1.2%) deviation from the optimal solution value for the Lagrangian lower bound (upper bound) on average. It has found an optimal solution 10 times (out of 18) and ran less than 1 minute on average. The greedy heuristics also run very fast but their solutions deviate from the optimal solution values by 20.8% (169.6%) for *Latest* (*Every*) on average. LR outperforms the solution approach by MIP and greedy heuristics on average for small instances.

11.4.2 Large Instances

In the second part of our experiments, we use 80 basic test instances with our modifications (i.e. adding fixed order cost $f_{0t} = \sum_{i \in N} h_0 d_i$ and fixed vehicle dispatching costs at two levels, $f_t = 0.8 * \sum_{i \in N} h_i d_i$ and $f_t = 0.2 * \sum_{i \in N} h_i d_i$). Thus, we solve 160 instances in total. We consider a relaxed variant of LR to run so that instead of finding an optimal integral solution to the subproblem $DIST$ in LR, we seek its LP relaxation solution. The implementation setting of this variant is tested within a set of preliminary experiments and it is decided to halve the scalar a every 10 consecutive non-improving iterations and to halt after 50 iterations. We did not report any results regarding the feasible solution to the MIP formulation of IRO because CPLEX could not attain a feasible solution for any instances at hand within two-hour time limit, but we report the percentage gap between its LP relaxation solution value (LPV) and the lower bound found by LR (LB), i.e. $100 * (LB - LPV)/LB$. We compare the performance of the LR upper bounding with the two greedy heuristics solutions and report the percent deviation of approximate solutions from the lower bound found by LR. We also report running times for all procedures. Computational results for instances with high and low f_t are averaged over 10 instances with respect to each combination of parameters (c_{ij}, h_i, h_0) and presented

Table 11.1 Results for small instances.

Instance*	CPU (in sec)		% deviation from optimal solution value				
	MIP	LR	LP	LB	UB	$Latest$	$Every$
bps1	1263	64	71.78	29.14	3.25	18.34	97.29
bps2	78	67	73.69	27.67	4.39	12.78	101.07
bps19	973	82	64.47	28.14	0.00	16.93	88.12
bps28	1838	59	77.95	34.73	0.04	16.46	242.57
bps31	1084	28	73.63	38.96	1.01	27.98	167.59
bps32	63	176	79.13	31.57	0.00	23.98	174.09
bps3	23	113	66.76	23.00	0.00	12.03	167.50
bps8	468	72	74.73	28.79	0.00	11.19	131.18
bps15	1421	36	63.26	34.99	0.91	18.33	96.29
bps24	159	30	70.40	36.43	0.00	39.81	189.47
bps36	74550	38	73.45	42.76	0.00	26.93	209.07
bps37	357	20	68.15	41.29	4.95	25.21	232.90
mbps3	13	67	66.27	18.80	0.00	8.94	206.25
mbps8	680	58	75.66	27.41	0.00	8.82	155.95
mbps15	22889	29	61.35	32.56	1.01	16.84	115.87
mbps24	625	21	71.13	36.21	0.00	37.95	198.42
mbps36	14753	35	73.29	45.54	0.00	26.60	225.06
mbps37	342	22	67.07	38.92	5.26	24.69	253.82
Average	6754.4	56.45	70.68	33.16	1.16	20.77	169.58

* Instances are derived from the same instances in [Bertazzi, Paletta and Speranza (2002)]. bps (mbps) denotes instances with m=1 and 0.8 (m=0.2).

in Table 11.2. Table 11.2 shows that the gap between the LR upper and lower bounds is not tight (on the average 53.6%) for large instances. Nevertheless, lower bounds generated by the LR are still far better than the LP relaxation of the IRO (approximately 38.9%). Almost all the results with low f_t are slightly better than those with high f_t. According to the results, LR is far superior to $Every$. On the other hand, the results generated by $Latest$ are very close to those obtained by LR. These results contradict with the results of small instances, where LR significantly surpass Latest. To shed light on this contradiction, we should investigate the features of $Latest$ and the instances at hand. $Latest$ tends to minimize inventory holding cost at retailers by replenishing them at the last possible moment. In our problem settings, we treat vehicle capacity as uncapacitated and have the opportunity to determine order amounts as well as their timings at the supplier. Furthermore, we solve routing of retailers to be replenished in each period (T many TSPs) as well as order decisions at supplier (an uncapacitated lot-sizing problem) to optimality. Hence, we expect $Latest$

Table 11.2 Results for large instances.

f_t	c_{ij}	h_i	h_o	CPU (in sec)					% deviation from LB			
				LP	LR	Latest	Every	LP	UB	Latest	Every
high	[0,500]	[0.6,1]	0.3	633.57	2791.84	9.18	0.70	40.04	27.34	27.71	205.00
			0.8	641.40	3072.08	9.09	0.72	38.76	34.28	35.01	217.19
		[0.1,0.5]	0.3	658.02	2777.85	9.10	0.76	38.93	56.02	55.26	703.55
			0.8	638.02	3071.11	9.08	0.72	37.84	73.21	74.56	730.59
	[0,1000]	[0.6,1]	0.3	625.48	2642.84	9.15	0.72	39.73	39.11	39.15	233.56
			0.8	642.29	2721.39	9.14	0.71	38.53	45.71	46.23	245.23
		[0.1,0.5]	0.3	649.78	2630.11	9.14	0.73	38.12	87.38	86.28	779.10
			0.8	640.45	2814.76	9.01	0.69	37.17	104.53	104.77	803.45
	Average(high)			641.12	2815.25	9.11	0.72	38.64	58.45	58.62	489.71
low	[0,500]	[0.6,1]	0.3	634.39	2640.86	8.77	1.00	40.53	18.66	18.97	194.07
			0.8	655.40	2950.36	9.12	0.72	39.30	25.11	26.24	206.11
		[0.1,0.5]	0.3	646.42	2688.62	9.15	0.69	39.45	46.74	46.63	696.08
			0.8	651.66	3057.58	9.11	0.73	38.35	64.79	66.14	723.21
	[0,1000]	[0.6,1]	0.3	626.64	2528.78	9.05	0.69	40.25	30.47	30.47	222.75
			0.8	649.64	2711.41	9.11	0.70	39.03	37.22	37.65	234.57
		[0.1,0.5]	0.3	657.45	2573.24	9.04	0.71	38.64	78.83	77.83	771.85
			0.8	644.79	2757.47	9.05	0.73	37.51	96.29	97.02	798.60
	Average(low)			645.80	2738.54	9.05	0.75	39.13	49.76	50.12	480.91

to perform well in our problem. However, we know that our algorithm is far superior to *Latest* on smaller test instances while performances are almost equivalent in the instances we considered in this section. We explain this situation by investigating the weights of the cost components on the total cost of the problem instances. For the small instances, supplier's cost, retailers' cost and transportation cost are 9.5%, 25.6% and 64.9%, respectively. On the other hand, those cost components are 10.1%, 68.5% and 21.4% for the large instances. Those figures show that inventory holding cost at retailers predominates other cost terms in large instances, while transportation cost has the largest weight in small instances. In both small and large instances, supplier's cost is the lowest cost term, thus insignificant with respect to other costs. For the large instances, percentage of retailers' cost on total cost is 68.5% on average, which means that minimizing that cost approximately minimizes the total cost. Thus, considering that *Latest* tends to minimize inventory holding cost at retailers, we can claim that *Latest* finds good quality solutions in large instances.

Table 11.2 also shows the sensitivity of solutions generated by LR with respect to different parameters. For instance, we can easily state that LR is more successful than *Latest* in instances, where unit holding cost at supplier is greater than unit holding cost at retailers. This conclusion is valid for instances with both high and low dispatching cost. For the LR algorithm on instances with high dispatching cost, increasing one of the cost parameters (except h_i) and keeping the rest constant lead to increase in gaps between LR upper and lower bounds. The smallest gap has been achieved when h_0 and c_{ij} are at their lowest value and h_i at its highest value. In contrast, the worst gap has been obtained when h_0 and c_{ij} are at their highest value and h_i at its lowest value. Furthermore, when dispatching cost decreases, gaps between LR upper and lower bounds improve and all the comments regarding sensitivity of solutions to unit holding cost at supplier and retailers as well as transportation cost are the same.

As far as the CPU times are concerned, one can observe from Table 11.2 that CPU time needed to solve LP relaxation of the IRO is over ten minutes, which shows the difficulty of solving the IRO to optimality. As expected, CPU times needed by *Latest* and *Every* are very low. Since *Latest* finds good quality solutions in short time on those instances, it can be a good alternative to solve IRO whenever holding cost at retailers is a predominating component of the total cost. On the other hand, our LR algorithm finds good quality solutions in reasonable times considering the length of horizon. In addition, LR also produces lower bounds in that time.

11.5 Conclusions

In this study, we addressed the integrated inventory management and vehicle routing problem with deterministic order-up-to level policy IRO. We have proposed a mathematical formulation and a Lagrangian relaxation based approach LR to find a lower bound on the objective function of the problem and to develop a heuristic for solving it. LR disaggregates the problem into N single retailer replenishment problems, VT distribution planning problems and a simple supplier's order and inventory problem. Computational results revealed that our LR algorithm produces good feasible solutions as we could not find a feasible solution to the large instances using the MIP solvers within hours and yields lower bounds better than the LP relaxation of the MIP formulation of IRO. We have adapted two greedy heuristics in the literature, called *Every* and *Latest*, to our context and compared our approach with them. According to our computational results, our LR is significantly superior to *Every* and comparable with *Latest*.

In this study, we have considered that the supplier can decide on its order policy. A future research issue may be to consider production planning issues with capacity limitations on production at the supplier (plant). Another possible research issue is to extend our work to involve multiple items as the model can easily be extended to account for multiple items if replenishment of each item can be made independent of others.

Bibliography

Archetti, C., Bertazzi, L., Laporte, G. and Speranza, M. G. (2007). A branch-and-cut algorithm for a vendor-managed inventory routing problem, *Transportation Science* **41**, pp. 382–391.

Baita, F., Ukovich, W., Pesenti, R. and Favaretto, D.(1998). Dynamic routing-and-inventory problems: A review, *Transportation Research A* **32**, pp. 585–598.

Berman, O. and Larson, R. C. (2001). Deliveries in an inventory/routing problem using stochastic dynamic programming, *Transportation Science* **35**, pp. 192–213.

Bertazzi, L., Paletta, G. and Speranza, M. G. (2002). Deterministic order-up-to level policies in an inventory routing problem, *Transportation Science* **36**, pp. 119–132.

Bertazzi, L., Paletta, G. and Speranza, M. G. (2005). Minimizing the total cost in an integrated vendor-managed inventory system, *Journal of Heuristics* **11**, pp. 393–419.

Campbell, A., Clarke, L. and Savelsbergh, M. W. P. (2002). Inventory routing in practice, in *The Vehicle Routing Problem*, eds: P.Toth and D.Vigo, *SIAM Monographs on Discrete Mathematics and Applications* pp. 309–330.

Campbell, A. and Savelsbergh, M. W. P. (2004). A decomposition approach for the inventory-routing problem, *Transportation Science* **38**, pp. 488–502.

Cetinkaya, S. and Lee, C. Y. (2000). Stock replenishment and shipment scheduling for vendor-managed inventory systems, *Management Science* **46**, pp. 217–232.

Chandra, P. (1993). A dynamic distribution model with warehouse and customer replenishment requirements, *Journal of the Operational Research Society* **44**, pp. 681–692.

Chandra, P. and Fisher, M. L. (1994). Coordination of production and distribution planning, *European Journal of Operational Research* **72**, pp. 503–517.

Fisher, M. L. (1981). The Lagrangian relaxation method for solving integer programming problems, *Management Science* **27**, pp. 1–18.

Fisher, M. L. (1985). An applications oriented guide to Lagrangian relaxation, *Interfaces* **15**, pp. 10–21.

Fumero, F. and Vercellis, C. (1999). Synchronized development of production, inventory and distribution schedules, *Transportation Science* **33**, pp. 330–340.

Georgia Institute of Technology (2007), *Concorde TSP solver*, http://www.tsp.gatech.edu/concorde.

Kleywegt, A. J., Nori, V. S. and Savelsbergh, M. W.P. (2002). The stochastic inventory routing problem with direct deliveries, *Transportation Science* **36**, pp. 94–118.

Kleywegt, A. J., Nori, V. S. and Savelsbergh, M. W.P. (2004). Dynamic programming approximation for a stochastic inventory routing problem, *Transportation Science* **38**, pp. 42–70.

Lawler, E. L., Lenstra, J. K., Rinnooy Kan, A. H.G. and Shmoys, D. B. (1985). *The traveling salesman prblem: A guided tour of combinatorial optimization*, (John Wiley & Sons).

Moin, N. H. and Salhi, S. (2007). Inventory routing problems: a logistical overview, *Journal of the Operational Research Society* **56**, pp. 345–356.

Pınar,Ö. and Süral, H. (2006). Coordinating inventory and transportation in a vendor managed system, *Proceedings of the Material Handling Conference*.

Sarmiento, A. M. and Nagi, R. (1999). A review of integrated analysis of production-distribution systems, *IIE Transactions* **31**, pp. 1061–1074.

Solyalı, O. and Süral, H. (2008). A single supplier-single retailer system with order-up-to level inventory policy, *Operations Research Letters* **36**, pp. 543–546.

Thomas, D. J. and Griffin, P. M. (1996). Coordinated supply chain management, *European Journal of Operational Research* **94**, pp. 1–15.

Wagner, H. M. and Whitin, T. M. (1958). Dynamic version of the economic lot size model, *Management Science* **5**, pp. 89–96.

Chapter 12

The Distribution of Deficit at Ruin on a Renewal Risk Model

K. K. Thampi
Department of Statistics, SNMC, M.G.University, Kerala-683516, India
e-mail: thampisnm@yahoo.co.in
M. J. Jacob
Department of Mathematics, NITC, Calicut-673601, India
e-mail: mjj@nitc.ac.in

Abstract

In this chapter, we consider the probability and severity of ruin for a renewal class of risk process in which the claim inter occurrence times is Generalized Exponential. We have obtained closed form expression for the distribution of the deficit at ruin. We illustrate the application of this result with several examples.

Key Words: Renewal process, severity of ruin, probability of ruin, generalized exponential distribution, Laplace transform

12.1 Introduction

Recently, many researchers are paying attention to the distribution of the severity of ruin in the general renewal risk model. The severity of ruin in the classical risk model was first discussed in a paper by Gerber et al.(1987). They introduced the function $G(u, y)$, defined as the probability that ruin occurs from initial surplus u with deficit at ruin no greater than y. Using an integral equation for $G(u, y)$, they have obtained a closed form solutions for $G(u, y)$ when the claim amount distribution is either an exponential mixture or gamma mixture. Dickson(1998) discussed the probability and severity of ruin when the claim arrival is Erlang(2) with claim amount distribution being Erlang(2). Lin and Willmot (2000) studied the joint and marginal moments of the time of ruin, the surplus before ruin and deficit at ruin. Avram and Usabel (2003) generalized Thorin's formula (1971) for the double Laplace transform of the finite time ruin probability, by considering the deficit at ruin in a renewal risk model with phase type inter-arrival times. Drekic et al. (2004) obtained a fairly robust result regarding the distribution of the deficit at ruin in the Sparre Andersen model.

The purpose of this chapter is to obtain some explicit expressions for the distribution of severity of ruin for a renewal risk model in which the inter-claim distribution is Generalized exponential (1991, 2001). We have also considered the probability that ruin occurs and the deficit at ruin exceeds a preassigned value.

In the rest of this section, we give some basic definitions and notations used in this chapter. Section 12.2 gives expression for the Laplace transform of the ruin probability. Section 12.3 introduces the severity of ruin and how the distribution of the severity of ruin is expressed in terms of Laplace transforms. In Section 12.4, we obtain $g(u, y)$, the density function of the severity of ruin for each of the different claim amounts.

Consider a risk process in which claims occur as renewal process. Let $T_1, T_2 \ldots$ be a sequence of i.i.d random variables where T_1 being the time until the first claim, for $i > 1$, T_i is the time between the $(i-1)^{th}$ and i^{th} claims. We assume that T_i has a Generalized exponential distribution with density function

$$h(t) = \gamma\lambda(1 - e^{-\lambda t})^{\gamma-1} e^{-\lambda t}, \quad \gamma, \ \lambda > 0, \ t > 0 \tag{12.1}$$

We illustrate ideas by restricting our attention to the case where $\gamma = 2$, but results obtained in this chapter can be generalized to any integer value of γ.

The insurer's surplus at time t, which we denote by $U(t)$, is given by

$$U(t) = u + ct - \sum_{i=1}^{N(t)} X_i \tag{12.2}$$

The claim surplus process $\{S(t), t \geq 0\}$ is

$$S(t) = \sum_{i=1}^{N(t)} X_i - ct \tag{12.3}$$

where $U(0) = u$ is the initial surplus. The counting process $\{N(t), \ t \geq 0\}$ whose inter-claim times $\{T_1, T_2...\}$ are distributed according to Generalized exponential with common mean $\mu = E(T_i)$. $\{X_i\}_{i=1}^{\infty}$ is a sequence of independent and identically distributed random variables, independent of $\{N(t), t \geq 0\}$, where X_i denotes the claim amount of ith claim. Let $f(x)$ denotes the density functions of X_I and let $m = E(X_I)$. Let $c = (1+\theta)\frac{E(X_i)}{E(T_i)}$, where $\theta > 0$ is the loading factor, denotes the insurer's premium income per unit time. We assume that

$$c\mu > m \qquad (12.4)$$

to ensure $U(t)$ has a positive drift.

Define the probability of ultimate ruin for this surplus process as

$$\psi(u) = P\{U(t) < 0 \text{ for some } t > 0\}$$

or we can define $\psi(u) = P\{\tau(u) < \infty\}$ where $\tau(u)$ denotes the time of ruin defined as $\tau(u) = \inf_{t \geq 0}\{t : U(t) < 0\}$. Let $\phi(u) = 1 - \psi(u)$ denote the survival probability.

Let $\hat{\psi}(s)$, $\hat{f}(s)$ and $\hat{h}(s)$ be the Laplace transforms of the functions $\psi(u)$, f(x) and h(t) respectively. If the moment generating function, $\hat{a}(s)$ of X_i exist, then the Lundberg inequality holds with $\psi(u) \leq e^{-Ru}$, where R is the unique positive value of s such that

$$\hat{a}(s)\hat{h}(cs) = 1 \qquad (12.5)$$

R is called the Lundberg exponent(adjustment coefficient).

12.2 Ruin Probability $\psi(u)$

We derive an integro-differential equation for $\psi(u)$ by conditioning on the time and amount of the first claim, we can write

$$\psi(u) = \int_0^\infty h(t) \int_0^{u+ct} f(x)\psi(u+ct-x)dxdt + \int_0^\infty h(t) \int_{u+ct}^\infty f(x)dxdt \qquad (12.6)$$

Substituting $z = u + ct$, $\psi(u)$ is differentiable and using the condition $g(0) = 0$, we get

$$c\psi''(u) = 3\lambda\psi'(u) - \frac{2\lambda^2}{c}\psi(u) + \frac{2\lambda^2}{c}\int_0^u \psi(u-x)f(x)dx + \frac{2\lambda^2}{c}\int_u^\infty f(x)dx$$

$$c^2\psi''(u) - 3\lambda c\psi'(u) + 2\lambda^2\psi(u) = 2\lambda^2 \int_0^u \psi(u-x)f(x)dx + 2\lambda^2(1-F(u))$$

Taking Laplace transform and rearranging

$$\hat{\psi}(s) = \frac{c^2s^2\psi(0) + c^2s\psi'(0) - 3\lambda cs\psi(0) + 2\lambda^2(1 - \hat{f}(s))}{s(c^2s^2 - 3\lambda cs + 2\lambda^2(1 - \hat{f}(s)))} \qquad (12.7)$$

We can eliminate $\psi'(0)$ from the equation (12.7) by considering the maximal aggregate loss process defined by

$$\max_{t>0}\{X(t) - ct\}$$

Since $S(t) = 0$ for $t = 0$, it follows that L is a non-negative random variable. We can relate the random variable L to the probability of ruin $\psi(u)$, as

$$1 - \psi(u) = P\{U(t) \geq 0, \text{ for all } t > 0\}$$
$$= P\{X(t) - ct \leq u \text{ for all } t > 0\}$$
$$= P(L \leq u)$$

Further, since $1 - \psi(0) = P\{L = 0\}$, the maximum loss attained at time $t = 0$, L has a mixed type distribution with a mass of $(1-\psi(0))$ at zero and the remaining probability is distributed continuously over positive values of L. Define $\hat{L}(s)$ to be the Laplace transform of the random variable L, so that

$$\hat{L}(s) = 1 - \psi(0) + \int_0^\infty e^{-su}(-\psi'(u))du$$
$$= 1 - s\hat{\psi}(s)$$
$$= \frac{c^2 s^2 - 3\lambda cs + 2\lambda^2(1 - \hat{f}(s))}{c^2 s^2 - 3\lambda cs + 2\lambda^2(1 - \hat{f}(s))} -$$
$$\frac{(c^2 s^2 \psi(0) + c^2 s \psi'(0) - 3\lambda cs \psi(0) + 2\lambda^2(1 - \hat{f}(s)))}{c^2 s^2 - 3\lambda cs + 2\lambda^2(1 - \hat{f}(s))}$$
$$= \frac{3\lambda cs \psi(0) - 3\lambda cs - c^2 s \psi'(0) + c^2 s^2(1 - \psi(0))}{c^2 s^2 - 3\lambda cs + 2\lambda^2(1 - \hat{f}(s))}$$
$$= \frac{c^2 s^2 (1 - \psi(0)) - 3\lambda cs(1 - \psi(0)) - c^2 s \psi'(0)}{c^2 s^2 - 3\lambda cs + 2\lambda^2(1 - \hat{f}(s))} \tag{12.8}$$

Using the condition $\hat{L}(s) = 1$ as $s = 0$, and applying L'Hospital rule in the right hand side of the expression (12.8) to obtain the limit as $s \to 0$, we get

$$\hat{\psi}(s) = \frac{c^2 s^2 \psi(0) - 2\lambda^2 ms + 2\lambda^2(1 - \hat{f}(s))}{s(c^2 s^2 - 3\lambda cs + 2\lambda^2(1 - \hat{f}(s)))} \tag{12.9}$$

Define $\hat{\phi}(s)$ be the Laplace transform of $\phi(u)$. Then, using standard properties of Laplace transforms, we have

$$\hat{\phi}(s) = \frac{1}{s} - \hat{\psi}(s)$$
$$= \frac{c^2 s \phi(0) + 2\lambda^2 m - 3\lambda c}{c^2 s^2 - 3\lambda cs + 2\lambda^2(1 - \hat{f}(s))}. \tag{12.10}$$

But the condition (12.4) stipulates $3c > 2\lambda m$, hence the numerator of expression (12.10) will have a positive zero $k = \frac{3\lambda c - 2\lambda^2 m}{c^2 \phi(0)}$, which is a solution of the denominator also. But the denominator has a unique positive zero as it is the defining equation for the adjustment coefficient. Hence $\phi(0)$ can be identified:

$$\phi(0) = \frac{3\lambda c - 2\lambda^2 m}{c^2 k}, \quad k > 0$$

12.3 Severity of Ruin

Our main concern in this chapter is what happens if ruin occurs and how serious this situation. We want to analyze the insurer's deficit at the time of ruin (Rolski (1999)). Let $\tau(u)$ be the time of ruin, the overshoot above the level u of the random process $\{S(t), t \geq 0\}$ crossing this level for the first time is defined by

$$Y^+(u) = \begin{cases} S(\tau(u)) - u & \text{if } \tau(u) < \infty \\ \infty & \text{if } \tau(u) = \infty \end{cases}$$

It is possible to express $Y^+(u)$ in terms of the surplus process,

$$Y^+(u) = \begin{cases} -U(\tau(u)) & \text{if } \tau(u) < \infty, \\ \infty & \text{if } \tau(u) = \infty \end{cases}$$

$Y^+(u)$ can be interpreted as the severity of ruin at time $\tau(u)$. The probability of ruin with initial surplus u and deficit immediately after claim causing ruin is at most y, denoted by $G(u, y)$, is

$$G(u, y) = P\{\tau(u) < \infty, Y^+(u) \leq y\}$$
$$= P\{\tau(u) < \infty, -y \leq U(\tau(u)) < 0\} \quad u \geq 0, \; y \geq 0 \quad (12.11)$$

It can be seen that $\psi(u) = \lim_{y \to \infty} G(u, y)$, with $\frac{G(u,y)}{\psi(u)} = P\{|U\tau(u)| \leq y/\tau(u) < \infty\}$ is a proper distribution function. Therefore $G(u, y)$ is a defective distribution function with corresponding density function $g(u, y)$. But in general, we are not even able to find an explicit expression for $\psi(u)$ in the renewal risk model, so there is no scope for achieving this goal for $G(u, y)$. But it is possible to obtain integral equation for $G(u, y)$ and we will be able to find $G(u, y)$ explicitly in special cases. Some these special cases are discussed in this chapter.

Let $g(0, y)$ denotes the probability that ruin occurs from the initial surplus 0 and deficit at the time of ruin is y. Now conditioning on the value of the first claim that takes the surplus process strictly below its initial capital u, we can write

$$\psi(u) = \int_0^u g(0,y)\psi(u-y)dy + \int_u^\infty g(0,y)dy \qquad (12.12)$$

Also $\quad \psi(0) = \int_0^\infty g(0,y)dy$

$$= \lim_{y\to\infty} G(0,y)$$

Therefore, we have

$$\psi(u) = \psi(0) + \int_0^u g(0,y)\psi(u-y)dy - \int_0^u g(0,y)dy$$

Defining $\hat{g}(s) = \int_0^\infty e^{-sy} g(0,y)dy$, then

$$\hat{\psi}(s) = \frac{1}{s}\psi(0) + \hat{g}(s)\hat{\psi}(s) - \frac{1}{s}\hat{g}(s)$$

$$\hat{g}(s) = \frac{s\hat{\psi}(s) - \psi(0)}{s\psi(s) - 1} \qquad (12.13)$$

Therefore, we can find $g(0,y)$ by inverting $\hat{g}(s)$.

An integral equation for $G(u,y)$ will be obtained by considering the first occasion in which the surplus falls below its initial level u either the surplus falls to $u - x \geq 0$, so that ruin subsequently occurs with a deficit of at most y, or ruin occurs with a deficit at most y. Hence

$$G(u,y) = \int_0^u g(0,x)G(u-x,y)dx + \int_u^{u+y} g(0,x)dx \qquad (12.14)$$

By successive substitution, we can write

$$G(u,y) = \int_0^u \sum_{n=0}^\infty g^{*n}(0,x) \int_{u-x}^{u-x+y} g(0,z)dzdx \qquad (12.15)$$

$g^{*n}(0,x) \int_{u-x}^{u-x+y} g(0,z)dzdx$ is the probability of the event that the surplus process is between $u-x$ and $u-x+dx$ at the nth record low and ruin occurs with the following record low such that the deficit is less than y.

Taking the derivative of (12.15), we get

$$g(u,y) = \int_0^u \sum_{n=0}^\infty g^{*n}(0,x)g(0, u+y-x)dx$$

Let $\hat{g}(s,y)$ is the Laplace transforms w.r.t u of $g(u,y)$, we obtain

$$\hat{g}(s,y) = \sum_{n=0}^\infty (\hat{g}(s))^n \hat{\xi}(s,y)$$

$$= \frac{\hat{\xi}(s,y)}{1 - \hat{g}(s)} \qquad (12.16)$$

where $\hat{\xi}(s,y) = \int_0^\infty e^{-su} \frac{\delta}{\delta y} G(0, u+y)$. The next task is to invert $\hat{g}(s,y)$ to obtain $g(u,y)$. We shall look at different claim amount distribution in which this can be done explicitly.

Now, we are interested in another probability function,

$$\psi(u,y) = P\{\tau(u) < \infty, U(\tau(u)) < -y\} \qquad (12.17)$$

which is the probability that ruin occurs and the deficit at ruin exceeds y.

Clearly, $\psi(u,y) = \psi(u) - G(u,y)$. This probability function $\psi(u,y)$ is important to the point of view of an insurer.

12.4 Different Claim Amount Distributions

In this section, we consider different claim amount distributions to illustrate the results of the previous sections. In all our numerical examples, the procedure is as follows: $\hat{\psi}(s)$ is obtained from (12.9), then $\hat{g}(s)$ is found from (12.13) which is inverted to get $g(0,y)$. Note that a simplified form of $\hat{g}(s)$ is again obtained to find $\hat{\xi}(s,y)$ and then $\hat{g}(s,y)$. Finally $\hat{g}(s,y)$ is inverted to yield $\frac{\delta}{\delta y} G(u,y)$. Numerical calculations of $G(u,y)$ at selected points (u,y) are also presented.

12.4.1 *Exponential Distribution*

Let us assume that the claim amount distribution is exponential with mean $1/\beta$, then

$$\hat{g}(s) = \frac{(s+\beta)(2m\lambda^2 - 3c\lambda + 3c\lambda\phi(0)) - 2\lambda^2\phi(0)}{(s+\beta)(2m\lambda^2 - 3c\lambda + c^2\beta\phi(0))} \qquad (12.18)$$

Inverting the Laplace transform of (12.18), and after eliminating the removable singularities (Mathews (2001)), we can write

$$g(0,y) = ke^{-\beta y} \qquad (12.19)$$

where $k = \frac{2\lambda^2 \phi(0)}{3c\lambda - 2\lambda^2 m + \beta c^2 \phi(0)}$

then

$$\hat{g}(s) = \frac{k}{s+\beta}$$

and

$$g(0, u+y) = ke^{-\beta(u+y)}, \quad \text{thus} \quad \hat{\xi}(s,y) = \frac{k}{\beta+s} e^{-\beta y}$$

Hence
$$\hat{g}(s,y) = \frac{k}{s+\beta-k}e^{-\beta y} \qquad (12.20)$$

Taking the inverse Laplace transform of (12.20), we get
$$g(u,y) = ke^{u(k-\beta)-\beta y}$$
which can be simplified to
$$g(u,y) = (\beta-r)e^{-ru-\beta y}$$
where r ($= R$) is the adjustment coefficient.

12.4.2 Gamma Distribution

If a single claim amount has Gamma distribution of the type $\Gamma(2,\beta)$, then we have
$$\hat{g}(s) = \frac{(s+\beta)^2(2m\lambda^2 - 3c\lambda + 3c\lambda\phi(0)) - 2\lambda^2\phi(0)(s+2\beta)}{(s+\beta)^2(2m\lambda^2 - 3c\lambda + c^2\beta\phi(0))} \qquad (12.21)$$

Taking the inverse Laplace transform of (12.21), we get
$$g(0,y) = k_1 e^{-\beta y} + k_2 y e^{-\beta y} \qquad (12.22)$$

$$\text{where } k_1 = \frac{2\lambda^2\phi(0)(3c\lambda - 2m\lambda^2 + 2c^2\beta\phi(0))}{(3c\lambda - 2m\lambda^2 + c^2\beta\phi(0))^2}$$

$$\text{and } k_2 = \frac{2\lambda^2\beta\phi(0)}{(3c\lambda - 2m\lambda^2 + c^2\beta\phi(0))}$$

then
$$\hat{g}(s) = \frac{k_1}{(s+\beta)} + \frac{k_2}{(s+\beta)^2}$$

Also
$$g(0,u+y) = k_1 e^{-\beta(u+y)} + k_2(u+y)e^{-\beta(u+y)}$$

Hence
$$\hat{g}(0,s+y) = \frac{k_1}{\beta+s}e^{-\beta y} + \frac{k_2}{\beta+s}ye^{-\beta y} + \frac{k_2}{(\beta+s)^2}e^{-\beta y}$$

Therefore,
$$\hat{g}(s,y) = \frac{1 - \frac{k_1}{(s+\beta)} - \frac{k_2}{(s+\beta)^2}}{\frac{k_1}{\beta+s}e^{-\beta y} + \frac{k_2}{\beta+s}ye^{-\beta y} + \frac{k_2}{(\beta+s)^2}e^{-\beta y}} \qquad (12.23)$$

Table 12.1 $\beta = 2, \lambda = 2, c = 1.1$

u \ y	1	2	3	4	5
0	0.444844	0.505048	0.513195	0.514298	0.514447
1	0.168455	0.191252	0.194338	0.194755	0.194812
2	0.063791	0.072424	0.073592	0.073750	0.073772
3	0.024156	0.027425	0.027868	0.027928	0.027936
4	0.009148	0.010385	0.010553	0.010576	0.010579
5	0.003464	0.003933	0.003996	0.004005	0.004006

Table 12.2 $\beta = 3, \lambda = 2, c = 1.1$

u \ y	1	2	3	4	5
0	0.663798	0.740657	0.746665	0.747073	0.747099
1	0.402647	0.440183	0.442922	0.443102	0.443113
2	0.232498	0.254034	0.255602	0.255705	0.255711
3	0.134071	0.146488	0.147392	0.147451	0.147455
4	0.077310	0.084469	0.084991	0.085025	0.085027
5	0.044579	0.048708	0.049008	0.049028	0.049029

The inverse Laplace transform of (12.23), and simplifying

$$g(u,y) = \left\{ \frac{(\beta-r_1)^2}{(r_2-r_1)} + \frac{(\beta-r_1)^2(\beta-r_2)y}{(r_1-r_2)} \right\} e^{-r_1 u - \beta y} + \left\{ \frac{(\beta-r_2)^2}{(r_1-r_2)} + \frac{(\beta-r_1)(\beta-r_2)^2 y}{(r_2-r_1)} \right\} e^{-r_2 u - \beta y}$$

where $r_1 (= R)$ and r_2 are the positive roots of the equation (12.5).

Tables 12.1 and 12.2 show the probability and severity of ruin $G(u, y)$ when the individual claim amount distributions are exponential and gamma respectively.

12.4.3 *Mixed Exponential Distribution*

Assuming that a single claim amount has a Mixed Exponential distribution with

$$f(x) = p_1 \beta_1 e^{-\beta_1 x} + p_2 \beta_2 e^{-\beta_2 x}, \quad x > 0, \ 0 < \beta_1 < \beta_2, \ p_1 + p_2 = 1$$

Therefore,

$$\hat{g}(s) = \frac{(s+\beta_1)(s+\beta_2)(2m\lambda^2 - 3c\lambda + 3c\lambda\phi(0)) - 2\lambda^2\phi(0)(\beta_1 p_2 + \beta_2 p_1)}{(s+\beta_1)(s+\beta_2)(2m\lambda^2 - 3c\lambda + c^2\beta\phi(0))} \quad (12.24)$$

Hence the inverse Laplace transform of (12.24)

$$g(0,y) = k_1 e^{-\beta_1 y} + k_2 e^{-\beta_2 y} \quad (12.25)$$

$$\text{where } k_1 = \frac{2p_1 \lambda^2 \phi(0)}{3c\lambda - 2m\lambda^2 + c^2 \beta_1 \phi(0)}$$

$$\text{and } k_2 = \frac{2p_2 \lambda^2 \phi(0)}{3c\lambda - 2m\lambda^2 + c^2 \beta_2 \phi(0)}$$

then

$$\hat{g}(s) = \frac{k_1}{s+\beta_1} + \frac{k_1}{s+\beta_1}$$

$$g(0, u+y) = k_1 e^{-\beta_1(u+y)} + k_2 e^{-\beta_2(u+y)}$$

and

$$\hat{\xi}(s,y) = \frac{k_1}{s+\beta_1} e^{-\beta_1 y} + \frac{k_2}{s+\beta_2} e^{-\beta_2 y}$$

Hence, we have

$$\hat{g}(s,y) = \frac{\frac{k_1}{s+\beta_1} e^{-\beta_1 y} + \frac{k_2}{s+\beta_2} e^{-\beta_2 y}}{1 - \frac{k_1}{s+\beta_1} - \frac{k_2}{s+\beta_2}} \quad (12.26)$$

Taking the inverse Laplace transform and after simplification

$$g(u,y) = \frac{(\beta_1-r_1)(\beta_1-r_2)(\beta_2-r_1)}{(\beta_1-\beta_2)(r_2-r_1)} e^{-r_1 u - \beta_1 y} +$$

$$\frac{(\beta_1-r_1)(\beta_1-r_2)(\beta_2-r_2)}{(\beta_1-\beta_2)(r_1-r_2)} e^{-r_2 u - \beta_1 y} +$$

$$\frac{(\beta_2-r_1)(\beta_1-r_1)(\beta_2-r_2)}{(\beta_1-\beta_2)(r_1-r_2)} e^{-r_1 u - \beta_2 y} +$$

$$\frac{(\beta_2-r_1)(\beta_2-r_2)(\beta_1-r_2)}{(\beta_1-\beta_2)(r_2-r_1)} e^{-r_2 u - \beta_2 y}$$

where $r_1 (= R)$ and r_2 are the positive roots of the equation (12.5).

12.4.4 Generalized Exponential Distribution

If the individual claim amount has a GE distribution with $GE(2,\beta)$ then

$$\hat{g}(s) = \frac{(s+\beta)(s+2\beta)(2m\lambda^2 - 3c\lambda + 3c\lambda\phi(0)) - 2\lambda^2\phi(0)(s+3\beta)}{(s+\beta)(s+2\beta)(2m\lambda^2 - 3c\lambda + c^2\beta\phi(0))} \tag{12.27}$$

Taking the inverse Laplace transform of (12.27)

$$g(0,y) = k_1 e^{-\beta y} - k_2 e^{-2\beta y} \tag{12.28}$$

$$\text{where } k_1 = \frac{4\lambda^2 \phi(0)}{3c\lambda - 2m\lambda^2 + c^2\beta\phi(0)}$$

$$\text{and } k_2 = \frac{2\lambda^2 \phi(0)}{3c\lambda - 2m\lambda^2 + 2c^2\beta\phi(0)}$$

then

$$\hat{g}(s) = \frac{k_1}{s+\beta} - \frac{k_2}{s+2\beta}$$

$$g(0, u+y) = k_1 e^{-\beta(u+y)} - k_2 e^{-2\beta(u+y)}$$

$$\hat{\xi}(s,y) = \frac{k_1}{s+\beta} e^{-\beta y} - \frac{k_2}{s+2\beta} e^{-2\beta y}$$

Hence, we have

$$\hat{g}(s,y) = \frac{\frac{k_1}{s+\beta} e^{-\beta y} - \frac{k_2}{s+2\beta} e^{-2\beta y}}{1 - \frac{k_1}{s+\beta} + \frac{k_2}{s+2\beta}} \tag{12.29}$$

Taking the inverse Laplace transform of the above, and then simplified

$$g(u,y) = \frac{(\beta - r_1)(\beta - r_2)(2\beta - r_1)}{\beta(r_1 - r_2)} e^{-r_1 u - \beta y} +$$
$$\frac{(\beta - r_1)(\beta - r_2)(2\beta - r_2)}{\beta(r_2 - r_1)} e^{-r_2 u - \beta y} +$$
$$\frac{(\beta - r_1)(2\beta - r_1)(2\beta - r_2)}{\beta(r_2 - r_1)} e^{-r_1 u - 2\beta y} +$$
$$\frac{(2\beta - r_1)(2\beta - r_2)(\beta - r_2)}{\beta(r_1 - r_2)} e^{-r_2 u - 2\beta y}$$

where $r_1 (= R)$ and r_2 are the positive roots of the equation (12.5).

Table 12.3 and 12.4 give the values of $G(u,y)$ for the claim size distributions mixed exponential and generalized exponential. Also figure 1 and 2 display graph of $G(u,y)$ for different values of u and y.

Table 12.3 $\beta_1 = 0.5$, $\beta_2 = 2$, $\lambda = 1$, $p_1 = 1/3$, $c = 1.1$

u \ y	1	2	3	4	5	10
0	0.268064	0.378087	0.437706	0.472904	0.494122	0.524098
1	0.162899	0.247954	0.297682	0.327592	0.345699	0.371307
2	0.117621	0.182917	0.221702	0.245116	0.259302	0.279370
3	0.088902	0.138952	0.168784	0.186808	0.197730	0.213181
4	0.067909	0.106260	0.129136	0.142959	0.151337	0.163188
5	0.051995	0.081379	0.098910	0.109504	0.115924	0.125006

Table 12.4 $\beta = 3$, $\lambda = 2$, $c = 1.1$

u \ y	1	2	3	4	5
0	0.736325	0.859406	0.876492	0.878812	0.879127
1	0.607585	0.697941	0.710318	0.711996	0.712223
2	0.488700	0.561246	0.571182	0.572529	0.572711
3	0.392931	0.451258	0.459247	0.460329	0.460476
4	0.315927	0.362824	0.369247	0.370117	0.370235
5	0.254014	0.291721	0.296885	0.297585	0.297679

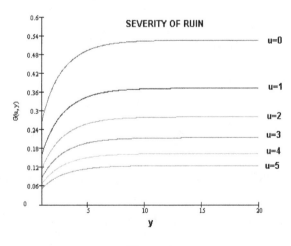

Figure 1

From the graph we note that for a given value of u, the distribution function $G(u, y)$ is increasing in y and tend to a limiting value after a certain limit of y. Hence as y is tending to infinity, the distribution $G(u, y)$ tends towards the ruin probability $\psi(u)$. See Tables 12.5 and 12.6 also.

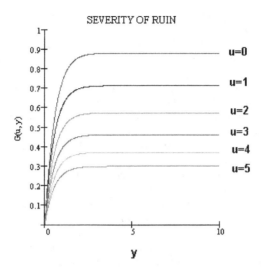

Figure 2

Table 12.5   Ruin Probabilities for mixed exponential claims

u	0	1	2	3	4	5
$\lim_{y \to \infty} G(u,y)$	0.526778	0.373597	0.281164	0.214562	0.164227	0.125818

Table 12.6   Ruin Probabilities for generalized exponential claims

u	0	1	2	3	4	5
$\lim_{y \to \infty} G(u,y)$	0.879178	0.712263	0.572745	0.460505	0.370260	0.297700

Acknowledgements. Both authors would like to thank the anonymous referees for their valuable and constructive suggestions.

Bibliography

Avram, F. and Usabel, M. (2003). Ruin probabilities and deficit for the renewal risk model with phase-type interarrival times, *Astin Bulletin* **34**, pp. 315–332.

Dickson, D. C. M. (1998). On a class of Renewal risk Processes, *North American Actuarial Journal* **2**, pp. 60–73.

Drekic, S., Stanford, D. A and Willmot, G. E. (2004). On the Distribution of Deficit at Ruin when Claims are Phase-type, *Scand. Actuarial Journal*, **2**, pp. 105–120.

Gupta, R. D. and Kundu, D. (1999). Generalized Exponential Distribution, *Astrl. & New Zealand J. Statist.* **41**, pp. 173–188.

Gupta, R. D. and Kundu, D. (2001). Generalized Exponential Distribution: Different Method of Estimation, *J. Statist. Comput. Simul.* **69**, pp. 315–338.

Gerber, H. U , Goovaerts, M. J and Kaas, R. (1987). On the probability and severity of ruin, *Astin Bulletin* **17**, pp. 151–163.

Mathews, J. H and Howell, R. W. (2001). *Complex Analysis for Mathematics and Engineering*, (Jones and Bartlett Publishers, Boston).

Lin, X. S and Willmot, G. E. (2000). The moments of the time of ruin, the surplus before ruin, and deficit at ruin, *Insurance:Mathematics and Economics* **27**, pp. 19–44.

Rolski, T, Schmidli, H, Schmidt, V. and Tuegels, J. (1999). *Stochastic Processes for Insurance and Finance*, (John Wiley & Sons).

Chapter 13

Fair Allocations to Discourage Free Riding Behavior

Kim Hang Pham Do
Department of Economics and Finance
Massey University, New Zealand
e-mail: K.H.Phamdo@massey.ac.nz
Henk Folmer
Department of Spatial Sciences
University of Groningen and
Economics of Consumers and Households Group
Wageningen University, the Netherlands
e-mail: Henk.Folmer@wur.nl

Abstract

This chapter presents the feasible proportional allocation rule to discourage free riding for a special class of free riding problems. Some theoretical and practical properties of the rule are discussed. Applications to the management of the Baltic Sea cod fishery and the Norwegian spring-spawning herring fishery are presented.

Key Words: Partition function form game, feasible allocation, free rider game, regional fishery management, proportional allocation rule

13.1 Introduction

Free riding occurs when players benefit from the actions and efforts of others without contributing to the costs incurred in generating the benefits. Typical situations where free riding may occur in the international arena are global warming abatement and management of international fish resources.

Free riding can be seen as a prisoner's dilemma. Common resource management, as in the case of international fish stocks, may take this form.

Consider for instance one of the main problems for an international fishery, the new member problem. The 1995 UN Fish Stocks Agreement (UNFSA) allows any nation to fish outside the Exclusive Economic Zone. Although the agreement mandates that a Regional Fishery Management Organization (RFMO) should manage such an international fish stock in a sustainable manner, distant water fishing nations may decide not to join a RFMO, but rather, to harvest in an individually optimal fashion (Bjørndal and Munro, 2003; Munro et al., 2004). This creates an incentive for all fishing nations not to join the RFMO and for incumbents to leave the RFMO which may lead to the breakdown of the fishery. (Observe the similarity to the tragedy of the commons (Hardin, 1968).)

This chapter discusses the proportional allocation[1] rule to discourage free riding. Assuming that the players can freely merge or break apart and are farsighted[2], we formulate a free rider problem as a game in partition function form (Thrall and Lucas, 1963). Using the principle of distributive fairness "equals should be treated equally, and unequals unequally, in proportion to relevant similarities and differences" (Moulin, 2003), we propose proportional allocation as a solution concept to achieve stable coalition structures. We also analyze the feasible set of coalitions, their values and how application of the proportional rule discourages free riding in the case of international fish resources.

Our approach is an extension of the work by Pham Do and Folmer (2006) and Kronbak and Lindroos (2005) in a search for fair solutions to discourage free riding in RFMOs. Pham Do and Folmer (2006) apply the Shapley value[3] for a special class of fishery games and Kronbak and Lindroos (2005) apply the satisfactory nucleolus. Although these sharing rules can be considered as "fair" for distributing the total positive gains from the grand coalition and stimulate the players to join the grand coalition, they are not sufficient to discourage free-riding, since for certain coalition structures free riding may result in a higher payoff (see Pham Do and Folmer (2006) and section 13.4 of this chapter). Our approach is also related to Eyckmans and Finus (2004) and Pham Do et al (2006). However, Eyckmans and Finus (2004) use the concept of internal and external stability of d'Aspremont et al. (1983) to propose a sharing scheme for the distribution

---

[1] Proportional allocation is not a new idea but it "is deeply rooted in law and custom as a norm of distributed justice" (Young, 1994).

[2] That is, it is the final and not the immediate payoffs that matter to the coalitions (Chander, 2003).

[3] The Shapley value in that paper refers to the modified Shapley value for the class of partition function form games (see Pham Do and Norde (2007) for details).

of the gains from cooperation for the particular class of games with only one non-trivial coalition, all other players are singletons. Pham Do et al. (2006) consider the population monotonic allocation schemes for a special class of fishery games with transferable technologies. This chapter, on the other hand, focuses on a feasible solution concept for a class of free rider games.

The next section presents some preliminaries, particularly the basic concepts of games in partition function form and free rider games. Section 13.3 deals with with the notion of feasible allocation and its properties for free rider games. Section 13.4 presents two applications. Concluding remarks follow in the last section.

13.2 Notation and Preliminaries

Let $N = \{1, 2, ..., n\}$ be a finite set of players. Subsets of N are called *coalitions*. It will be convenient to economize on brackets and suppress the commas between elements of the same coalition. Thus, we will write, for example, 124 instead of $\{1, 2, 4\}$. A *partition* κ of N is a set of pairwise disjoint nonempty coalitions, $\kappa = \{S_1, ..., S_m\}$, such that their union is N. Let $\mathcal{P}(N)$ be the set of all partitions of N. For a partition $\kappa \in \mathcal{P}(N)$ and $i \in N$, let $S(i, \kappa)$ be the coalition $S \in \kappa$ to which player i belongs.

A partition of a subset $S \subset N$ is denoted by κ_S; a singleton coalition by $\{i\}$, the coalition structure consisting of all *singleton coalitions* by $[N]$; the grand coalition by N and the coalition structure consisting of the grand coalition only by $\{N\}$. Finally, let $|S|$ be the number of players in S and $|\kappa|$ the number of coalitions in κ.

For $i \in N$ and $\kappa \in \mathcal{P}(N)$, we define a coalition structure arranged by *affiliating* $S(i, \kappa)$ to T ($\neq S$) $\in \kappa$, denoted by $\kappa_{+i}(T)$, as :

$$\kappa_{+i}(T) = \{(\kappa\backslash\{S, T\}) \cup (S\backslash\{i\}) \cup (T \cup \{i\})\}. \tag{13.1}$$

A coalition structure $\kappa_{+i}(T)$ is simply a merge of player i into T. (Observe that the number of coalitions does not increase, i.e. $|\kappa| \geq |\kappa_{+i}(T)|$).

A coalition structure $\kappa_{-i}(S)$ arranged by *withdrawing* i from S is defined as follows:

$$\kappa_{-i}(S) = \{(\kappa\backslash S) \cup \{i\} \cup (S\backslash\{i\})\} \tag{13.2}$$

A coalition structure $\kappa_{-i}(S)$ is simply a split of player i from S. (Again, the number of coalitions does not decrease, i.e $|\kappa_{-i}(S)| \geq |\kappa|$.)

For each $i \in N$, let κ^i denote a coalition structure where player i plays as a singleton, and $\kappa^i(N)$ be the set $\{\kappa \in \mathcal{P}(N) | \{i\} \in \kappa\}$, i.e the set of all coalition structures where player i plays as a singleton. So, $\kappa^i = \{\{i\}, \kappa_{N \setminus i}\} \in \kappa^i(N)$.

Example 13.1. Consider the coalition structure $\kappa = \{123, 45\}$. For $i = 1, T = \{123\}$ and $S = \{45\}$, then $\kappa_{+1}(S) = \{23, 145\}$, $\kappa_{-1}(T) = \{1, 23, 45\}$, and κ^1 can be one of the following coalitions $\{1, 23, 45\}$, $\{1, 25, 34\}$, $\{1, 24, 35\}$, $\{1, 2345\}$, $\{1, 2, 345\}$, $\{1, 3, 245\}$, $\{1, 4, 235\}$, $\{1, 5, 234\}$, $\{1, 23, 4, 5\}$, $\{1, 2, 34, 5\}$, $\{1, 24, 3, 5\}$, $\{1, 25, 3, 4\}$, $\{1, 35, 2, 4\}$, $\{1, 2, 3, 45\}$, $\{1, 2, 3, 4, 5\}$.

A pair (S, κ) which consists of a coalition S and a partition κ of N to which S belongs is called an *embedded coalition*. Let $E(N)$ denote the set of all embedded coalitions, i.e.

$$E(N) = \{(S, \kappa) \in 2^N \times \mathcal{P}(N) | \ S \in \kappa\}.$$

Definition 13.1. A mapping

$$w : E(N) \longrightarrow R$$

that assigns a real value $w(S, \kappa)$ to each embedded coalition (S, κ) is called a *partition function*. The ordered pair (N, w) is called a *partition function form game*[4] (pffg).

The value $w(S, \kappa)$ represents the payoff of coalition S given that coalition structure κ forms. For a given partition $\kappa = \{S_1, S_2, ..., S_m\}$ and partition function w, let $\overline{w}(S_1, S_2, ..., S_m)$ denote the m-vector $(w(S_i, \kappa))_{i=1}^m$. The set of partition function form games with player set N is denoted by $\Gamma(N)$. For convenience, we write w as a pffg, instead of (N, w).

Definition 13.2. Let $w \in \Gamma(N)$. We call player $j \in N$ a *free rider* when it expects to benefit from a merger of the other players by staying outside the coalition. Formally, j is a free rider if for $\kappa^j \neq [N], w(j, \kappa^j) > w(j, [N])$.

Example 13.2. Consider the partition function form game w defined by:

$\overline{w}(1, 2, 3) = (0, 0, 0)$, $\overline{w}(12, 3) = (2, 0)$, $\overline{w}(23, 1) = (3, 2)$, $\overline{w}(13, 2) = (2, 1)$, $\overline{w}(123) = 10$. This game has two free riders: player 1 and player

---
[4]For an application of partition function form games to fisheries see Pintassilgo (2003).

2, since $\min_{\kappa^i \in \kappa^i(N)} \{w(i,\kappa^i)\} = 0$, $\forall i \in N$, $\max_{\kappa^1 \in \kappa^1(N)} \{w(1,\kappa^1)\} = 2$ and $\max_{\kappa^2 \in \kappa^2(N)} \{w(2,\kappa^2)\} = 1$, whereas $\max_{\kappa^3 \in \kappa^3(N)} \{w(3,\kappa^3)\} = 0$.

Definition 13.3. Let $w \in \Gamma(N)$. w is called a free rider game if the two following conditions hold

(i) $w(N,\{N\}) \geq \sum_{S \in \kappa \in \Gamma(N)} w(S,\kappa)$, and
(ii) $\forall i \in N, \forall S,T \in \kappa^i \backslash \{i\}$, $w(i, \kappa^i \backslash \{S,T\} \cup \{S \cup T\}) \geq w(i,\kappa^i)$.

Condition (i) implies that the grand coalition is the most efficient coalition, while condition (ii) implies that player i expects to benefit from the merger of coalitions by not joining the merger.

The set of free rider games is denoted by $\mathcal{FRG}(\mathcal{N})$.

Let $w \in \mathcal{FRG}(\mathcal{N})$ and $i \in N$. We define the minimum and maximum payoffs for player i as follows:

$$\eta_i = \min_{\kappa^i \in \kappa^i(N)} \{w(i,\kappa^i)\}, \quad (13.3)$$

$$\theta_i = \max_{\kappa^i \in \kappa^i(N)} \{w(i,\kappa^i)\}.$$

The value η_i is the payoff guaranteed to player i if it stays alone regardless of what the partition of $N\backslash\{i\}$ does, whereas θ_i is the maximum payoff that player i can expect when all others cooperate. The interval $[\eta_i, \theta_i]$ is called a *feasible right* for each player $i \in N$. For any free rider $i \in N$, $\theta_i > \eta_i$.

One can easily see that for every $w \in \mathcal{FRG}(\mathcal{N})$,

$$\eta_i = w(i,[N]), \text{ and} \quad (13.4)$$

$$\theta_i = w(i, \kappa_{-i}(N)). \quad (13.5)$$

Note that (13.4) implies that the worst payoff is obtained when all players behave non-cooperativelly (act as singletons) whereas (13.5) indicates an incentive for players to free ride since a free rider expects to get the highest payoff if it is the only outsider of the grand coalition.

Definition 13.4. Let $w \in \mathcal{FRG}(\mathcal{N})$. A coalition S is called stable if no subcoalition can improve its payoff by breaking up from the coalition, *ceteris paribus*, i.e. $w(S,\kappa) \geq \sum_{S_b \in \kappa_S} w(S_b, \kappa_S \cup (\kappa\backslash S))$.

The 1995 UN Agreement calls for cooperative management through RFMOs. This implies that the extension of a RFMOs and its stability are crucial features. This translates into the following necessary requirements for a RFMO (as a stable coalition in the coalition structure κ with $|\kappa| < N$).

(C1) Feasibility: $w(S,\kappa) \geq \sum_{i \in S} w(i,[N])$;
(C2) Potential stability: $w(S,\kappa) \geq \sum_{i \in S} w(i,\kappa_{-i}(S))$;
(C3) Strong stability: S is potentially stable and

$$w(S,\kappa) = \max_{\kappa_{N \backslash S}} w(S, \kappa_{N \backslash S} \cup S)$$

The conditions (C1) and (C2) are necessary for forming a coalition, while (C3) implies the stability of coalition S under a coalition structure κ. Note that if (C2) holds for all coalitions, the coalition structure can be considered as a potentially stable coalition structure.

Definition 13.5. A game $w \in \mathcal{FRG}(\mathcal{N})$ is called potentially stable if there exists a coalition structure κ such that $w(S,\kappa) \geq \sum_{i \in S} w(i, \kappa_{-i}(S))$ for every $S \in \kappa$.

A potentially stable game implies the existence of a potentially stable coalition structure in the sense that no player is interested in leaving its coalition to adopt free rider behavior. Moreover, if the grand coalition is stable, then no player is interested to leave it.

13.3 Feasible Allocations

We now turn to the notion of feasible allocation to induce free riders to give up their behaviour. We make use of Myerson (1978) who points out that the basic requirement of a fair solution is that its allocation is feasible. Moreover, he observes that the construction of fair allocations (settlements) should be based on the expected payoffs in all feasible coalitions, particularly the grand coalition, taking into account threats. Below we focus on the construction of a feasible allocation. We shall pay attention to the question whether the coalition is profitable and how profit should be divided so as to induce the players to form a coalitions such that a free rider has an incentive to cooperate.

For every coalition S, a *reasonable allocation* (with respect to S in κ) is defined as a vector $x = (x_i)_{i \in S} \in R^{|S|}$ satisfying $w(S,\kappa) \geq x(S) = \sum_{i \in S} x_i$

and $x_i \geq w(i, [N])$ for every $i \in S$. Reasonable allocation implies that for every coalition, the sum of its allocation values (awards) does not exceed the worth of the coalition, whereas on the other hand the payoff of player i exceeds its payoff if the coalition structure consists of singletons only. The set of all reasonable payoffs for S in w is denoted by $X(S, w)$.

Definition 13.6. The semi-stable set of w is defined by

$$SemS(N, w) = \{x \in X(N, w) | \forall S \in \kappa, \ x(S) = w(S, \kappa)\}. \tag{13.6}$$

Semi-stability implies that all players can form a coalition structure in such a way that every player can find a coalition for itself that meets the demand of all members, exactly divides total payoff and that the payoff for each $i \in S$ is individually rational. The semi-stable set exists for every free rider game w, as this coalition structure consists of all singletons satisfying all conditions of Definition 13.6. (Note that the semi-stable set differs from the imputation set known from the characteristic function (TU) game. An imputation set is the payoff vector for the grand coalition, whereas a semi-stable set assigns a vector to every possible coalition (structure), specifying individual payoffs to coalition members and outsiders).

Definition 13.7. A weighted scheme of coalition S is a collection of real numbers $\lambda_S = (\lambda_{S,i})_{i \in S} \in R^{|S|}$ satisfying $\sum_{i \in S} \lambda_{S,i} = 1$ and $\lambda_{S,i} \in [0, 1]$.

For example, an *upper weighted value* is defined as the collection of

$$\lambda_S = (\frac{\theta_i}{\sum_{j \in S} \theta_j})_{i \in S}; \tag{13.7}$$

and a *lower weighted value* is the collection of

$$\lambda_S = (\frac{\eta_i}{\sum_{j \in S} \eta_j})_{i \in S}, \tag{13.8}$$

where $\theta_i = \max_{\kappa^i \in \kappa^i(N)} \{w(i, \kappa^i)\}$ and $\eta_i = \min_{\kappa^i \in \kappa^i(N)} \{w(i, \kappa^i)\}$.

Definition 13.8. A *valuation*[5] *is a mapping Ψ which associates to each coalition structure $\kappa \in P(N)$ a vector of individual payoffs in R^N.*

---
[5] The notion "valuation" indicates that each player is able to evaluate directly the payoff it obtains in different coalition structures. Valuations thus emerge when the rule of division of the payoffs between coalition members is fixed (for further details, see Bloch, 2003).

Definition 13.9. *A weighted valuation is a valuation* Ψ *such that*

$$\Psi_i(S, w) = a_i + \lambda_{S,i} G(S, \kappa), \tag{13.9}$$

for every coalition $S(i, \kappa)$, *where* $a_i \in [\eta_i, \theta_i]$, $G(S, \kappa) = w(S, \kappa) - \sum_{i \in S} a_i$, *and* $\lambda_S = (\lambda_{S,i})_{i \in S}$ *is a weighted scheme of* S.

A weighted valuation gives an expected value to each player with respect to the distribution among the players of all the free rider values and the gain from cooperation. A weighted valuation is called *proportional valuation* if λ_S is chosen such that $\lambda_{S,i} = \frac{\lambda_i}{\sum_{i \in S} \lambda_i}$, where $\lambda_i \in R^+$.

Let $\Psi : \mathcal{FRG}(\mathcal{N}) \to \mathbb{R}^N$ be a valuation. The weighted valuation Ψ

(i) is *individually rational (IR)* if $\Psi_i(w) \geq w(i, [N])$ for all $i \in N$.
(ii) is *relatively efficient (RE)* if for $w \in \Gamma(N)$

$$\sum_{i \in S \in \kappa} \Psi_i(S, w) = w(S, \kappa) \text{ for all } S \in \kappa.$$

(iii) satisfies *fair ranking (FR)* if for players $i, j \in N$, $\theta_i \geq \theta_j$, for $S(i, \kappa)$ and $S(j, \kappa)$, then $\Psi_i(S, w) \geq \Psi_j(S, w)$.
(iv) satisfies *claim right (CR)* if for player $i \in N$, $\min_{\kappa^i \in \kappa^i(N)} \{w(i, \kappa^i)\} = \max_{\kappa^i \in \kappa^i(N)} \{w(i, \kappa^i)\}$, then $\Psi_i(S, w) \geq \max_{\kappa^i \in \kappa^i(N)} w(i, \kappa^i)$.
(v) is *relatively proportional (RP)* if for every player $i \in N$ and $S(\kappa, i)$, $w(i, \kappa_{-i}(S)) = \lambda_i \sum_{j \in S} w(j, \kappa_{-j}(S))$, then $\Psi_i(S, w) = \lambda_i w(S, \kappa)$ where $\lambda_i \in [0, 1]$ and $\sum_{i \in S} \lambda_i = 1$.

Below we shall pay attention to proportional valuation where λ_S is an upper weighted value of S. Note that this valuation splits the surplus (if $G(S, \kappa) > 0$) or loss (if $G(S, \kappa) < 0$) proportionally to what could be obtained by each player as an outsider.

Proposition 13.1. *For every potentially stable game* $w \in \mathcal{FRG}(\mathcal{N})$, *there exists a proportional allocation that satisfies the five properties IR, RE, FR, CR and RP.*

Proof. For every $S \in \kappa$, define $\Psi_i(S, w) = w(\{i\}, \kappa_{-i}(S)) + \lambda_i G(S, \kappa)$, where $G(S, \kappa) = w(S, \kappa) - \sum_{i \in S} w(\{i\}, \kappa_{-i}(S))$, $\lambda_i = \frac{\theta_i}{\sum_{j \in S} \theta_j}$, and $\theta_i = \max_{\kappa^i \in \kappa^i(N)} w(i, \kappa^i))$. Since $G(S, \kappa) \geq 0$, it follows that $\Psi_i(S, w) \geq w(i, [N])$,

$\Psi(S,w) = \sum_{i \in S} \Psi i(S,w) = w(S,\kappa)$. Thus, if $w(i,\kappa_{-i}(S)) \geq w(j,\kappa_{-j}(S))$ implies that $\theta_i \geq \theta_j$ then $\Psi_i(S,w) \geq \Psi_j(S,w)$. Now let $i \in N$ be a player such that $\min_{\kappa^i \in \kappa^i(N)} \{w(i,\kappa^i)\} = \max_{\kappa^i \in \kappa^i(N)} \{w(i,\kappa^i)\}$. Thus, $\eta_i = \theta_i = w(i,\kappa^i) \leq w(i,\kappa_{-i}(S)) + \lambda_i G(S,\kappa) = \Psi_i(S,w)$. □

Example 13.3. Consider the oligopoly game defined as $\overline{w}(1,2,3) = (36,16,9)$; $\overline{w}(12,3) = (57.78, 18.78)$; $\overline{w}(13,2) = (49,25)$; $\overline{w}(23,1) = (25,49)$ and $\overline{w}(123) = 90.25$. In this game, $\eta = (\eta_i)_{i=1,2,3} = (36,16,9)$ and $\theta = (\theta_i)_{i=1,2,3} = (49,25,18)$. The proportional rule with *upper weighted value* λ_N leads to $\Psi(N,w) = (47.66, 24.32, 18.27)$.

The modified Shapley value[6] for this game would lead to $Sh(w) = (46.70, 24.71, 18.83)$.

This example shows that the modified Shapley values assigns more value to players 2 and 3 than to player 1, while the proportional valuation assigns more value to player 1 and less value to players 2 and 3.

We define an *adjustment of proportional allocation APV(w)* as

$$APV(w) = \{\Psi(S,w) | \forall S \in \kappa, \forall \kappa \in \mathcal{P}(N) \text{ and } w \in \Gamma(N)\}, \text{ where}$$

$$\Psi_i(S,w) = \begin{cases} w(i,\kappa_{-i}(S)) + \lambda_i G(S,\kappa), & \text{if } S \text{ is potentially stable} \\ w(i,[N]) + \lambda_i(w(S,\kappa) - \sum_{i \in S} w(i,[N])), & \text{otherwise} \end{cases}$$
(13.10)

Proposition 13.2. Let $w \in \mathcal{FRG}(\mathcal{N})$, then $APV(w) \subset SemS(N,w)$.

Proof. Let $\lambda_S = (\lambda_i)_{i \in S}$ be a weighted scheme. Since $w \in \mathcal{FRG}(\mathcal{N})$, it follows that $w(S,\kappa) \geq \sum_{i \in S} w(i,[N])$. Therefore,

(i) if S is not potentially stable, then $\Psi_i(S,w) = w(i,[N]) + \lambda_i(w(S,\kappa) - \sum_{i \in S} w(i,[N])) \geq w(i,[N])$

(ii) if S is potentially stable, then $w(S,\kappa) \geq \sum_{i \in S} w(i,\kappa_{-i}(S))$ implies that $G(S,\kappa) \geq 0 \Rightarrow \Psi_i(S,w) = w(i,\kappa_{-i}(S)) + \lambda_i G(S,\kappa) \geq w(i,\kappa_{-i}(S)) \geq w(\{i\},[N])$.

Since $\Psi(S) = \sum_{i \in S} \Psi_i(S,w) = w(S,\kappa) \Rightarrow APV(w) \subset SemS(N,w)$. □

---
[6] Recall that the modified Shapley value (Pham Do and Norde, 2007) is the Shapley value for the class of partition function form games. It is calculated as the average of the marginal contributions for each player in all coalition structures consisting of one non-trivial coalition and others as singletons.

The propositions above lead to the following Theorem.

Theorem 13.1. *For every free rider game, there exists a feasible allocation that satisfies individual rationality, relative efficiency, fair ranking and claim right. Moreover, if this game is potentially stable then this allocation is relatively proportional.*

Remark 13.1. Let $w \in \mathcal{FRG}(\mathcal{N})$. The grand coalition is strong stable if $w(N, \{N\}) = \max_{\kappa \in P(N)} \sum_{S \in \kappa} w(S, \kappa) \geq \sum_{i \in N} w(i, \kappa_{-i}(N))$.

13.4 Applications

This section presents applications of the feasible allocation rule to the Baltic Sea cod fishery and the Norwegian spring-spawning herring fishery. The underlying bioeconomic models[7] and calculations are adopted from Kronbak and Lindroos (2005) and Lindroos and Kaitala (2000).

13.4.1 *The Baltic Sea cod fishery*

In the Baltic Sea cod fishery[8] there are three participants: four "old" EU member states (Denmark, Finland, Germany and Sweden), four "new" EU member states (Estonia, Latvia, Lithuania, Poland) and the Russian Federation. The International Baltic Sea Fishery Commission (IBSFC)[9] manages the Baltic Sea cod fishery. The countries participating in the Baltic Sea cod fishery are represented in the IBSFC by their coalitions (1: old EU member states, 2: new EU member states, and 3: Russian Federation). The optimal strategy of each coalition is to maximize its net present value, given the behavior of the non-members.

There are five possible coalition structures: $[N] = \{1, 2, 3\}$, $\{N\} = \{123\}, \{12, 3\}, \{13, 2\}$, and $\{23, 1\}$. Table 1 show the payoffs of the coalition structures (Kronbak and Lindroos, 2005).

---

[7]The specific biological recruitment of the stocks are applied within the framework of the Beverton-Holt model. We adopt the net present values for each possible coalition (depending on several parameters such as mortality, stock recruitment, costs, etc) from these authors to construct the free rider games.

[8]The Baltic Sea fishery is not a high sea fishery and is not facing the problem of new members. Nevertheless, there could be a problem of free riding, a situation that we analyze below.

[9]The IBSFC was abolished in January 2007 when the EU member states withdrew from it in 2006.

Table 1. *The possible benefits (Dkr (mil.)) from five coalition structures in the Baltic Sea cod fishery*

Coalition	Net benefit	Free rider value
1	23069	-
2	16738	-
3	15608	-
12	42562	20276
13	41250	21094
23	33544	28456
123	74717	-

Source: Adjusted from Kronbak and Lindroos (2005)

From Table 1, the free rider game w is obtained[10] as follows:

$\overline{w}(1,2,3) = (23069, 16738, 15608); \overline{w}(12,3) = (42562, 20276);$
$\overline{w}(13,2) = (41250, 21094); \overline{w}(1,23) = (28456, 33544); \overline{w}(123) = 74717.$

This game is potentially stable since $w(S, \kappa) \geq \sum_{i \in S} w(i, \kappa_{-i}(S))$, for all S, and all κ. Moreover, $w(N, \{N\}) = \max_{\kappa \in \mathcal{P}(N)} \sum_{S \in \kappa} w(S, \kappa) = 74717 \geq \sum_{i \in N} w(i, \kappa_{-i}(N)) = 69826$ implies the stability of the grand coalition.

In Table 2 the outcome of the proportional allocation rule (Proposition 13.1) is presented. We also present the outcomes of the alternative sharing rules modified Shapley value (Pham Do and Norde, 2007) and satisfactory nucleolus[11] (Kronbak and Lindroos, 2005) for comparison.

Table 2. *The feasible allocations in the Baltic Sea cod fishery (value in Dkr (mil.), percentages in brackets)*

Player	Free rider value	Modified Shapley value	Satisfactory nucleolus	Proportional allocation
1	28456 (40.8)	29962 (40.1)	30111 (40.3)	30451 (40.8)
2	21094 (30.2)	23013 (30.8)	22714 (30.4)	22571 (30.2)
3	20276 (29.0)	21743 (29.1)	21892 (29.3)	21694 (29.0)

[10] We adopt the Chander (2003) approach, where the player (country) outside the coalition will play non-cooperatively against the coalition, to construct the partition function.
[11] The satisfactory nucleolus is a modified imputation calculated in a similar fashion as the nucleolus (for details, see Kronbak and Lindroos, 2005).

From Table 2 it follows that the proportional allocation rule preserves each coalition's share under free riding behavior. Moreover, in absolute terms each coalition is better off than under free riding.

13.4.2 The Norwegian spring-spawning herring fishery

In the Norwegian spring-spawning herring fishery the following nations participate: Norway, Iceland, The Russian Federation, Faeroe Islands and some members of the EU. The latter is a distant water fishing nation. Lindroos and Kaitala (2000) argue that on the basis of historical developments the following coalitions are involved in the fishery: coalition 1 (Norway and the Russian Federation), coalition 2 (Iceland and the Faeroe Islands) and coalition 3 (EU). Table 3 shows the values of possible coalition structures.

From Table 3 the following free rider game w is obtained:
$\overline{w}(1,2,3) = (4878, 2313, 986); \overline{w}(12,3) = (19562, 14534);$
$\overline{w}(13,2) = (18141, 17544); \overline{w}(23,1) = (17544, 18141); \overline{w}(123) = 44494.$

Table 3. The possible benefits (Dkr (mil.)) for five coalition structures

Coalition	Net benefit	Free rider value
1	4878	-
2	2313	-
3	896	-
12	19562	14534
13	18141	17544
23	17544	18141
123	44494	-

Source: Lindroos and Kaitala (2000)

Observe that the grand coalition is not potentially stable, as $w(N, \{N\}) = 44494 \leq \sum_{i \in N} w(i, \kappa_{-i}(N)) = 50219$. However, it is efficient, as $w(N, \{N\}) \geq \sum_{S \in \kappa} w(S, \kappa)$ for all $\kappa \in \mathcal{P}(\mathcal{N})$. Since it is not potentially stable, Lindroos and Kaitala (2000) conclude that a multilateral agreement is not feasible. However, since

(1) the RFMO can freely accept new members and that members can break apart;

(2) the players are farsighted and aware of the fact that free riding ultimately will lead to the worst case scenario of a break-down of the fishery;

it makes sense to consider efficient allocation. Particularly, Chander (2003) argues that there are two alternative coalition structures in the long run under the assumption that all players are farsighted: full cooperation and no-cooperation (other outcomes can be considered as intermediate outcomes that will ultimately lead to either full cooperation or no-cooperation). The ultimate outcome depends on how each player evaluates its share from the final surplus. We apply the proportional allocation rule to calculate feasible allocations. For comparison we also present the outcome for the modified Shapely value[12]. The results are presented in Table 4.

Table 4. Allocations in the Norwegian fishery (value in Dkr (mil.), percentages in brackets).

Player	Free rider	Modified Shapley value	Proportional allocation
1	18141 (36.1)	16030 (36.7)	16074 (36.1)
2	17544 (35.0)	14816 (33.3)	15540 (35.0)
3	14534 (28.9)	13348 (30.0)	12880 (28.9)

We observe that the outcome for each coalition under each allocation rule is smaller than the outcome under free riding. Moreover, the modified Shapley value is relatively more beneficial for player 1 (36.7 % vs 36.1) whereas players 2 and 3 are relatively worse off. Finally, proportional allocation is the only rule that preserves the proportional shares under free riding. Therefore, its outcome is most likely to be accepted.

13.5 Concluding Remarks

The purpose of this chapter is to analyze the properties of the proportional allocation rule as "fair" sharing rule for a special class of free rider games and shows how this rule can be applied to stimulate cooperation and to discourage free riding. We present five conditions that a reasonable and fair sharing rule should meet: individual rationality, relative efficiency, fair

---

[12] Since the satisfactory nucleolus (Kronbak and Lindroos (2005)) is applicable to a stable game only, we do not use it in this comparison.

ranking, claim right and relative proportionaly. We show that the proportional rule satisfies all five properties.

We also point out that alternative allocation rules, particularly the modified Shapley value and satisfactory nucleolus, do not satisfy the relative proportional characteristic. Furthermore, we compare the proportional allocations to other sharing devices, notably the modified Shapley value. Two applications to international fisheries are presented. We have shown that if all players are free to merge or break apart and are farsighted, then all players have an incentive to cooperative, since they are ultimately better off than in a non-cooperative outcome. In particular, the proportional rule is the only one that preserves the proportional shares under free riding. Therefore, it's most likely to be accepted in a real world policy context.

Bibliography

Bjørndal, T. and Munro, G. (2003). The management of high seas fisheries and the implementation of the UN fish stock agreement of 1995, in *The International Yearbook of Environmental and Resource Economics 2003/2004*, Folmer and Tietenberg (eds), Edward Elgar.

Bloch, F. (2003). Noncooperative models of coalition formation in games with spillovers, in *Endogenous Formation of Economic Coalitions*, Carraro (eds.), chapter 2. Edward Elgard.

Chander, P. (2003). The $\gamma-$core and coalition formation. *CORE Discussion paper 2003/46*. Université catholique de Louvain, (forthcoming in *International Journal of Game Theory*).

Eyckmans, J. and Finus, M. (2004). An Almost Ideal Sharing Scheme for Coalition Games with Externalities, FEEM Working Paper No. 155.04.

Hardin, G. (1968). The tragedy of the commons, *Science* **163**, pp. 191–211.

Kronbak, L. G. and Lindroos, M. (2005). *Sharing rule and stability in coalition game with exteranalities: the case of the Baltic Sea Cod fishery*, Discussion paper N ^{0} 7, Dept. of Economics and Management, University of Helsinki.

Lindroos, M. and Kaitala, V. (2000). Nash equilibrium in a coalition game of Norwegian Spring-spawning herring fishery, *Marine Resource Economics* **15**, pp. 321–340.

Monlin, H. (2003). *Fair Division and Collective Welfare*, (The MIT Press).

Munro, G., Van Houtte, A. and Willmann, R. (2004). *The Conservation and Management of Shared Fish Stocks: Legal and Economic Aspects*, FAO Fisheries Technical Paper, 465, Rome.

Myerson, R. (1978). Threat equilibria and fair settlements in coopeartive games, *Mathematics of Operations Research* **3**, pp. 265–274.

Pham Do, K. H. and Folmer, H. (2006). International Fisheries Agreements: the feasibility and impacts of partial cooperation. In *"The Theory And Practice Of Environmental And Resource Economics, Essays in Honour of Karl-Gustaf Löfgren"* (Eds) Aronsson, Axelsson and Brännlund, Edward Elgar.

Pham Do, K. H., Folmer, H. and Norde, H. (2006). *Fishery management games: How to reduce effort and admit of new members"*, Discussion paper, AIE 06.10, Massey University (forthcoming in *International Game Theory Review*).

Pham Do, K. H. and Norde, H. (2007). The Shapley value for partition function form games, *International Game Theory Review* **9**, pp. 353–360.

Pintassilgo, P. (2003). A coalition approach to the management of high sea fisheries in the presence of externalities, *Natural Resource Modeling* **16**, pp. 175-197.

Thrall, R. M. and Lucas, W. F. (1963). n-person games in partition function form, *Naval Research Logistic Quarterly* **10**, pp. 281-293.

Young, H. P. (1994). *Equity: In Theory and Practice*, (Princeton University Press).

Chapter 14

A Stackelberg Differential Game with Overlapping Generations for the Management of a Renewable Resource

Luca Grilli[1,2]
*Dipartimento di Scienze Economiche, Matematiche e Statistiche
Università degli Studi di Foggia
Largo Papa Giovanni Paolo II, 1 - 71100 - Foggia, Italy
e-mail: l.grilli@unifg.it*

Abstract

In this chapter we study a differential game, for the extraction activity of a renewable good, in which players are overlapping generations. The framework of overlapping generations allows us to consider intragenerational (players in the same generation) and intergenerational (players in different generations) game equilibria. We consider the case in which players, even if identical, face competition in an asymmetric way. Since we consider overlapping generations, players have asynchronous time horizons, in contrast with a number of studies in intertemporal exploitation of resources in which players have identical time horizons. We conclude by considering the case in which players compete in a leader-follower way. We introduce a Stackelberg differential game with asynchronous time horizons and non fixed role structure. The overlapping generations' framework results in the presence of two different behaviours, the myopic and the non-myopic behaviour. We present a possible solution for the myopic case.

Key Words: Feedback Nash Equilibrium, resource extraction, overlapping generations, asynchronous horizon, asymmetric players, Stackelberg differential game

---

[1] Financially supported by University of Foggia, Local Project 2006-2007 and National Project PRIN.
[2] The author deeply thanks Steffen Jørgensen for his useful suggestions and comments.

14.1 Introduction

The problem of resource extraction activity is of great interest and it is studied by mean of many different approaches. From the pioneering work by [Levhari and Mirman (1980)] the game theoretic approach counts a number of studies such as [Kaitala (1993)], [Plourde and Yeung (1989)], [Jørgensen and Sorger (1990); Jørgensen and Yeung (1996); Jørgensen and Yeung (1999)], [Fischer and Mirman (1992)], [Clark (1980)], [Dockner, Feichtinger and Mehlmann (1989); Dockner and Kaitala (1989); Dockner, Jørgensen, Long and Sorger (2000)](this is an absolutely incomplete list).

Dealing with the problem of resource extraction of a natural resource in presence of different extractors it is necessary to consider at least the two following features: competition among extractors and intergenerational equity. In fact actions of present generations (in terms of extraction rates) influence the choices of future generation in an obvious way.

Competition among extractors can be modelled with a game, but the framework of agents with identical (finite of infinite) time horizon, that is the most used in literature, cannot deal with the problem of intergenerational equity.

It is well known, among economists, that if extractors do not care about the effects of their extraction policies on future levels of stock of resource this can result in the so called "tragedy of commons"[Hotelling (1931); Pigou (1932); Gordon (1954)].

In the chapter by [Burton (1993)] is studied the problem of intertemporal preferences and intergenerational equity in a context of resource extraction, [Mourmouras (1993)] consider a model with overlapping generations. Therefore, none of these papers use a game theoretical approach.

[Jørgensen and Yeung (1999)], for the first time, study a differential game in which players are overlapping generations of extractors of a renewable good.

The framework of overlapping generations allows to consider the problem of intergenerational equity and to introduce a differential game with asynchronous horizons.

It is necessary to define in a more details the concept of "intergenerational equity". [Rawls (1971)] claims:

"...consider the question of justice between generations. There is no need to stress the difficulties that this problem raises. It subjects any ethical theory to severe if not impossible lest"

[Solow (1986)] consider the problem of a natural resource:
" *the current generation is always entitled to take as much resource out of the common intertemporal pool as it can, provided only that it leaves behind the possibility to each succeeding generation can be as well off as this one* "

It seems that, in an environmental context, with "intergenerational equity" we can mean that the extraction strategies of present generations can satisfy their own needs and in the same time guarantee to future generations to do, at least, the same.

On the other hand, the overlapping generations framework introduces in the model the presence of asynchronous horizons since players in different generation have different time horizons.

In this chapter, we consider the case in which players are not perfectly symmetric. It seems natural to suppose that players in different generations, even if identical, do not compete in a symmetric way. We suppose that old extractors have an advantage in extraction activity, as a consequence, for instance, of a better technology (or more experience). We want to stress that this is only one of the possible choices, the following model can be applied also to the reverse case.

As we will show, the advantage in extraction activity compensate partially the difference in marginal rent of young generations and old generations.

In the first case we introduce the asymmetry in the model considering that extractors in different generations face different costs (see [Grilli (2008, 2003)]), in particular old generations face lower costs than young generations.

The chapter presents also the case in which the asymmetry is introduced under the hypothesis that competition among players happens in a leader-follower way. The framework of overlapping generations introduces a "dynamical" structure in the Stackelberg game. Each extractor starts her economic life as a follower and terminates as a leader; in our knowledge this is a novelty in differential games and it seems to be a fruitful field of investigation both for theory and applications.

It is possible to consider the behaviour of the players in two different ways: myopic or non-myopic. The myopic player solves the game in the first part of her life (as a follower) without taking into account that she will act as a leader in the second part of her economic life (and the leader knows it). The non-myopic case, that is the most intersting and needs to be studied in more details, the follower solves her problem considering that she will become the leader in the last part of the game (and the leader knows it).

In the first section we present the model with asymmetric players with different costs presented by [Grilli (2008, 2003)], we show the main features of the FNE and their implications. This section is a starting point for the following one.

The last section provides the theoretical framework of a Stackelberg differential game with overlapping generations. We discuss the model and, using the main results presented in the previous section, we present a solution to the myopic case. Remarks and comments are discussed at the end of this section.

Conclusions and future developments are proposed at the end of the chapter.

14.2 Differential Game and Overlapping Generations

In this section we present the basic ideas of the model presented by [Grilli (2008, 2003)] as an extension of the model by [Jørgensen and Yeung (1999)] in the case of asymmetric players. We will enter in some details because this model is strictly connected with the case of Leader-Follower competition, that will be presented in the following section. In particular we will show that the myopic case is completly tractable by mean of this approach. Consider an economy with a homogeneous good that is a single renewable resource and a sequence of $J+1$ overlapping generations of extractors. Suppose that in each generation there are $n \geq 2$ identical individuals. The problem of extractor i is to determine her extraction rate $h_i^j(s)$, where $i \in N = \{1, 2, \ldots, n\}$ and $j \in \{0, 1, \ldots, J\}$. The set of admissible extractions rates for extractor i in generation j is Ω_i^j. Let $x(s)$ indicate the resource stock at time s and let suppose that $\Omega_i^j = \mathbb{R}^+$ if $x > 0$ and $\Omega_i^j = 0$ if $x = 0$.

Each extractor has an economic life span T (for the initial generation is $T/2$). Let t_j indicate the time in which generation $j \in \{, 1, \ldots, J-1\}$ is born; in the time interval $[t_j, t_j + T/2]$ players in generation j face competition with players in generation $j-1$ that are in the last part of their economic life, in the time interval $[t_j + T/2, t_j + T]$ they compete, as an old generation, with players (young) in generation $j+1$.

The following equalities can make this dynamic more clear and will be useful later: $[t_j, t_j + T/2] = [t_{j-1} + T/2, t_{j-1} + T]$ and $[t_j + T/2, t_j + T] =$

$[t_{j+1}, t_{j+1}+T/2]$. Generation J, in the last part of her life, when generation $J-1$ exit the game, is the only one generation in the game. For the rest of the chapter let consider only the case $j \in \{1, \ldots, J-1\}$ that is the most relevant.

In order to introduce an asymmetry in the game let suppose that players face different costs according to which part of their economic life they are living in. We define the cost functions $C_i^{*j}(h_i^j(s), x(s))$ for $s \in [t_j, t_j + T/2)$ and $C_i^{**j}(h_i^j(s), x(s))$ for $s \in [t_j + T/2, t_j + T]$ (see [Grilli (2008)]).

If $C_i^{**j} < C_i^{*j}$ extractors in old generation face lower cost than young competitors.

The resource extracted is a homogeneous good in a market with a price given by the inverse demand curve $P(s)$.

The payoff of extractor i in generation j is:

$$J_i^j =$$

$$\int_{t_j}^{t_j+T/2} \left\{ h_i^j(s) P(s) - C_i^{*j}(h_i^j(s), x(s)) \right\} e^{-r(s-t_j)} ds$$

$$+ \int_{t_j+T/2}^{t_j+T} \left\{ h_i^j(s) P(s) - C_i^{**j}(h_i^j(s), x(s)) \right\} e^{-r(s-t_j-T/2)} ds \quad (14.1)$$

where r is a discount rate. We recall that the game begins in $t_1 = t_0 + T/2$ when generation 1 and generation 0 co-exists.

Let x_0 be the stock size at time $t_1 = 0$, the dynamics of the resource stock is ruled by:

$$\dot{x}(s) = G(x(s)) \quad \text{and} \quad x(t_0) = x_0 > 0. \quad (14.2)$$

Where function $G(x)$ describe the dynamics of a renewable good. In general, pure compensation hypothesis implies that the relative growth rate $G(x)/x$ is decreasing in x, $G(0) = G(x^*) = 0$ for some $x^* > 0$.

In presence of extraction activity equation (14.9) becomes:

$$\dot{x}(s) = G(x(s)) - \sum_{k=1}^{n} h_k^{j-1}(s) - \sum_{k=1}^{n} h_k^j(s) \quad \text{for } s \in [t_j, t_j + T/2), \quad (14.3)$$

and

$$\dot{x}(s) = G(x(s)) - \sum_{k=1}^{n} h_k^j(s) - \sum_{k=1}^{n} h_k^{j+1}(s) \quad \text{for } s \in [t_j + T/2, t_j + T]. \quad (14.4)$$

By considering the overlapping generation framework and the functional form of cost function the game that successive generations have to solve can be divided into the following two differential games $\Gamma_1(x_0, T)$:

$$\max_{h_i^j \in \Omega_i^j} J_i^j =$$

$$\max \int_{t_1}^{T_f} \left\{ h_i^j(s) P(s) - C_i^{*j}(h_i^j(s), x(s)) \right\} e^{-r(s-t_j)} ds \quad (14.5)$$

$$\text{s.t. } \dot{x}(s) = G(x(s)) - \sum_{k=1}^{n} h_k^{j-1}(s) - \sum_{k=1}^{n} h_k^j(s),$$

where $T_f = t_J + T/2$ and $h_i^j(s) = 0$ for $s \notin [t_j, t_j + T/2]$, and the game $\Gamma_2(x_0, T)$:

$$\max_{h_i^j \in \Omega_i^j} J_i^j =$$

$$\max \int_{t_1}^{T_f} \left\{ h_i^j(s) P(s) - C_i^{**j}(h_i^j(s), x(s)) \right\} e^{-r(s-t_j-T/2)} ds \quad (14.6)$$

$$\text{s.t. } \dot{x}(s) = G(x(s)) - \sum_{k=1}^{n} h_k^{j+1}(s) - \sum_{k=1}^{n} h_k^j(s),$$

where $T_f = t_J + T$ and $h_i^j(s) = 0$ for $s \notin [t_j + T/2, t_j + T]$.

The games $\Gamma_1(x_0, T)$ and $\Gamma_2(x_0, T)$, as we will show, are connected and cannot be solved independently.

Under the assumption that extractors cannot use precommitted strategies since they do not communicate among one another; each player has to adopt a decision rule that depends on the current level of stock resource, so we look for Feedback Nash Equilibrium for the games $\Gamma_1(x_0, T)$ and $\Gamma_2(x_0, T)$.

Extractor i in generation j has to compete, in the first part of her economic life, as a young extractor, with players in her generation and with players in generation $j - 1$; otherwise, in the second part of her life she competes, as an old extractor, with players in generation $j + 1$ (young generation). In this model the competition among players in different generation is not perfectly symmetric as a consequence of different costs.

Each player adopts the feedback strategies $\phi_i^j(x, s)$ for the game $\Gamma_1(x_0, T)$ and for the game $\Gamma_2(x_0, T)$. These strategies are continuous in s and uniformly Lipschitz in x for each s.

Let denote the value functions, if they exist, with $V_i^{j*}(x,t)$ for $t \in [t_j, t_j + T/2]$ and $V_i^{j**}(x,t)$ for $t \in [t_j + T/2, t_j + T]$, with terminal conditions $V_i^{j*}(x, t_j + T/2) = V_i^{j**}(x, t_j + T/2)$ and $V_i^{j**}(x, t_j + T) = 0$. The hypothesis $V_i^{j**}(x, t_j + T) = 0$ is a natural condition since in that moment the old extractor exit the game. In the time interval $[t_j, t_j + T/2] = [t_{j-1} + T/2, t_{j-1} + T]$, there are n extractors in generation j and n extractors in generation $j-1$, they all have to solve the problem:

$$\max_{h_i^j \in \Omega_i^j} J_i^j =$$

$$\int_{t_j}^{t_j+T/2} \left\{ h_i^j(s) P(s) - C_i^{*j}(h_i^j(s), x(s)) \right\} e^{-r(s-t_j)} ds$$

$$+ V_i^{j**}(x, t_j + T/2)$$

$$\max_{h_i^{j-1} \in \Omega_i^{j-1}} J_i^{j-1} =$$

$$\int_{t_{j-1}+T/2}^{t_{j-1}+T} \left\{ h_i^{j-1}(s) P(s) - C_i^{**j-1}(h_i^{j-1}(s), x(s)) \right\} e^{-r(s-t_j)} ds$$

s.t. $\dot{x}(s) = G(x(s)) - \sum_{k=1}^{n} h_i^{j-1}(s) - \sum_{k=1}^{n} h_i^j(s).$

In order to find an analytical solution of this problem it is necessary to assume some functional form for $C_i^{*j}(h_i^j(t), x(t))$, $C_i^{**}(h_i^j(t), x(t))$, $P(s)$ and $G(x(t))$. In the paper by [Grilli (2008)] the optimal strategies $\phi_i^{j*}(x,t)$ and $\phi_i^{j**}(x,t)$ for the two games are derived in closed form. In this case we have the following functions:

- Cost function (with $l < c$):

$$C_i^j(h_i^j(s), x(s)) = \begin{cases} \frac{c}{x(s)^{1/2}} h_i^j(s) & s \in [t_j, t_j + T/2] \\ \frac{l}{x(s)^{1/2}} h_i^j(s) & s \in [t_j + T/2, t_j + T] \end{cases} \quad (14.7)$$

- The inverse demand curve:

$$P(s) = Q(s)^{-1/2} = \left(\sum_{i \in N} h_i^j(s)\right)^{-1/2}. \qquad (14.8)$$

- The dynamics of the resource stock is ruled by:

$$\dot{x}(s) = a\,x(s)^{1/2} - b\,x(s) \quad \text{and} \quad x(t_0) = x_0 > 0. \qquad (14.9)$$

The optimal strategies for extractor i in generation j both for $t \in [t_j, t_j + T/2]$ and for $t \in [t_j + T/2, t_j + T]$ are:

$$\phi_i^{j*}(x,t) =$$

$$\frac{\frac{1}{2}(4n-1)^2 x \left\{ n\left(\frac{A^{**}(t-t_{j-1}-T/2)}{2}\right) - (n-\frac{1}{2})\left(\frac{A^*(t-t_j)}{2}\right) + n(l-c) + \frac{c}{2} \right\}}{\left[n(l + \frac{A^{**}(t-t_j-T/2)}{2}) + n(c + \frac{A^*(t-t_j)}{2}) \right]^3} \qquad (14.10)$$

for $t \in [t_j, t_j + T/2]$;

$$\phi_i^{j**}(x,t) =$$

$$\frac{\frac{1}{2}(4n-1)^2 x \left\{ n\left(\frac{A^*(t-t_{j+1})}{2}\right) - (n-\frac{1}{2})\left(\frac{A^{**}(t-t_j-T/2)}{2}\right) + n(c-l) + \frac{l}{2} \right\}}{\left[n(c + \frac{A^*(t-t_{j+1})}{2}) + n(l + \frac{A^{**}(t-t_j-T/2)}{2}) \right]^3} \qquad (14.11)$$

for $t \in [t_j + T/2, t_j + T]$.

The optimal strategies present the following topics:

- Equations (14.10) and (14.11) are linear in the state variable x (this is a key point for the following section).
- Strategies $\phi_i^{j*}(x,t)$ depend on the value function of their competitors in the same generation (intragenerational) $V_k^{j*}(x,t)$, on the value function of the previous generation (intergenerational) $V_k^{j-1**}(x,t)$
- The strategies of "young" generation are negatively affected by a factor $n(l-c)$
- The strategies of "old" generation have again influence both intragenerational and intergenerational, the advantage from lower costs that is proportional to $n(c-l)$ compensates partially their marginal resource rent that is decreased
- $\phi_i^{j*}(x,t) \geq 0$ requires that $A^* - A^{**} \leq \frac{1}{n}(c + \frac{A^*}{2}) - \frac{c-l}{2}$

14.3 The Leader-Follower Approach

In the previous section the asymmetry in the game was due to a different cost structure for each generation, in this section we discuss and present the case in which the problem of asymmetry in the game is introduced directly by the rules of the game. Let us suppose that competition among generations follows a Leader-Follower structure. We suppose, for sake of simplicity, that in each generation there is only one player. We have a Stackelberg differential game in which players have asynchronous time horizons (as a consequence of the overlapping generations) and this is, in our opinion, a novelty in differential games. If we suppose that the Leader and the Follower in this game is respectively the old and the young player, by mean of the overlapping generation framework, these roles are not fixed but depend on the time period each player is living in, this is another novelty in differential game theory. So we can suppose that young player competes as follower with player in old generation that is the leader (it is also possible to assume the reverse). From the point of view of player in generation j, if $s \in [t_j, t_j + T/2]$ she competes as a follower with player in generation $j-1$ that is the leader; if $s \in [t_j + T/2, T_j + T]$, she competes as a leader with the player in generation $j+1$ that is the follower.

The behaviour of the players can be of two types: myopic or not myopic. We consider the myopic behaviour as the behaviour of a player who solves the two games separately, in other words she solve the game in the first part of her life (as a follower) without considering that in the future she will be a leader, otherwise, she acts in a non-myopic way if she solves the two games as an unique game, and so when she solve the game as a follower she knows and takes into due account that in the second part of her life she will be a leader.

14.3.1 *The Model*

In order to model the Stackelberg setup, if $s \in [t_j, t_j + T/2]$ we indicate with h_F^j the extraction rate for player in generation j (she is young and competes as a follower) and with h_L^{j-1} the extraction rate for player in generation $j-1$ (the leader). On the other side, if $s \in [t_j + T/2, t_j + T]$ than h_L^j is the extraction rate for player in generation j and h_F^{j+1} is the control of player in the just born generation $j+1$. We consider the case in which the information structure is markovian.

The payoff of the extractor in generation j is ruled by:

$$J^j =$$

$$\int_{t_j}^{t_j+T/2} \left\{\pi_F(s, x(s), h_L^{j-1})\right\} e^{-r(s-t_j)} ds$$

$$+ \int_{t_j+T/2}^{t_j+T} \left\{\pi_L(s, x(s), h_F^{j+1})\right\} e^{-r(s-t_j-T/2)} ds \qquad (14.12)$$

where the functions π_F and π_L are respectively the instantaneous profit of the player when she is follower and leader.

The dynamics of the resource in presence of extraction activity is:

$$\dot{x}(s) = G(x(s)) - h_F^j(s) - h_L^{j-1}(s) \text{ for } s \in [t_j, t_j+T/2), \qquad (14.13)$$

and

$$\dot{x}(s) = G(x(s)) - h_L^j(s) - h_F^{j+1}(s) \text{ for } s \in [t_j+T/2, t_j+T]. \qquad (14.14)$$

If we consider the myopic case, the players solve their optimization problem separating their economic life time into two parts and solve the two problems independently. In this case we can apply the same arguments presented in the previous section (see [Grilli (2008, 2003)] for details) and consider two differential games $\Gamma_1^S(x_0, T)$ and $\Gamma_2^S(x_0, T)$ (apex S in order to remind that these are the games in the Stackelberg case) and obtain (at least in principle) the optimal strategies.

In this case $\Gamma_1^S(x_0, T)$ becomes:

$$\max_{h_F^j \in \Omega_F^j} J_F^j =$$

$$\max \int_{t_1}^{T_f} \left\{\pi_F(s, x(s), h_L^{j-1})\right\} e^{-r(s-t_j)} ds \qquad (14.15)$$

$$\text{s.t. } \dot{x}(s) = G(x(s)) - h_F^j(s) - h_L^{j-1}(s),$$

where $T_f = t_J + T/2$ and $h_F^j(s) = 0$ for $s \notin [t_j, t_j+T/2]$,

and $\Gamma_2^S(x_0, T)$ is:

$$\max_{h_L^j \in \Omega_L^j} J_L^j =$$

$$\max \int_{t_1}^{T_f} \left\{ \pi_L(s, x(s), h_F^{j+1}) \right\} e^{-r(s-t_j-T/2)} ds \qquad (14.16)$$

$$\text{s.t. } \dot{x}(s) = G(x(s)) - h_L^j(s) - h_F^{j+1}(s),$$

where $T_f = t_J + T$ and $h_L^j(s) = 0$ for $s \notin [t_j + T/2, t_j + T]$.

In each game standard arguments can be applied for the solution of the Markovian Stackelberg Equilibria ([Başar and Olsder (1999)]). At least in principle it is possible to obtain the follower's best reply $h_F^j = R(t, x(t), h_L^{j-1})$ in $\Gamma_1^S(x_0, T)$ ($h_F^{j+1} = R(t, x(t), h_L^j)$ in $\Gamma_2^S(x_0, T)$), and the leader determines her strategy $\phi_L^{j-1}(t, x(t))$ ($\phi_L^j(t, x(t))$ in $\Gamma_2^S(x_0, T)$) as the solution of her maximization problem under the hypothesis that the follower plays $R(t, x(t), h_L^{j-1})$ ($R(t, x(t), h_L^j)$ in $\Gamma_2^S(x_0, T)$).

It is well known the reason why the best reply function can be obtained "at least in principle", in fact the problem of the follower is a non-standard optimal control problem since h_L^{j-1} can be any extraction rate in the admissible set. In order to avoid this problem a possible solution is to limit the possible strategies for the leader to a subset, and, in particular, one that has been often proposed in literature is the subset of linear and stationary strategies of the form

$$h_L^{j-1}(t, x) = a + bx, \qquad (14.17)$$

where a and b are constants.

In our case this assumption is supported by the results presented in the previous section. We recall that in [Grilli (2008)] the FNE strategies are obtained in closed form and they are linear in the state variable. As a consequence it seems a reasonable assumption for the solution of this class of games to restrict the set of strategies for the leader to the linear case (eq. 14.17).

We have shown that, by following this approach, the myopic case is completely tractable and solvable by applying the technique presented in [Grilli (2008)] so to obtain the solutions in each one of the two subgames.

On the other hand it is not clear how it is possible to consider and study the case in which the two games are not solved separately, that is the non-myopic case. In this direction we hope to stimulate further investigations.

14.3.2 Remarks

(1) How many players?

In the Stackelberg setup presented in the previous section there is one leader and one follower, that is one extractor in each generation. This extractor is a follower while she is young (first half of her economic life time), and a leader when she is old (second half of her economic life time). The Stackelberg setup also works, in general, if there is one leader and n followers. However, in the Overlapping Generations framework this does not make sense (the number of players in each generation must be the same). What does happen to the Stackelberg setup if we consider the case of more than one leader? It seems that the model cannot be applied to such a situation because if there is more than one leader the follower could not design a rational reaction. As a consequence it seems that the only acceptable framework for the Stackelberg setup is the one presented in this chapter.

(2) Who is the Leader?

In the model presented above we have supposed that each extractor is a follower while she is young and a leader when she is old. The main reason for it is to give an advantage to old extractors following the ideas of the model presented in [Grilli (2008)]. It is easy to prove that, in that model, the reverse case is an easy exercise. Is it the same in the Stackelberg setup? The answer in this case requires a serious investigation and the reason is the following. As we have proved in the previous section, in the myopic case where the young extractor is the follower, the leader (old) plays a standard Stackelberg game and selects a strategy, knowing that the follower will play his best response to her announced strategy. The conclusion is that we have two standard Stackelberg games, played sequentially (this is the reason why the model in [Grilli (2008)] provides us a possible solution).

The situation is different if we reverse the roles of extractors and assume that they are leader when young and follower when old. In this case the simple myopic solution is not satisfactory since each extractor has the option, when designing the strategy as a leader in the first subinterval, to take into account that she will be the follower in the second subinterval. This is what we have called the non-myopic approach which requests further investigation. The general principle would be to design the leader strategy in anticipation of the fact that she will be follower in the next subinterval. A degree of freedom is endogenous in a dynamic game since the state variable can be affected by the players. Moreover, the leader strategy could be chosen as an incentive strategy (e.g., a linear one), designed to obtain some advantage in the game where she is the follower. If we apply the hypothesis of linear strategies (eq. 14.17) in this case it could lead to incentives that are strategically unsatisfactory since their determination just requires to fix two parameters once and for all. As a result precommitment sneaks can appear.

14.4 Conclusions

In this chapter we have presented two different approaches in order to introduce an asymmetry in the differential game for resource extraction activity in which players are overlapping generations of extractors. From one side we consider, following [Grilli (2008)], the asymmetry in the competition among players as a consequence of a differentiate cost structure. We present the model and give some details about the FNE, in particular we show how the solutions depend on both intragenerational and intergenerational factors, and that the difference in marginal resource rent is reduced for old extractors.

In the last section we introduce a leader-follower structure in the game. This is a novelty in differential game theory since we have, for the first time, asynchronous time horizons (as a consequence of the overlapping generations) and a Leader-Follower structure that is not fixed but depends on the time period each player is living in.

We underline the presence of two different behaviours, the myopic and the non-myopic behaviour. We show that, using a similar approach as in [Grilli (2008)], it is possible to model and solve the myopic case also by mean of the approach of linear, stationary strategies for the leader (supported by results in the FNE solutions). The non-myopic case is another interesting case that needs to be studied and defined in a more general framework. We hope that this chapter can stimulated studies in this direction that seems to be interesting both for theory and practice in differential games.

Bibliography

Başar, T. and Olsder, G. J. (1999). *Dynamic Noncooperative Game Theory*, (SIAM,Philadelphia, 2nd edition).

Burton, P. (1993). Intertemporal preferences and intergenerational equity considerations in optimal resource harvesting, *Journal of Environmental Economics and Management* **24**, pp. 119–132.

Carrera, C. and Moran, M. (1995). General dynamics in overlapping generations models, *Journal of Economic Dynamics and Control* **19**, pp. 813–830.

Chiarella, C., Kemp, M. C., Long, N. V. and Okuguchi, K. (1984). On the Economics of international fisheries, *International Economic Review* **25**, pp. 85–92.

Clark, C. W. (1976). *Mathematical Bioeconomics: The Optimal Management of Renewable Resources*, (Wiley, New York).

Clark, C. W. (1980). *Restricted access to common-property fishery resources: A game theoretic analysis* in *Dynamic Optimization and Mathematical Economics*, ed. P.T.Liu, Plenum, New York, pp. 117–132.

Clemhout, S. and Wan Jr., H. (1985). Dynamic common property resources and environmental problems, *Journal of Optimization Theory and Applications* **46**, pp. 471–481.

Diamond, P. (1965). National Debt in a neoclassical growth model, *American Economic Review* **55**, pp. 1126–1150.

Dockner, E., Feichtinger, G. and Mehlmann, A. (1989). Noncooperative solutions for a differential game model of a fishery, *Journal of Economics Dynamics and Control* **13**, pp. 1–20.

Dockner, E. and Kaitala, V. (1989). On efficient equilibrium solutions in dynamic game of resource management, *Resource and Energy* **11**, pp. 23–34.

Dockner, E., Jørgensen, S., Long, N. V. and Sorger, G. (2000). *Differential games in economics and management science*, (Cambridge University Press, Cambridge).

Fischer, R. and Mirman, L. (1992). Strategic dynamic interactions: Fish wars, *Journal of Economic Dynamics and Control* **16**, pp. 267–287.

Gordon, H. S. (1954). The Economic Theory of a Common Property Resource, *Journal of Political Economy* **62**, pp. 124–142.

Grilli, L. (2008). Resource extraction activity: an intergenerational approach, *Game Theory and Applications* **13**.

Grilli, L. (2003). *Giochi Differenziali con Orizzonti Asincroni e Applicazioni alla Gestione delle Risorse Rinnovabili*, Phd Thesis 11/2003, Università degli Studi di Napoli "Federico II".

Hotelling, H. (1931). The Economics of Exhaustible Resource, *Journal of Political Economy* **39**, pp. 137–175.

Kaitala, V. (1993). Equilibria in a stochastic resource management game under imperfect information, *European Journal of Operation Research* **71**, pp. 439–453.

John, A. and Pecchenino, R. (1994). An Overlapping Generations Model of Growth and the Environment, *Economic Journal* **104**, pp. 1393–1410.

Jørgensen, S. and Sorger, G. (1990). Feedback Nash equilibrium in a problem of optimal fishery management, *Journal of Optimization Theory and Applications* **64**, pp. 293–310.

Jørgensen, S. and Yeung, D. W. K. (1996). Stochastic differential game model of a common property fishery, *Journal of Optimization Theory and Applications* **90**, pp. 391–403.

Jørgensen, S. and Yeung, D. W. K. (1999). Inter- and intragenerational renewable resource extraction, *Annals of Operation Research* **88**, pp. 275–289.

Jørgensen, S. and Yeung, D. W. K. (2001). Intergenerational cooperative solution of a renewable resource extraction game, *Game Theory and Applications* **6**, pp. 53–72.

Levhari, D. and Mirman, L. J. (1980). The great fish war: An example using a Cournot Nash solution, *Journal of Economics* **11**, pp. 322–334.

Mourmouras, A. (1993). Conservationist government policies and intergenerational equity in an overlapping generations model with renewable resources, *Journal of Public Economics* **51**, pp. 249–268.

Pigou, A. C. (1932). The Economics of Welfare, (London, Macmillan).

Plourde, C. and Yeung, D. W. K. (1989). Harvesting of a transboundary replenishable fish stock: A noncooperative game solution, *Marine Resource Economics* **6**, pp. 57–71.

Rawls, J. (1971). *A theory of Justice*, (Cambridge, Harvard University Press).

Riley, J. G. (1980). The Just Rate of Depletion of a Natural Resource, *Journal of Environmental Economics and Management* **7**, pp. 291–307.

Solow, R. M. (1974). Intergenerational equity and exhaustible resources, *Review of Economic Studies Symposium* pp. 29–46.

Solow, R. M. (1986). On the intergenerational allocation of natural resource, *Scandinavian Journal of Economics* **88**, pp. 141–149.

Stephan, G. Muller-Furstenberger, G. and Previdoli, P. (1997). Overlapping Generations or Infinitely-Lived Agents, *Environmental and Resource Economics* **10**, pp. 27–40.

Weibull, J. (1995). Evolutionary Game Theory, (MA:The M.I.T. Press, Cambridge).

Yeung, D. W. K. (2000). Feedback solution of a Class of Differential Games with Endogenous Horizons, *Journal of Optimization Theory and Applications* **106**, pp. 657–675.

Yeung, D. W. K. (2001). Infinite-Horizon Stochastic Differential Games with Branching Payoffs, *Journal of Optimization Theory and Applications* **111**, pp. 445–460.

Yoshida, M. (1998). Nash Equilibrium Dynamics of Environmental and Human Capital, *International Tax and Public Finance* **5**, pp. 357–377.

Chapter 15

A Benders' Partitioning Approach for Solving the Optimal Communication Spanning Tree Problem

Yogesh K. Agarwal
Decision Sciences Group
Indian Institute of Management, Lucknow 226013, India
e-mail: yka@iiml.ac.in
Prabha Sharma
Department of Mathematics and Statistics
Indian Institute of Technology Kanpur, India
e-mail: prabha@iitk.ac.in

Abstract

Optimal Communication Spanning Tree Problem (OCSTP) is formulated as a mixed integer programming problem and Benders' Partitioning approach is applied for solving it. It is shown that after fixing the values of integer variables for defining a given tree, the dual of the resulting problem is very easy to solve. This dual solution is used to generate a cut for the Benders' master problem. Rather than solving the master problem directly as an integer program, we use the standard local search algorithm to obtain an approximate solution. The algorithm proposed by us evaluates the Benders' objective function at each neighboring tree, and moves to the neighbor that minimizes this objective function. A cut for the master problem is generated from the new solution and added to the master problem. In order to achieve faster convergence, some constrains are imposed on the search. We present computational results on randomly generated 100-node problems, and compare the standard local search with Benders' search. The results show that starting from randomly selected initial trees, average solution quality produced by Benders' search is significantly better than that produced by standard local search.

Key Words: Local search, locally optimal solution, communication spanning tree, Benders' partitioning, mixed integer programming

15.1 Introduction and Problem Definition

Given a graph $G = (V, E)$ with flow costs c_{ij} for each edge (i, j), and the traffic demands between various pairs of vertices, it is required to design a spanning tree network which will minimize the total cost of flow for all traffic demands. This problem is well-known in the literature as the Optimal Communication Spanning Tree Problem (OCSTP). In this chapter we formulate the problem as a mixed integer program and develop a heuristic approach based on Benders' Partitioning for solving it.

We use the following notation to formulate the problem.

i, j node indices

x_{ij} binary variable,

x_{ij} $= 1$ if edge (i, j) is present in the solution,

$= 0$ otherwise

k commodity (node-pair) index

$s(k), t(k)$ source, sink nodes of commodity k

d_k traffic demand of commodity k

$D = \sum_k d_k$, total demand of all commodities.

y_{ij}^k flow of commodity k on link (i, j) from node i to node j.

c_{ij} cost of unit flow on link (i, j) in either direction.

Using the above notation, the problem can be formulated as the following Mixed Integer Program.

Minimize $\sum_i \sum_j \sum_k c_{ij} y_{ij}^k$

Subject to

$\sum_j y_{ji}^k - \sum_j y_{ij}^k = -d_k,$ for $i = s(k)$ $\forall k$

$\sum_j y_{ji}^k - \sum_j y_{ij}^k = d_k$ for $i = t(k)$ $\forall k$ I

$\sum_j y_{ji}^k - \sum_j y_{ij}^k = 0$ for $i \neq s(k), t(k)$ $\forall k$ (MIP)

$\sum_k (y_{ij}^k + y_{ji}^k) \leq D.x_{ij}$ $\forall (i, j)$ II

$x \in T$

$y_{ij}^k \geq 0$

Constraint set I are the standard flow conservation constraints for each commodity. Constraints II are akin to the capacity constraints in a network design problem, even though there is no capacity constraint in this problem. The purpose of these constraints is to ensure that the flow of any commodity is not permitted on a link which is not included in the tree represented by the binary tree vector x. If x_{ij} is 1, the RHS of this constraint is D, which is the total demand of all commodities, and does not impose any constraint on the flow on this link. On the other hand, when x_{ij} is zero, the RHS is zero, and no flow is permitted on the link. Here T is the set of indicator vectors of all trees. Note that the constraint $x \in T$ is implicit in nature, and can be replaced by explicit linear constraints that the vector x must satisfy in order to be a tree vector. Since our algorithm is a neighborhood search approach which moves from one tree to a neighboring tree, these constraints are always implicitly satisfied.

In the present work, we apply Bender's partitioning to the above mixed integer formulation, and develop a neighborhood search based heuristic procedure to solve the Benders' Master Problem. Computational experiments show that this heuristic can yield significantly improved results compared to standard local search approach.

The chapter is organized as follows. After reviewing the literature in Section 15.2, we briefly review the Benders' Partitioning technique in Section 15.3, and go on to develop its specific application to OCSTP in Section 15.4. Section 15.5 gives the details of local search approach as directly applied to OCSTP, and Section 15.6 extends this approach to solve the Benders' Master Problem. It also discusses various implementation related issues. Section 15.7 develops a methodology for comparing the performance of local search algorithms, and Section 15.8 provides the computational results. Some concluding remarks are made in Section 15.9.

15.2 Literature Review

OCSTP can be viewed as a generalization of the Network Design Problem (NDP) on graph $G = (V, E)$ which is defined as follows. For all $(i, j) \in E$, let c_{ij} denote the cost of traversing the edge (i, j), B a budget and C a threshold criterion, where B and C are both positive integers. The problem is to find a sub-graph $G' = (V, E')$ of G with cost $\sum c_{ij} \leq B$ and the criterion value $F(G') \leq C$, where $F(G')$ denotes the sum of the costs of the shortest paths in G' between all vertex pairs. Johnson, Lenstra and Rinnooy

Kan (1978) have shown that NDP is NP-complete. Hence the optimization version of the problem, in which the graph G' should minimize the function $F(G')$ is NP-hard. They also showed that the simplified version of NDP in which all $c_{ij} = 1$ is also NP-complete.

The OCSTP was first introduced by Hu (1974). It can be shown that OCSTP is a generalization of NDP, since by choosing B large enough the budget constraint can be made ineffective, and the criterion function is generalized to the cost of communication by taking traffic volume into account. Therefore, OCSTP is also NP-hard.

Maffioli proposed a special case of OCSTP in which all demands $d_k = 1$. Hu calls this problem the Optimum Distance Spanning Tree Problem (ODSTP), which is in fact the optimization version of NDP with the additional requirement that G' be a spanning tree. This problem is also NP-hard. Hu considered another special case of the problem in which all $c_{ij} = 1$ and d_k values are arbitrary. He called this problem the Optimum Requirement Spanning Tree Problem, and was able to solve it polynomially. Another special case considered by Hu is one in which the c_{ij} values satisfy the generalized triangle inequality. Under this condition, he showed that the optimum tree will be a star tree.

Ahuja and Murty (1978) gave a branch and bound algorithm and a two-phase local search heuristic for OCSTP, but both their algorithms are suited for small and moderate size problems. Sharma (2006) has used local search and parametrization of demands to construct a pseudo-polynomial algorithm for obtaining approximate solutions. In case the c_{ij} values satisfy the generalized triangle inequality, she has given a polynomial time algorithm which combines local search with parametrization of c_{ij}.

15.3 A Brief Overview of the Benders' Partitioning Approach

We consider solving this problem by Benders' Partitioning approach. A brief overview of the Benders' Partitioning approach is given below. Consider the following general MIP problem.

$$\begin{aligned}
\text{Minimize} \quad & cx + dy \\
\text{Subject to:} \quad & Ax + By \geq b \\
& x \geq 0, \text{ integer} \\
& y \geq 0
\end{aligned} \tag{M}$$

For a given integer vector x, the problem reduces to:

$$cx + \begin{array}{ll} \text{Minimize} & dy \\ \text{Subject to:} & By \geq b - Ax \\ & y \geq 0 \end{array} \qquad \text{(L)}$$

The dual of this problem is:

$$cx + \begin{array}{ll} \text{Maximize} & u(b - Ax) \\ \text{Subject to:} & uB \leq d \\ & u \geq 0 \end{array} \qquad \text{(DL)}$$

For a given integer vector x, the optimal solution of DL will occur at an extreme point u_p of polyhedron $P = \{uB \leq d, u \geq 0\}$. For any feasible solution x, the value $cx + u_p(b - Ax)$ is an upper bound on the optimal solution of the original problem M. We therefore wish to find an integer vector x, for which the corresponding optimal solution u_p produces the least possible value of $cx + u_p(b - Ax)$. Let S_P be the set of all extreme points of P. Then, this problem can be posed as the following master problem, called the Benders' Master Problem (BMP), which is a pure integer program:

$$\begin{array}{ll} \text{Minimize} & z \\ \text{Subject to} & z \geq cx + u_p(b - Ax) \quad \forall p \in S_P \\ & x \geq 0, \text{ integer} \end{array} \qquad \text{(BMP)}$$

The number of constraints in the master problem is very large, but only a small number are active at the optimal solution. These constraints are generated iteratively in the Benders' Partitioning Approach, which works as follows. We start with any feasible vector x (in this case any tree), and solve problem DL to generate the optimal extreme point u_p. We then generate the corresponding constraint (called Benders' cut), and define problem BMP with a single constraint. The next step is to solve the BMP to get a new solution x, and repeat this process. The solution to problem DL is an upper bound, while the solution of BMP is a lower bound on the original MIP. As the process proceeds, the gap between the two bounds decreases, and eventually reaches zero when the optimal solution to the problem is reached.

15.4 Applying Benders' Partitioning to OCSTP

For OCSTP, given a tree vector x, the solution of problem L is trivial, as each demand is routed along the one and only path available in the tree.

The corresponding dual solution to problem DL can also be computed easily by utilizing the complementary slackness conditions between problems L and DL. The details of this computation are given in the next section. Thus, given a tree vector x, a new constraint of the master problem can be easily generated. As mentioned before, in addition to the generated constraints, the master problem has additional implicit constraints to ensure that the solution vector x is a tree.

The main challenge in applying this approach to OCSTP is to solve the master problem, which is an integer program. Due to the Benders' cuts added to master problem, the LP relaxed solution need not be integral even if all the facets of the tree polytope are present in the BMP.

Rather than solving the BMP optimally, we propose to use a neighborhood search heuristic for solving this problem. This solution, though not necessarily optimal, can be used to generate a fresh constraint for the BMP. This basic step can be repeatedly used within a heuristic search strategy to arrive at a solution to the problem. Before going into the details of this heuristic approach, we discuss the method of computing the solution of problem DL for a given tree vector x.

15.4.1 *Computation of the Dual Solution*

The MIP formulation of the OCSTP is restated below.

$$\text{Minimize} \quad \sum_i \sum_j \sum_k c_{ij} y_{ij}^k$$

$$\begin{aligned}
\text{Subject to} \quad & \sum_j y_{ji}^k - \sum_j y_{ij}^k = -d_k, & \text{for } i = s(k) & \quad \forall k \\
& \sum_j y_{ji}^k - \sum_j y_{ij}^k = d_k & \text{for } i = t(k) & \quad \forall k \\
& \sum_j y_{ji}^k - \sum_j y_{ij}^k = 0 & \text{for } i \neq s(k), t(k) & \quad \forall k \\
& \sum_k (y_{ij}^k + y_{ji}^k) \leq D x_{ij} & & \quad \forall (i,j) \\
& x \in T \\
& y_{ij}^k \geq 0
\end{aligned}$$

For a given binary vector $x \in T$, problem reduces to

Minimize $\sum_i \sum_j \sum_k c_{ij} y_{ij}^k$

Subject to: $\sum_j y_{ji}^k - \sum_j y_{ij}^k = -d_k$ for $i = s(k)$ $\forall k$

$\sum_j y_{ji}^k - \sum_j y_{ij}^k = d_k$ for $i = t(k)$ $\forall k$ (I)

$\sum_j y_{ij}^k - \sum_j y_{ji}^k = 0$ for $i \neq s(k), t(k)$ $\forall k$ (L)

$\sum_k (y_{ij}^k + y_{ji}^k) \leq D x_{ij}$ $\forall (i,j)$ (II)

$y_{ij}^k \geq 0$

Note that each x_{ij} is not a variable but a parameter in the above problem. Associating the dual variables u_i^k with constraints (I) and v_{ij} with constraints (II), the dual of the above problem is:

Maximize $\sum_k d_k(u_{t(k)}^k - u_{s(k)}^k) + D \sum_{(i,j)} v_{ij} . x_{ij}$

Subject to: $u_j^k - u_i^k \leq c_{ij} + v_{ij}$ $\forall i,j,k$ (DL)

$v_{ij} \geq 0$

u_i^k unrestricted

For a given tree, we now address the issue of computing the optimal dual solution. Since constraints II have a slack for each of the tree arcs, we must have $v_{ij} = 0$ for all tree arcs. This implies that for tree arcs we must have $u_j^k - u_i^k \leq c_{ij}$. Moreover, if y_{ij}^k is positive, this condition must hold with equality due to complimentary slackness, i.e. $u_j^k = u_i^k + c_{ij}$. When y_{ij}^k is zero on some edge of the tree, we let $u_j^k = u_i^k$ for such edges, so that the LHS of the above constraint is zero, and it is automatically satisfied, assuming that $c_{ij} \geq 0 \; \forall \; (i,j)$.

The above method of computing u_i^k implies that if node i is on the path of commodity k, then u_i^k is equal to the length of the path from $s(k)$ to i. On the other hand, if node i is not on the path of commodity k, then u_i^k is equal to u_j^k, where j is the node at which the path from $s(k)$ to i diverges from the commodity path.

Having computed the u_i^k values, the v_{ij} values for non-tree edges must satisfy the dual constraint

$$v_{ij} \geq (u_j^k - u_i^k) - c_{ij} \; \forall \; k$$

Note that in defining v_{ij}, (i,j) is treated as an undirected edge, while the variables y_{ij}^k are directed. Since the above condition corresponds to variable y_{ij}^k, a similar condition also exists for the variable y_{ji}^k, i.e., $v_{ij} \geq (u_i^k - u_j^k) - c_{ij}$. These two conditions can be combined into a single condition:

$$v_{ij} \geq |u_i^k - u_j^k| - c_{ij} \; \forall \; k$$

Note that if both nodes i and j happen to be on the path of commodity k in the tree, then $|u_i^k - u_j^k|$ is simply equal to the distance between nodes i and j in the tree, which we denote by δ_{ij}. If one or both of nodes i and j are not on the commodity path, then our method of computing u_i^k implies that $|u_i^k - u_j^k| \leq \delta_{ij}$. Note that the above inequality must hold for every commodity, including the commodity whose end-nodes are nodes i and j. Clearly, the value of $|u_i^k - u_j^k|$ will be maximum for this commodity, and it will be equal to δ_{ij}. Thus, if we set $v_{ij} = \delta_{ij} - c_{ij}$, it will be a valid dual solution consistent with the u_i^k values computed as described earlier. This makes the computation of v_{ij} extremely efficient.

15.4.2 Computation of the Benders' Cut

In case of a general MIP, for a given integer solution x, having computed the optimal dual solution u, the Benders' cut generated for the Master problem is $z \geq cx + u(b - Ax)$. For the MIP formulation of OCSTP, c is a null vector, reducing the cut to $z \geq u(b - Ax)$. Let $b - Ax$ be split into two components:

- those corresponding to the flow conservation constraints, i.e. dual variables u_i^k, and
- those corresponding to capacity constraints, i.e. dual variables v_{ij}.

In other words $z \geq (u, v)[b_1 - A_1 x, b_2 - A_2 x]$. But A_1 is a null matrix, and b_2 is a null vector. This reduces the constraint to:

$$z \geq ub_1 - vA_2 x$$

It is easy to see that $ub_1 = \sum_k d_k(u_{t(k)}^k - u_{s(k)}^k)$. Due to the way u_i^k values were computed in the last section, this expression is equal to the total cost of flow on tree T represented by vector x, i.e. the cost of the OCSTP objective function for the solution represented by vector x. Let us call this value α. It is easy to verify that in the expression $vA_2 x$, the coefficient of variable x_{ij} is Dv_{ij}. Defining $\beta_{ij} = Dv_{ij}$, the Benders' cut generated by a given tree solution x is:

$$z + \sum_{(i,j)} \beta_{ij} x_{ij} \geq \alpha$$

where α and β_{ij} values are computed from the dual solution (u, v) as described above.

15.4.3 The Benders' Master Problem and the Solution Approach

Having defined the form of the Benders' cut for OCSTP, the master problem can be rewritten as follows.

$$\begin{aligned}
&\text{Minimize } z \\
&\text{Subject to} \quad z + \sum_{(i,j)} \beta_{ij}^p \cdot x_{ij} \geq \alpha^p \quad \forall x_p \in T \\
&\qquad\qquad\quad x \in T \\
&\qquad\qquad\quad x_{ij} = 0, 1 \\
&\qquad\qquad\quad z \geq 0
\end{aligned} \qquad \text{(BMP)}$$

Here β^p, and α^p are the coefficients of the Benders' cut generated from the tree vector x_p, and T is the collection of trees from which the Benders' cuts have been generated so far. The condition $x \in T$ can be handled implicitly in case of neighborhood search heuristic by moving from one tree vector to a neighboring tree vector.

As per the Benders' Partitioning algorithm, the problem can be solved as follows. Start with an arbitrary tree vector x_p, and compute the corresponding dual solution (u, v) of problem DL. Generate the Benders' cut and add it to the master problem, BMP. Solve BMP to get a new tree vector, and repeat the process. Note that the solution of DL provides an upper bound on the original problem, and solution of BMP a lower bound. As more cuts are added to the BMP, the two bounds come closer. The optimal solution is reached when they converge to the same value.

15.5 Neighborhood Search Heuristic for OCSTP

The spanning tree problem lends itself to a very natural definition of neighborhood for applying local search techniques. Several researchers have used such neighborhood search based approaches to develop heuristics for OCSTP (see Ahuja and Murty (1987) and Sharma (2006)). Given a tree T, consider adding an edge e not in T to the tree. This edge forms a unique cycle with a subset of edges in T. Removing one of the edges in this cycle leads to a new spanning tree, which is defined as a neighbor of the current solution. The set of all trees that can be obtained in this manner from the current tree constitutes the neighborhood of the current solution.

Let Z_P be the cost of OCSTP solution corresponding to a given tree. Then using the above definition of neighborhood, a local search heuristic for the OCSTP is given below.

Local Serach Heuristic (H_L)
Step 0: Start with a randomly selected tree vector x.
Step 1: Examine all neighboring trees of the current solution x and evaluate their costs Z_P.
Step 2: If none of the neighbors has a lower cost than the current solution, report current solution as locally optimal solution, and STOP. Otherwise, replace x with the neighbor with the least Z_P value, and go back to Step 1.

We use this heuristic as a benchmark to evaluate the performance of the Benders' Partitioning based heuristic described later.

Along the lines of the above heuristic, we shall develop the a heuristic for solving the BMP which is described in Section 15.6. Next we address some computational issues related to implementation of the above heuristic.

15.5.1 *Implementation Issues*

Computing the cost of a neighboring tree from scratch is computationally expensive, and there are efficient shortcuts for such computation which have been proposed by other researchers also (see Ahuja and Murty (1987)). Here we briefly present such a computational shortcut, which is also utilized later to improve the efficiency of the Benders Heuristic proposed by us.

Suppose that edge (p, q) of the current tree is removed. Then tree T is divided into two components $T_1 = (N_1, E_1)$ and $T_2 = (N_2, E_2)$. Let $E_{12} = \{(i, j) : i \in N_1 \text{ and } j \in N_2\}$. The total flow cost for a tree may be viewed as the sum of three cost components: C_1, C_2 and C_{12}. Here C_1 is the total cost of flow for demands whose both end points are in N_1, C_2 the total cost for demands with both end-points in N_2, and C_{12} is the total cost of inter-component traffic, i.e. the flow cost for those demands whose one end-point belongs to N_1, and the other to N_2. It is noteworthy that for a neighboring tree obtained by replacing edge (p, q) by and edge (r, s), where $r \in N_1$ and $s \in N_2$, the components C_1 and C_2 for the new tree remain unchanged, only component C_{12} changes. The shortcut proposed below exploits this observation, while proposing an efficient method for computing the revised value of C_{12} for each edge (r, s).

Let θ_{ij} be the distance of the unique path between nodes i and j in the original tree T. Then, we compute $C_1 = \sum_{i \in N1} \sum_{j \in N1} \theta_{ij} \cdot d_{ij}$ and $C_2 = \sum_{i \in N2} \sum_{j \in N2} \theta_{ij} \cdot d_{ij}$, where d_{ij} is the demand between nodes i and j. Suppose that we wish to know the specific edge (r, s) that will result in minimum value of the OCSTP objective function when it replaces the edge (p, q). Let

$$D_i^2 = \sum_{j \in N2} d_{ij} \quad \forall\, i \in N_1 \text{ total traffic from } i \in N_1 \text{ to } N_2,$$

$$D_i^1 = \sum_{j \in N1} d_{ij} \quad \forall\, i \in N_2 \text{ total traffic from } i \in N_2 \text{ to } N_1 \text{ and}$$

$$D^{12} = \sum_{i \in N1} \sum_{j \in N2} d_{ij} \quad \text{total traffic between } N_1 \text{ and } N_2$$

We now compute $C_r = \sum_{i \in N1} D_i^2 \cdot \alpha_{ir}$, the total cost of transporting to node r the traffic originating in T_1 but destined to T_2. Similarly, compute $C_s = \sum_{i \in N2} D_i^2 \cdot \alpha_{is}$, the total cost of transporting to node s the traffic originating in T_2 but destined to T_1.

Then it is clear that the OCSTP objective function value of the neighboring tree obtained by removing edge (p, q) and adding edge (r, s) is

$$C_{pqrs} = C_1 + C_2 + C_r + C_s + D^{12}.c_{rs}$$

In order to find the edge (r, s) that minimizes the cost, we simply have to minimize $C_r + C_s + D^{12}.c_{rs}$ over all edges (r, s) with $r \in T_1$ and $s \in T_2$. If the C_i values are computed for all i in advance, this requires only one multiplication and two additions to evaluate the cost of neighboring tree for each candidate edge (r, s). In order to find the least cost neighbor, this computation has to be repeated $(n - 1)$ times, i.e. once for each edge $(p, q) \in T$. This computational shortcut leads to considerable improvement in the computational efficiency of the Local Search heuristic. We later utilize this idea to boost the efficiency of Benders' heuristic also.

15.6 Neighborhood Search Using Benders' Partitioning Approach

As pointed out earlier, rather than solving the Benders' Master Problem optimally, we propose to solve it using the neighborhood search heuristic. Now we present such a heuristic based on the neighborhood defined in the last section. This heuristic moves from one tree solution to the next while adding a Benders' cut to the BMP at each step.

Benders' Heuristic (H_B)

Step 0: Start with a randomly selected tree vector x_0.

Step 1: Generate a Benders' cut from solution x_0, and add it to the BMP.

Step 2: For each neighboring solution x, let $Z_B(x)$ be the BMP objective function. Let S be the set of solutions with $Z_B(x) < Z_B(x_0)$. If S is empty, report the current solution as locally optimal and STOP. Otherwise go to Step 3.

Step 3: Let $x* \in S$ be the solution with the least Z_B value. Let $x_0 \leftarrow x*$, and go back to Step 1.

We refer to the sequence of solutions x_0 visited by the above algorithm as Benders' trajectory. Note that heuristic H_B pays no attention to the OCSTP objective function Z_P of the solution while shifting to a neighboring solution. As a result, the Z_P values of the solutions visited along the trajectory demonstrate large fluctuations in either direction at each step. Therefore, this heuristic demonstrates extremely poor convergence. Even on a small problem of only 9 nodes, algorithm H_B takes around 1000 iterations on average before terminating. The number of iterations increases to around 3000 for a 10-node problem, and to 8000 for a 12-node problem. It is clearly not pragmatic to use this algorithm on larger problems. However, it is important to note that on the problems tested, the algorithm almost always terminated either on the best known solution, or a solution very close to this solution. For the 9-node problem, we started H_B from 50 different randomly generated initial solutions. In 41 cases out of 50 the algorithm terminated at the best known solution (cost 1784), while in the remaining 9 cases it terminated at the second best known solution (cost 1790). On the other hand, algorithm H_L could produce the best known solution in only 8 cases out of 50, and the second best solution in 17 cases. The cost of the worst solution produced by H_L was 1873.

These observations suggest that if a computationally efficient way of implementing the Benders' heuristic can be found, it can yield results superior to ordinary Local Search heuristic. The modified heuristic proposed below attempts to overcome the computational drawbacks of heuristic H_B.

Modified Benders' Heuristic (H'_B)

Step 0: Start with a randomly selected tree vector x_0.

Step 1 : Generate a Benders' cut from solution x_0, and add it to the Problem BMP.

Step 2 : For each neighboring solution x, let $Z_P(x)$ and $Z_B(x)$, respec-

tively, be the OCSTP and BMP objective function values for that solution. Also, let $S' = \{x : Z_P(x) = Z_P(x_0) \text{ and } Z_B(x) < Z_B(x_0)\}$. If S is empty, go to Step 4. Otherwise, go to Step 3.
Step 3 : Let $x* \in S'$ be the solution with least Z_B value. Let $x_0 \leftarrow x*$, and go back to Step 1.
Step 4: Perform local search (Algorithm H_L) with x_0 as the initial solution, and report the resulting local optimum as the final solution. STOP.

15.6.1 Computational Issues Pertaining to Implementation of H'_B

A major computational issue pertaining to heuristics H'_B stems from the fact that evaluating $Z_P(x)$ for a neighboring solution is significantly more time consuming than evaluating $Z_B(x)$. Although the time to evaluate $Z_B(x)$ depends on the number of cuts added to the BMP, even with a relatively large number of cuts it is must faster to evaluate Z_B than Z_P. This is a major computational advantage of Hueristic H'_B over H_L. Although it may appear from the description of H'_B that Z_P needs to be evaluated at every neighboring solution, in reality, through proper implementation, it needs to be evaluated at a very small number of neighboring solutions.

This computational advantage of H'_B is further enhanced for larger problems, because the time needed to evaluate Z_P increases much faster than that for Z_B as the problem size increases. As the algorithm H'_B proceeds, more and more cuts are added to BMP, and the efficiency of evaluating Z_B will gradually decline. However, this problem is alleviated by keeping only a small number of most recently generated cuts in the BMP, and discarding those generated earlier. The computational experiments suggest that even if a very small number (say 5 to 10) most recently generated cuts are retained, there is no perceptible decline in the quality of solutions generated by H'_B.

15.6.2 Computational Enhancements in Implementation of H'_B

In algorithm H'_B, among all the neighboring solutions which improve Z_P, we select one that leads to maximum improvement in the Z_B value. In contrast to the local search algorithm, which is greedy and selects the neighbor with least Z_P value, we sacrifice part of improvement in Z_P value to ensure that we get the best possible improvement in Z_B subject to the condition that

Z_P does not get worse. Since the average improvement in the Z_P value in H'_B is less than that in H_L, it is natural that H'_B will require more iterations to converge to a local solution. Accordingly, the computational workload of H'_B will be higher than that of H_L. We devise a scheme that exploits the shortcut method of computing Z_P so as to overcome this disadvantage.

Note that for a given outgoing link (p,q), once we have computed the C_r and C_s values for all $r \in N_1$ and $s \in N_2$, it is extremely efficient to compute the cost of any neighbor that involves removal of the link (p,q). Our effort is to avoid searching the entire neighborhood and confine the search to a small number of outgoing links, whenever possible. A simplistic way to implement the algorithm H'_B is as follows. Compute Z_B for all neighbors, and sort them in increasing order of Z_B. Start computing the Z_P value for each neighbor in the sorted list until a neighbor is found with a better Z_P value than the current solution, and move to this neighbor. There are two computational bottlenecks in this approach.

First bottleneck is the sorting of Z_B values. Since the number of neighbors of a tree is $O(n^3)$, the sorting of all Z_B values is computationally quite expensive. Therefore, rather than physically sorting all Z_B values, we subdivide the range of Z_B values from Z_B^{min} to Z_B^{max} into k equal ranges, called "slices". Thus, the tth slice contains neighbors in the range:
$$Z_B^{min} + (Z_B^{max} - Z_B^{min}) * (t-1)/k \leq Z_B \leq Z_B^{min} + (Z_B^{max} - Z_B^{min}) * t/k$$
We compute the Z_P values for the neighbors in the first slice, and as soon as a neighbor is found with a better Z_P value than the current solution, we accept this as the new solution, and move to the next iteration. If no neighbor in the first slice has a better Z_P value, we process the next slice, and so on, until either a better solution is found, or all neighbors have been searched. Although this approach compromises the principle of moving the neighbor with the best Z_B value in the set S', the computational savings more than make up for this compromise. The number of slices k should be neither too small nor too large for obvious reasons. We decide to use $k=5$ in our computational experiments.

The second bottleneck is the computation of Z_P values. We would like to confine the computation of Z_P values to as few outgoing edges as possible to minimize the computational effort. We select the outgoing link which is involved in the largest number of neighbors and compute the Z_P values for these neighbors in the current slice. If a neighbor with better Z_P value is found, we accept it as the new solution and move on to the next iteration, otherwise we move to the next outgoing link with largest number of neighbors.

These computational strategies significantly improve the performance of the Benders' heuristic so as to overcome the effect of larger number of iteration needed as compared to those needed for local search.

15.7 Computational Experiments and Results

In this section we describe the computational experiments and their results to compare the performaces of algorithms H_L and H'_B. However, before presenting these results, we show in the next section that the superiority of one algorithm over another may not always be immediately obvious in such a comparison. We present a methodology to unambiguously show the superiority of one algorithm over another in such a case.

15.7.1 *A Methodology for Performance Comparison of Local Search Heuristics with Random Restart*

Consider two local-search type heuristic algorithms, say A and B, for solving a problem. We assume that both algorithms can begin from any randomly selected starting solution, and the final solution produced by the algorithm may differ for each starting solution. How do we compare the performance of the two algorithms? If on the average one algorithm produces better quality solution, and also requires less running time, it is clearly superior to the other one. However, when the average solution quality is better, but the running time is more, this decision requires a careful comparison methodology to decide as to which is a better algorithm. In this section we develop a systematic approach for answering this question, which is used to demonstrate the superiority of Benders' Heuristic to the ordinary Local Search Heuristic.

Since each initial solution produces a potentially different final solution, let us consider the objective function value of the final solution as a random variable, say X and Y for the two algorithms, A and B. Let T_A and T_B, respectively, be the average running times of the two algorithms. Let algorithm A run k times, producing solution X_i in the ith run, and take the best of these solutions, i.e. $Z^A(k) = Min_i\{X_i\}$ as the final solution. Let $Z^B(k)$ be similarly defined. Let $E(Z^A(k))$ be the expected value of $Z^A(k)$, and $k.T_A$ is the average time spent on these k runs. Consider plotting $E(Z^A(k))$ against $k.T_A$ for various values of k to get a graph, and also plotting a similar graph for algorithm B. This graph shows the quality of solution obtained as a function of cpu time spent. Plotting both the graphs together will clearly demonstrate the superiority of one algorithm over the other.

In order to get the $E(Z^A(k))$ values, we run the algorithm 100 times from different randomly generated starting points. Considering these 100 solution values as the universe of all solution, we can estimate the value of $E(Z^A(k))$ by taking many random samples of size k from these 100 solutions, determining the minimum in each sample, and taking the average of these minima as an estimate of $E(Z^A(k))$.

An illustrative sample of graphs for algorithm H_L, and H'_B is given in the Figure 1. It is clear from the graph that for reaching a desired solution quality, represented by a horizontal dotted line in the figure, the time required by Benders' heuristic is much smaller than that for the local search heuristic. These times are given by the intersection points of the horizontal line with the two curves.

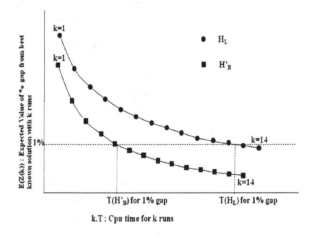

Figure 1 : Performance comparison of Heuristics H_L and H'_B

From these curves we can determine the average time needed by an algorithm to produce a solution of desired quality, e.g. within 1% of the best known solution. For example, Figure 1 shows that the expected gap of 1% can achieved with algorithm H'_B with only $k = 5$, while it takes $k = 12$ to achieve the same expected gap with algorithm H_L. The cpu times in the two cases are approximately 18 seconds and 46 seconds, respectively. We report the cpu times computed in this way in the computational results in the next section while comparing the performance of the two algorithms.

15.8 Computational Results

We tested the performance of algorithms H_L and H'_B on two sets of ten randomly generated problems of 100 nodes each defined on complete graphs. In the first set of problems, called Random Problems, the c_{ij} values were randomly generated from a uniform distribution $U(a, a+100)$. Here a may be viewed as a fixed cost of using an edge. We have conducted computational experiments with two values of a, i.e. $a = 20$ and $a = 50$. The traffic demands d_k for each node pair were also generated from a uniform distribution $U(1, 20)$. In the second set of problems, called Euclidian Problems, the c_{ij} values are based on Euclidian distances with a fixed cost element. We randomly generate the coordinates of the nodes on a 100 by 100 square, and take $c_{ij} = a + d_{ij}$, where d_{ij} is the Euclidian distance between points i and j. In this case we report results with $a = 2$ and $a = 5$.

In the context of fixed cost, it is important to point out that Ahuja and Murty (1987) have conducted their computational experiments with zero fixed cost, and their results show that the Local Search Heuristic produces the optimal solution most of the time. In our experiments we found that when the fixed cost is zero, the local search produces the best known solution with a very large number of random starting points. However, when the fixed cost is large, this is not the case, and only very few starting points produce the best known solution. This suggests that the problem is more difficult to solve by local search when fixed costs are larger. On the other hand, if the fixed cost is very high in comparison to the variable component, the problem becomes trivial to solve. This is because with a large enough fixed cost, the optimum solution almost always occurs at a star tree. In this case the c_{ij} values will satisfy the generalized triangle inequality of Hu (1974), and he has shown that in such a case the optimal solution is a star tree. For our computational experiments, we have chosen the fixed cost values so that the best star tree solution is significantly worse than the best-known solution. The fixed cost values selected above were chosen after verifying this.

Tables 1 and 2 list the results for Random Problems with fixed costs of 50 and 20 respectively, while Tables 3 and 4 list the results for Euclidian Problems with fixed costs of 5 and 2, respectively. For each instance in each set of problems, algorithms H_L and H'_B were applied from 100 randomly selected starting trees. The average percentage gap of these 100 solutions from the best known solution for each instance is reported in columns 7 and 8, respectively, for the two algorithms. Columns 1 and 2 list the frequency of solutions within one percent of the best known solution for the two algorithms, while columns 3 through 6 report these frequency comparisons for solutions falling within 2% and 3% of the best known solution, respectively. The average cpu times per run (in seconds on Pentium 2.0 GHz Core 2 Duo machine) are reported in columns 9 and 10 respectively. Expected times to reach within 1% of best known solution based on the methodology discussed in the previous section using algorithms H_L and H'_B, respectively, are listed in columns 11 and 12. A comparison of columns 11 and 12 shows that for Random Problems with fixed cost of 50 the Benders' search requires, on average, about 28 seconds to reach a solution within 1% of the best known solution, while the ordinary local search required about 64 seconds, i.e. a saving of about 56%. This percentage saving is comparable, i.e. 53% (from 47.42 to 22.36), in case of problems with fixed cost of 20. In case of Euclidian Problems we find that the advantage of Benders' search is not as high as in case of Random Problems. While for Euclidian Problems with fixed cost of 2, the percentage saving in time to reach within 1% of best-known solution is 43% (from 42.11 to 23.83), it declines to a negligible 9.4% (from 34.90 to 31.62) when the fixed cost is 5. This seems to suggest that the Benders' search is more advantageous for Random Problems than for Euclidian Problems.

Table 1 : Random Problems with Fixed Cost = 50

Pb. No.	1	2	3	4	5	6	7	8	9	10	11	12
	Within 1% of Best		Within 2% of Best		Within 3% of Best		Avg. gap %		CPU time per Run		Time to reach 1% of Best	
	LS	BS	LS	BS	LS	BS	LS	BS	LS	BS	LS	BS
1	1	25	1	25	1	30	8.0	4.7	3.05	6.98	Large	34.9
2	4	23	4	27	8	47	6.3	3.3	2.95	6.91	85.5	34.5
3	3	19	11	48	17	59	6.1	2.6	3.00	6.72	53.9	26.8
4	3	32	4	37	7	62	6.1	2.5	2.99	6.90	86.7	20.7
5	5	19	13	40	19	51	5.1	2.7	2.96	6.68	47.4	26.7
6	9	29	9	44	15	44	5.0	2.9	3.04	6.94	42.6	27.8
7	2	15	4	33	8	41	5.3	3.1	2.96	6.85	85.7	34.2
8	3	16	12	40	13	61	4.9	2.7	3.03	6.83	63.6	27.3
9	4	27	4	27	8	37	5.5	3.4	2.83	6.50	74	32.5
10	9	22	11	35	20	65	4.9	2.7	2.84	6.41	34	25.6
Avg	4.3	22.7	7.3	35.6	11.6	49.7	5.7	3.1	2.97	6.77	63.7	28.5

Table 2 : Random Problems with Fixed Cost = 20

Pb. No.	1	2	3	4	5	6	7	8	9	10	11	12
	Within 1% of Best		Within 2% of Best		Within 3% of Best		Avg. gap %		CPU time per Run		Time to reach 1% of Best	
	LS	BS	LS	BS	LS	BS	LS	BS	LS	BS	LS	BS
1	8	24	28	50	34	60	8.3	5.1	3.34	4.84	96.8	33.9
2	19	17	32	40	33	45	5.1	3.3	3.48	4.88	17.4	19.5
3	6	22	11	22	16	48	7.6	4.3	3.54	4.77	42.4	23.8
4	12	31	12	31	14	31	6.4	4.3	3.62	5.01	39.8	20
5	9	39	9	39	9	39	10.7	5.3	3.35	5.01	60.2	20
6	2	18	2	41	14	41	5.3	3.3	3.50	5.01	101.6	25
7	9	16	9	16	12	30	8.2	5.6	3.53	4.99	53	34.9
8	13	21	24	48	33	59	5.5	2.9	3.46	4.88	24.2	19.5
9	4	19	11	36	23	56	5.8	3.1	3.41	4.76	61.4	23.8
10	8	24	28	50	34	60	5.5	2.6	3.35	4.92	26.8	14.7
Avg	9	23.1	16.6	37.3	22.2	46.9	6.8	4.0	3.46	4.91	47.42	22.36

Table 3 : Euclidian Problems with Fixed Cost = 5

Pb. No.	1	2	3	4	5	6	7	8	9	10	11	12
	Within 1% of Best		Within 2% of Best		Within 3% of Best		Avg. gap %		CPU time per Run		Time to reach 1% of Best	
	LS	BS	LS	BS	LS	BS	LS	BS	LS	BS	LS	BS
1	13	41	20	70	33	85	8.9	2.0	3.09	8.75	21.65	17.5
2	20	40	33	63	40	79	7.6	2.2	3.01	9.46	18.06	27.14
3	8	26	16	37	26	62	9.4	2.5	3.25	9.29	32.54	27.88
4	2	18	15	59	65	82	5.5	2.1	2.98	8.02	77.52	32.01
5	5	19	20	55	36	71	8.9	2.5	3.19	8.18	44.72	49.06
6	10	29	16	56	33	81	8.9	2.2	3.21	8.16	38.53	24.47
7	15	26	35	66	38	78	10.1	2.2	3.03	8.94	18.17	26.82
8	9	31	14	56	20	59	10.3	2.9	3.03	8.02	33.35	32.09
9	20	34	29	70	36	83	7.7	2.0	3.15	8.06	18.89	24.17
10	8	18	10	36	31	67	7.4	2.5	3.23	8.19	32.34	40.93
Avg	11	28.2	20.8	56.8	35.8	74.7	8.5	2.3	3.12	8.52	34.90	31.62

Table 4 : Euclidian Problems with Fixed Cost = 2

Pb. No.	1	2	3	4	5	6	7	8	9	10	11	12
	Within 1% of Best		Within 2% of Best		Within 3% of Best		Avg. gap %		CPU time per Run		Time to reach 1% of Best	
	LS	BS	LS	BS	LS	BS	LS	BS	LS	BS	LS	BS
1	6	21	34	66	47	85	4.1	2.0	3.86	8.29	100.29	24.87
2	12	28	24	62	54	80	4.2	2.2	3.70	8.37	29.57	25.11
3	24	41	39	80	68	97	2.9	1.4	4.03	8.50	20.16	16.99
4	2	15	17	50	55	89	4.5	2.0	3.45	7.74	79.39	30.96
5	2	16	20	54	36	78	5.2	2.3	3.70	7.77	70.28	31.06
6	7	36	30	73	49	93	4.0	1.5	3.86	7.98	53.98	15.96
7	24	43	59	76	69	91	3.4	1.5	3.70	8.22	14.80	16.45
8	9	31	27	57	40	65	4.6	2.9	3.80	7.96	34.22	23.89
9	10	23	45	77	72	87	2.9	1.8	3.79	7.22	26.50	21.66
10	6	21	23	51	44	78	3.8	2.3	3.85	8.10	50.07	32.42
Avg	10	27.5	31.8	64.6	53.4	84.3	3.9	2.0	3.77	8.02	42.11	23.83

15.9 Concluding Remarks

A new Benders' Partitioning based search heuristic for solving the OCSTP has been presented in this chapter, which appears to perform significantly better than the ordinary local search. To the best of our knowledge, this is the first time the Benders' Partitioning approach has been used to develop a local search heuristic for solving a combinatorial optimization problem. In view of the significant improvement in the performance of local search algorithm with this approach, application of the same should also be attempted for other mixed integer programming problems which lend themselves to the application of Benders' partitioning approach.

Bibliography

Ahuja, R. K. and Murty, V. V. S. (1987). Exact and Heuristic Algorithms for the Optimum Communication Spanning Tree Problem, *Transportation Science* **21**, pp. 163–170.

Hu, T. C. (1974). Optimum Communication Spanning Trees, *SIAM J. of Compt.* **3**, pp. 188–195.

Johnson, D. S., Lenstra, J. K. and Rinnooy Kan, A. H. G. (1978). The complexity of the Network Design Problem, *Networks* **8**, pp. 279–285.

Peleg, D. and Reshef, F. (1998). *Deterministic Polylog Approximation for Minimum Communication Spanning Trees*, Lecture Notes in Computer Science, 1443, Springer Verlag, pp. 670-686.

Rothlauf, F., Gerstacker. J. and Heinz, A. (2003). *On the Optimal Communication Spanning Tree Problem*, Working Paper 10/2003, University of Mannheim, Dept. of Information Systems, Mannheim, Germany.

Sharma, P. (2006). Algorithms for the Optimum Communication Spanning Tree Problem, *Ann. Oper. Res.* **143**, pp. 203–209.

Chapter 16

Sperner's Lemma with Multiple Labels

R. B. Bapat
Indian Statistical Institute
New Delhi: 110016, India
e-mail: rbb@isid.ac.in

Abstract

An analogue of Sperner's lemma with multiple, unrestricted labels is proved. It extends a result due to Hochberg, McDiarmid and Saks obtained in the context of determining the bandwidth of the triangulated triangle.

Key Words: Sperner's lemma, triangulated triangle, multiple labels

16.1 Introduction

Sperner's lemma is a well-known result in combinatorial topology and is often used to provide a constructive proof of Brouwer's fixed point theorem. There have been several generalizations and extensions of this lemma (see, for example, [Bapat (1989)], [Gale (1984)], [Hochberg, McDiarmid and Saks (1995)], [Idzik (1999)], [De Loera, Peterson and Su (2002)], [Meunier (2008)]).

[Hochberg, McDiarmid and Saks (1995)] proved a result (see Theorem 16.3), using Sperner's lemma, and applied it to the problem of determining the bandwidth of the triangulated triangle. The present note was motivated by the observation that a small modification of Theorem 16.3 produces a result which is in fact equivalent to Sperner's lemma. The modification also has the attractive property that, unlike Sperner's lemma, it imposes no conditions on the labeling.

[Gale (1984)] gave a generalization of the Knaster-Kuratowski-

Mazurkiewicz (or the KKM) lemma, another result in combinatorial topology, which is equivalent to Sperner's lemma. A similar generalization of Sperner's lemma was obtained in [Bapat (1989)], which involves multiple labels at each vertex (see Theorem 16.4). In this chapter we work with multiple labels and obtain an analogue of Theorem 16.3, which turns out to be equivalent to Theorem 16.4.

It may be remarked that Theorem 16.5 remains valid, with appropriate modifications, in higher dimension. Furthermore we may have an arbitrary triangulation of the triangle. We have retained the two dimensional framework and consider the triangulated triangle for clarity. Theorem 16.3 which motivated this work, does not carry over to higher dimension.

16.2 Sperner's Lemma with Arbitrary Labels

Let k be a positive integer. The triangulated triangle T_k is the graph defined as follows. The vertex set of T_k is

$$V(T_k) = \{(x, y, z) : x, y, z \text{ are nonnegative integers, } x + y + z = k\}.$$

Two vertices (x, y, z) and (x', y', z') are joined by an edge if

$$|x - x'| + |y - y'| + |z - z'| = 1.$$

The graph T_5 is shown in Figure 1.

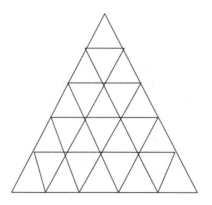

Figure 1: The graph T_5

For $i = 1, 2, 3$; the side P_i of T_k is defined as the subgraph induced by the vertices (x, y, z) with the i-th coordinate equal to zero. We set v_1, v_2, v_3 to be the vertices $(k, 0, 0), (0, k, 0), (0, 0, k)$ respectively. We now state Sperner's lemma. The proof is well-known, see, for example, [Bondy and Murty (1976)].

Theorem 16.1 (Sperner's Lemma). *Suppose the vertices of T_k are assigned labels from the set $\{1,2,3\}$ so that P_i has no vertex labeled i (and hence v_i is labeled i) for $i = 1,2,3$. Then there must be a triangular face of T_k whose vertices have three distinct labels.*

We now introduce some terminology. For basic definitions in Graph Theory, see [Bondy and Murty (1976)], [West, D. B. (2001)]. A subset S of $V(T_k)$ is said to form a connecting set if the subgraph induced by S has a connected component which contains vertices from each of P_1, P_2, P_3. As an example, note that $V(P_i)$ is a connecting set for each P_i.

Theorem 16.2. *Suppose each vertex of T_k is assigned an integer label. Then one of the following assertions must hold:*

(i) There is a triangular face of T_k whose vertices have three distinct labels
(ii) There exists an integer t such that the vertices of T_k labeled t form a connecting set.

Theorem 16.2 will be a consequence of the main result, Theorem 16.5. The following result was obtained in [Hochberg, McDiarmid and Saks (1995)] where it was applied to determine the bandwidth of the triangulated triangle. We obtain it as a consequence of Theorem 16.2.

Theorem 16.3. *Suppose each vertex of T_k is assigned one of the two colours, red or blue. Then either the vertices with red colour or the vertices with blue colour, but not both, form a connecting set.*

Proof. Label vertices with red colour by 1 and those with blue colour by 2. Then clearly (i) of Theorem 16.2 does not hold and therefore (ii) of Theorem 16.2 must hold. Thus either the vertices with red colour or the vertices with blue colour form a connecting set. If vertices with red colour form a connecting set then it can be seen that it precludes those of blue colour from forming a connecting set. (This is the feature which does not remain true in higher dimension.) □

16.3 Multiple Labels

The following "permutation-based" analogue of Sperner's lemma was proved in [Bapat (1989)].

Theorem 16.4. *Let k be a positive integer and suppose each vertex v of T_k is assigned a 3-tuple of labels $(\ell_1(v), \ell_2(v), \ell_3(v))$ where each label is from the set $\{1, 2, 3\}$. Suppose i does not appear in any label assigned to a vertex in $P_i, i = 1, 2, 3$. Then there exists a triangular face of T_k with vertices v_1, v_2, v_3 and a permutation σ of $\{1, 2, 3\}$ such that the labels $\ell_{\sigma(1)}(v_1), \ell_{\sigma(2)}(v_2)$ and $\ell_{\sigma(3)}(v_3)$ are distinct.*

Let k be a positive integer. The graph $T_k^{(3)}$ is defined as follows. The vertex set $V(T_k^{(3)}) = V(T_k) \times \{1, 2, 3\}$. Two vertices (u, i) and (v, j) of $T_k^{(3)}$ are joined by an edge if u and v are adjacent in T_k and $i \neq j$.

A subset S of $V(T_k^{(3)})$ is said to form a connecting set if the subgraph induced by S has a connected component which contains vertices from each of $V(P_i) \times \{1, 2, 3\}, i = 1, 2, 3$. As an example, note that $V(P_i) \times \{1, 2, 3\}$ is a connecting set for each i. The following is the main result.

Theorem 16.5. *Suppose each vertex of $T_k^{(3)}$ is assigned an integer label. Then one of the following assertions must hold:*

(i) There is a triangular face of T_k with vertices v_1, v_2, v_3 and a permutation σ of $\{1, 2, 3\}$ such that $(v_1, \sigma(1)), (v_2, \sigma(2)), (v_3, \sigma(3))$ are assigned three distinct labels

(ii) There exists an integer t such that the vertices of $T_k^{(3)}$ labeled t form a connecting set.

Proof. We assume that (i) and (ii) do not hold and get a contradiction. Consider the vertex (v, i) of $T_k^{(3)}$ and suppose it is assigned label s. Let G_s be the component of the subgraph of $T_k^{(3)}$ induced by the vertices labeled s, containing (v, i). Assign (v, i) the label w where w is the least integer such that G_s has no vertices from $P_w \times \{1, 2, 3\}$. We refer to this as the new labeling to distinguish it from the original one. Since (ii) does not hold, the new label is well-defined. Thus we obtain a labeling of vertices of $T_k^{(3)}$ by elements from the set $\{1, 2, 3\}$. Observe that a labeling of the vertices of $T_k^{(3)}$ may also be viewed as a labeling of $V(T_k)$ by 3-tuples and this observation is tacitly used in what follows. Clearly, in the new labeling any vertex from $P_i \times \{1, 2, 3\}$ cannot get label i and therefore the labeling satisfies the conditions of Theorem 16.4. Hence there exists a triangular face of T_k with vertices v_1, v_2, v_3 and a permutation σ of $\{1, 2, 3\}$ such that $(v_1, \sigma(1)), (v_2, \sigma(2))$ and $(v_3, \sigma(3))$ have all three labels $1, 2, 3$. Since (i) does not hold, two of the three vertices must get the same label in the original labeling. Suppose, without loss of generality, that the vertices labeled i and j, $i < j$, in the new labeling are both labeled s in the original labeling. Then from the vertex labeled j we can reach $P_i \times \{1, 2, 3\}$ via vertices labeled s while from the vertex labeled i we cannot do so. This is a contradiction since the vertices labeled i and j are adjacent in the graph $T_k^{(3)}$. That completes the proof. □

We remark that we may derive Theorem 16.4 from Theorem 16.5 and thus Theorem 16.4 is equivalent to Theorem 16.5. This is seen as follows. Suppose the vertices of T_k are assigned 3-tuples of labels from the set $\{1, 2, 3\}$ so that P_i has no vertex label containing i, $i = 1, 2, 3$. As before, we may view this as a labeling of the vertices of $T_k^{(3)}$. Then either (i) or (ii) of Theorem 16.5 holds. Clearly from a vertex of $T_k^{(3)}$ labeled i we cannot reach side $P_i \times \{1, 2, 3\}$, since label i is missing in $P_i \times \{1, 2, 3\}$. Thus (ii) of Theorem 16.5 cannot hold. Thus by (i), there must be a triangular face of T_k with vertices v_1, v_2, v_3 and a permutation σ of $\{1, 2, 3\}$ such that $(v_1, \sigma(1)), (v_2, \sigma(2))$ and $(v_3, \sigma(3))$ have all three labels. Therefore Theorem 16.4 is proved.

We conclude with the demonstration that Theorem 16.2 may be derived from Theorem 16.5. Suppose each vertex of T_k is assigned an integer label. Then we may define a labeling of the vertices of $T_k^{(3)}$ by setting the label of (v, i) equal to the label of v in the labeling of T_k for each $i = 1, 2, 3$. Now it is easy to see that an application of Theorem 16.5 with this labeling results in the proof of Theorem 16.2 for the given labeling.

Bibliography

Bapat, R. B. (1989). A constructive proof of a permutation-based generalization of Sperner's lemma, *Mathematical Programming*, **44**, pp. 113–120.

Bondy, J. A. and Murty, U. S. R. (1976). *Graph Theory with Applications*, (Macmillan).

Gale, D. (1984). Equilibrium in a discrete exchange economy with money, *International Journal of Game Theory* **13**, pp. 61–64.

Hochberg, R., McDiarmid, C. and Saks, M. (1995). On the bandwidth of triangulated triangles, *Discrete Mathematics* **138**, pp. 261–265.

Idzik, A. (1999). *The existence of the satisfactory point, Current Trends in Economic Theory and Applications, Studies in Economic Theory* Vol. 8, Ed. A. Alkan, C.D.Aliprantis, N.C.Yannelis, Springer, pp. 225–232.

De Loera, J. A., Peterson, E. and Su, F. E. (2002). A polytopal generalization of Sperner's lemma, *Journal of Combinatorial Theory, Series A* **100**, pp. 1-26.

Meunier, F. (2008). Combinatorial Stokes' formulae, *European Journal of Combinatorics* **29**, pp. 286–297.

West, D. B. (2001). *Introduction to Graph Theory*, 2nd Edition, (Prentice-Hall Inc.).

Chapter 17

Incremental Twin Support Vector Machines

Reshma Khemchandani
Department of Mathematics, Indian Institute of Technology
Hauz Khas, New Delhi-110016, India.
e-mail: reshmaiitd@gmail.com

Jayadeva
Department of Electrical Engineering, Indian Institute of Technology
Hauz Khas, New Delhi-110016, India
e-mail:jayadeva@ee.iitd.ac.in

Suresh Chandra
Department of Mathematics, Indian Institute of Technology
Hauz Khas, New Delhi-110016, India.
e-mail: chandras@maths.iitd.ac.in

Abstract

Support Vector Machines (SVMs) suffer from the problem of large memory requirements and CPU time when trained in batch mode on large data sets. Therefore incremental techniques have been developed to facilitate batch SVM learning. In this chapter we propose a new incremental technique called Incremental Twin Support Vector Machines for training in batch mode. This technique is based on a newly developed classifier, called Twin Support Vector Machines (TWSVM) classifier. The TWSVM classifier determines two non-parallel planes by solving two related support vector machines-type problems, each of which is smaller than in a conventional Incremental SVM. Numerical implementation on several benchmark datasets has shown that the Incremental Twin SVM is not only fast, but also has good generalization.

Key Words: Support vector machines, pattern classification, machine learning, incremental learning

17.1 Introduction

Support Vector Machines (SVMs) [Burges (1998); Gunn (1998); Vapnik (1995)] involve the solution of a quadratic programming problem subject to linear inequality constraints. Consequently, they suffer from the problem of a large memory and CPU time requirement when trained in batch mode on large data sets. Incremental learning techniques are one possible solution to tackle these problems as only a subset of the data is considered at each step of the learning process. This helps in managing the memory and time requirements of the learning algorithm.

Incremental learning can also be used when the whole data is not available a prior, i.e., we are working in an online scenario. Incremental techniques in the SVM framework have been proposed in [Cauwenberghs and Poggio (2000); Domeniconi and Gunopulos (2001); Syed, Liu, and Sung (1999)] to facilitate batch learning over very large data sets and stream data sets.

Incremental SVM techniques learn new data by discarding all past examples except for the support vectors and the misclassified samples, i.e., a new set of data is loaded in the memory along with the current set of support vectors and the misclassified samples to obtain the updated classifier [Cauwenberghs and Poggio (2000); Domeniconi and Gunopulos (2001); Syed, Liu, and Sung (1999)]. Hence, the support vectors obtained from this learning process are the new representation of the data considered so far, and they are kept in memory. The process is repeated till all the training data is used.

Recently, we had proposed Twin Support Vector Machines (TWSVMs) for binary data classification. The TWSVM is a non-parallel plane classifier where the data samples of each class are proximal to one of the two non-parallel planes. These non-parallel planes are obtained by solving a pair of small sized quadratic programming problems (QPPs), as compared to conventional SVMs where a large size QPP need to be solved. This strategy of solving two smaller sized QPPs rather than a large size QPP makes TWSVMs work almost four times faster than standard SVMs.

In this chapter, we deal with TWSVMs in an incremental learning scenario. The data set is presented to the algorithm in several batches instead of in a single training step. This helps to reduce the time and space requirements. Taking motivation from incremental SVMs, we define the concept of support vectors in the context of TWSVMs. We also make appropriate modifications in the training procedure, i.e., along with the new batch of data loaded in the memory, the current set of support vectors and the misclassfied samples obtained from the previous step of training are used for training. Experimental results are presented to demonstrate the efficacy of the proposed algorithm.

The chapter is organized as follows: Section 17.2 briefly dwells on linear Twin Support Vector Machines for binary data classification and extends to incremental learning scenario in Section 17.3. Section 17.4 deals with experimental results and Section 17.5 contains concluding remarks.

17.2 Twin Support Vector Machines

In this section, we give a brief outline of Twin Support Vector Machines (TWSVMs)[Jayadeva, Khemchandani and Chandra (2007)]. As mentioned earlier, TWSVM classifier is obtained by solving the two QPPs, which has the formulation of a typical SVM, except that not all patterns appear in the constraints of either problem at the same time.

Let the patterns to be classified be denoted by a set of m row vectors $P_i, (i = 1, 2, \ldots, m)$ in the n-dimensional real space $\mathbf{R}^n$, and let $y_i \in \{1, -1\}$ denote the class to which the ith pattern belongs. Matrices A and B represent data points belonging to classes 1 and -1, respectively. Let the number of patterns in classes 1 and -1 be given by m_1 and m_2, respectively. Therefore, the sizes of matrices A and B are of sizes $(m_1 \times n)$ and $(m_2 \times n)$, respectively.

The TWSVM classifier is obtained by solving the following pair of quadratic programming problems

(TWSVM1) $\underset{w^{(1)},\, b^{(1)},\, q}{Min} \; \frac{1}{2}(Aw^{(1)} + e_1 b^{(1)})^T(Aw^{(1)} + e_1 b^{(1)}) + c_1 e_2^T q$

subject to

$$-(Bw^{(1)} + e_2 b^{(1)}) + q \geq e_2,$$
$$q \geq 0, \tag{17.1}$$

and,

(TWSVM2) $\underset{w^{(2)},\, b^{(2)},\, q}{Min} \; \frac{1}{2}(Bw^{(2)} + e_2 b^{(2)})^T(Bw^{(2)} + e_2 b^{(2)}) + c_2 e_1^T q$

subject to

$$(Aw^{(2)} + e_1 b^{(2)}) + q \geq e_1,$$
$$q \geq 0, \tag{17.2}$$

where c_1, $c_2 > 0$ are parameters, q is the vector of error variables, and e_1 and e_2 are vectors of ones of appropriate dimensions.

In a nutshell, TWSVMs comprise of a pair of quadratic programming problems, such that in each QPP the objective function corresponds to a particular class, and the constraints are determined by patterns of the other class. Thus, TWSVMs give rise to two smaller sized QPPs. In (TWSVM1), patterns of class 1 are clustered around the plane $x^T w^{(1)} + b^{(1)} = 0$. Similarly in (TWSVM2), patterns of class -1 cluster around the plane $x^T w^{(2)} + b^{(2)} = 0$. We observe that TWSVM is approximately four times faster than the usual SVM. This is because the complexity of the usual SVM is no more than m^3, and TWSVM solves two problems viz. (17.1) and (17.2), each of which is roughly of size $\frac{m}{2}$. Thus, the ratio of runtimes is approximately $\frac{m^3}{2 \times (\frac{m}{2})^3} = 4$.

The Lagrangian corresponding to the problem (TWSVM1) (17.1), is given by

$$L(w^{(1)}, b^{(1)}, q, \alpha, \beta) = \frac{1}{2}(Aw^{(1)} + e_1 b^{(1)})^T (Aw^{(1)} + e_1 b^{(1)})$$
$$+ c_1 e_2^T q - \alpha^T(-(Bw^{(1)} + e_2 b^{(1)}) + q - e_2)) - \beta^T q \quad (17.3)$$

where $\alpha = (\alpha_1, \alpha_2 \ldots \alpha_{m_2})^T$, and $\beta = (\beta_1, \beta_2 \ldots \beta_{m_2})^T$ are the vectors of Lagrange multipliers. The Karush-Kuhn-Tucker (K. K. T.) necessary and sufficient optimality conditions [Mangasarian (1994)] for (TWSVM1) are given by

$$A^T(Aw^{(1)} + e_1 b^{(1)}) + B^T \alpha = 0 \quad (17.4)$$
$$e_1^T(Aw^{(1)} + e_1 b^{(1)}) + e_2^T \alpha = 0 \quad (17.5)$$
$$c_1 e_2 - \alpha - \beta = 0 \quad (17.6)$$
$$-(Bw^{(1)} + e_2 b^{(1)}) + q \geq e_2, \quad q \geq 0 \quad (17.7)$$
$$\alpha^T(-(Bw^{(1)} + e_2 b^{(1)}) + q - e_2) = 0, \quad \beta^T q = 0 \quad (17.8)$$
$$\alpha \geq 0, \quad \beta \geq 0. \quad (17.9)$$

Since $\beta \geq 0$, from (17.6) we have

$$0 \leq \alpha \leq c_1. \quad (17.10)$$

Next, combining (17.4) and (17.5) leads to

$$[A^T \ e_1^T][A \ e_1]\begin{bmatrix} w^{(1)} \\ b^{(1)} \end{bmatrix} + [B^T \ e_2^T]\alpha = 0. \quad (17.11)$$

We define

$$H = [A \ e_1], \quad G = [B \ e_2], \quad (17.12)$$

and the augmented vector $u = \begin{bmatrix} w^{(1)} \\ b^{(1)} \end{bmatrix}$. With these notations, (17.11) may be rewritten as

$$H^T H u + G^T \alpha = 0 \quad \text{i.e} \quad u = -(H^T H)^{-1} G^T \alpha. \tag{17.13}$$

Using (17.3) and the above K.K.T conditions, we obtain the Wolfe dual [Mangasarian (1994)] of (TWSVM1) as follows

$$(DTWSVM1) \underset{\alpha}{Max} \quad e_2^T \alpha - \tfrac{1}{2} \alpha^T G (H^T H)^{-1} G^T \alpha$$

subject to

$$0 \leq \alpha \leq c_1. \tag{17.14}$$

Similarly, we consider (TWSVM2) and obtain its dual as

$$(DTWSVM2) \underset{\gamma}{Max} \quad e_1^T \gamma - \tfrac{1}{2} \gamma^T P (Q^T Q)^{-1} P^T \gamma$$

subject to

$$0 \leq \gamma \leq c_2. \tag{17.15}$$

Here, $P = [A \quad e_1]$, $Q = [B \quad e_2]$, and the augmented vector $v = \begin{bmatrix} w^{(2)} \\ b^{(2)} \end{bmatrix}$, which is given by

$$v = (Q^T Q)^{-1} P^T \gamma. \tag{17.16}$$

In the above discussion, the matrices $H^T H$ and $Q^T Q$ are matrices of size $(n+1) \times (n+1)$, where in general, n is much smaller in comparison to the number of patterns of classes 1 and -1.

Once vectors u and v are known from (17.13) and (17.16), the separating planes

$$x^T w^{(1)} + b^{(1)} = 0 \quad \text{and} \quad x^T w^{(2)} + b^{(2)} = 0 \tag{17.17}$$

are obtained. A new data sample $x \in \mathbf{R}^n$ is assigned to class r ($r = 1, 2$), depending on which of the two planes given by (17.17) it lies closest to. Thus

$$\text{class}(x) = arg \min_{r=1,2} (d_r(x))$$

$$\text{where} \quad d_r(x) = \left(\frac{|x^T w^{(r)} + b^{(r)}|}{||w^{(r)}||} \right), \tag{17.18}$$

and $||w||$ is the L_2 norm of vector w.

From the Karush-Kuhn-Tucker conditions (17.4)-(17.10), we observe that patterns of class -1 for which $0 < \alpha_i < c_1$, ($i = 1, 2, \ldots, m_2$), lie on the hyperplane given by $x^T w^{(1)} + b^{(1)} = 0$. Taking motivation from standard SVMs, we can define such patterns of class -1 as support vectors of class 1 *with respect to* class -1, as they play an important role in determining the required plane. A similar observation holds for the problem TWSVM2.

17.3 Training Incremental Twin Support Vector Machines

In principle algorithm based on active set method for training SVMs can also be considered as incremental learning algorithm [Mangasarian and Musicant (2000); Osuna, Freund, and Girosi (1997)]. This is because only a small part of the data sample is used for training at every step. However, these approaches are not useful to incremental learning since none of the samples are discarded during the training. Thus, these samples have to be reconsidered at each step. Therefore, not much improvement in terms of space and time requirement can be expected from such algorithms.

In general, the incrementally built model will not be too far from the model built with the complete data set at once. This is because, at each incremental step, the support vectors represent the essential class boundary information. Further, this boundary information is utilized to generate the classifier in succeeding steps. A similar concept applies in the case of twin support vector machines with the added advantage that the size of each of the pair of QPPs is much smaller than the large size QPP that arise in the case of a conventional SVM.

Once a new batch of data is loaded into memory, there are different possibilities for updating of the current model [Campbell, Cristianini and Smola (2000); Mitra, Murty, and Pal (2000)]. In SVM based models, at each step the learned model from the previously considered data is preserved in the form of support vectors or erroneous samples. These support vectors or erroneous samples along with the new samples loaded in the memory are used for updating the classifier.

We seek a model for incremental learning that maintains an updated representation of recent batches of data, i.e., in our model we incorporate both the aforementioned techniques. We first partition the training data into batches of a fixed size. When the new batch of data is loaded into memory to obtain the updated TWSVM classifier, it is augmented with the support vectors and the erroneous samples from the previous batch of data. Thus, the model TWSVM$_t$ at time t (preserved in the form of support vectors and erroneous samples), is used to classify a given set of loaded data. If the trained data is misclassified, it is kept, otherwise it is discarded. The support vectors and misclassified samples of TWSVM$_t$ together with the new data sample loaded into the memory are used as training data to obtain the new classifier TWSVM$_{t+1}$. This procedure is continued till the complete data is exhausted. The algorithm is summarized below.

Given a dataset, we first split it into a training set TR and a test set TE. The training set TR is further split into k parts TR$_i$, where $i = 1, 2, \ldots, k$, with each part containing mutually exclusive $k\%$ of the data from TR. For incremental training the update procedure is described as follows

Set $i = 1$

Step 1. Train TWSVM$_i$ with TR$_i$ and preserve the set of support vectors (SV$_i$) and erroneous samples (E$_i$) obtained w.r.t the TWSVM$_i$ classifier.
Step 2. Set $i \leftarrow i + 1$.
Step 3. TR$_i \leftarrow$ TR$_i \cup$ (SV$_{i-1}$) $\cup$ (E$_{i-1}$)
Step 4. Use the updated TWSVM$_i$ classifier thus obtained to evaluate the prediction accuracy of the set TE.

Further test set accuracy of the data set is determined by following the standard ten fold cross-validation methodology. Therefore the above procedure is repeated for each of ten folds and then average over ten folds has been reported.

17.4 Experimental Results

Incremental TWSVM (Inc-TWSVM), TWSVM, Incremental SVM (Inc-SVM), and SVM data classification methods were implemented by using MATLAB verion 7 [Matlab] running on a PC with an Intel P4 processor (3 GHz) with 1 GB RAM. The methods were evaluated on datasets from the UCI Machine Learning Repository [Blake and Merz (1998)]. Test set accuracy was determined by following the standard ten fold cross-validation methodology [Duda, Hart and Strok (2001)] with the incremental training procedure explained in Section 17.3.

Table 1 summarizes Inc-TWSVM performance on some benchmark datasets available at the UCI machine learning repository [Blake and Merz

(1998)]. The table compares the (mean ± standard deviation) performance of the Inc-TWSVM classifier with that of TWSVM, Inc-SVM, and SVM. Optimal values of c_1 and c_2 were obtained by using a tuning set comprising of 10% of the dataset. Table 2 compares the training time for ten folds of Inc-TWSVM with that of Inc-SVM. The tables indicate that Inc-TWSVM is not just effective but also faster than the conventional Inc-SVM. This is because it solves two quadratic programming problems of a smaller size instead of a large size QPP.

Table 1: Test Set Accuracy with a Linear Kernel.

Data Set	Inc-TWSVM	TWSVM	Inc-SVM	SVM
Heart-statlog (270×14)	84.81±4.52	84.07±4.70	83.70±4.44	83.70±6.02
Heart-c (303×14)	84.14±5.71	83.80±5.53	84.44±5.27	82.48±5.62
Ionosphere (351×34)	87.46±6.42	86.05±7.70	87.74±6.02	86.88 ±5.62
Votes (435×16)	96.09±2.93	96.31±3.47	95.87±2.66	95.40±2.71
Australian (690×14)	85.94±2.83	86.96±2.67	85.51±2.67	85.51±2.67
Japanese Credit (653×14)	86.37±2.96	86.52±4.04	86.42±4.04	86.68±4.06

Accuracies have been indicated as percentages.

Table 2: Training Times (in seconds)

Data Set	Inc-TWSVM	Inc-SVM
Heart-statlog	14.14	52.60
Heart-c	24.90	68.61
Ionosphere	13.20	65.23
Votes	11.96	36.67
Australian	57.30	272.1
Japanese Credit	75.1	370.5

17.5 Concluding Remarks

In this chapter, we have proposed incremental learning with a twin support vector machines, proposed in [Jayadeva, Khemchandani and Chandra (2007)], consists of learning new data by discarding all past examples except support vectors and erroneous points.

In Incremental TWSVM, we solve two quadratic programming problems of a smaller size instead of a large sized one as we do in traditional SVMs. This makes Incremental TWSVM faster than a standard Incremental SVM classifier. Furthermore, in contrast to a single hyperplane as given by traditional SVMs, TWSVMs yield two non-parallel planes such that each plane is close to one of the two datasets, and is distant from the other dataset. In terms of generalization, Incremental TWSVM compares favourably with TWSVM and SVM.

Acknowledgements. The first author (Reshma Khemchandani) acknowledges the financial support of the Council of Scientific and Industrial Research (India) in the form of a scholarship for pursuing her Ph.D.

Bibliography

Blake, C. L. and Merz, C. J. (1998). *UCI Repository for Machine Learning databases*, Irvine, CA: University of California, Department of Information and Computer Sciences. On-line at http://www.ics.uci.edu/ mlearn /MLRepository.html

Burges, C. A. (1998). Tutorial on Support Vector Machines for Pattern Recognition, *Data Mining and Knowledge Discovery* **2**, pp. 1–43.

Campbell, C., Cristianini, N. and Smola, A. (2000). *Query learning with large margin classifiers*, In Proceedings, 17th International Conference on Machine Learning, Eds. Morgan Kaufmann, San Francisco, CA.

Cauwenberghs, G. and Poggio, T. (2000). *Incremental and decremental support vector machine learning*, NIPS 2000, USA: MIT Press, pp. 409- 415.

Domeniconi, C. and Gunopulos, D. (2001). *Incremental support vector machine construction*, Nick Cercone, Tsau Young Lin, Xindong Wu (Eds.): Proceedings of the 2001 IEEE International Conference on Data Mining, San Jose, California, USA. IEEE Computer Society, pp. 589-592.

Gunn, S. R. (1998). *Support Vector Machines for Classification and Regression*, Technical Report, School of Electronics and Computer Science, University of Southampton, Southampton, U.K, On-line at http://www.isis.ecs.soton.ac.uk/resources/svminfo/

Duda, R. O., Hart, P. R. and Stork, D. G. (2001). *Pattern Classification*, 2nd edition, (John Wiley and Sons, Inc, New York).

Jayadeva, Khemchandani, R. and Chandra, S. (2007). Twin Support Vector Machines for Pattern Classification, *IEEE Trans. on PAMI* **29**, pp. 901–911.

Liu, P., He, Q. and Chen, Q. (2004). *Incremental Batch Learning with Support Vector Machines*, Proceedings of the 5th World Congress on Intelligent Control and Automation, Hangzhou, P.R. China, June 15-19.

Mitra, P., Murthy, C. A. and Pal, S. K. (2000). *Data Condensation in Large Databases by Incremental Learning with Support Vector Machines*, International Conference on Pattern Recognition.

Mangasarian, O. L. (1994). *Nonlinear Programming*, (SIAM, Philadelphia, PA).

Mangasarian, O. L. and Musicant, D. R. (2000). *Active Support Vector Machine Classification*, Technical Report 00-04, Data Mining Institute, Computer Sciences Department, University of Wisconsin, Madison, Wisconsin.

Osuna, E., Freund, R. and Girosi, F. (1997). *An improved training algorithm for support vector machines*, In Proceedings, *IEEE Workshop on Neural Networks for Signal Processing*, Amelia Island Florida, pp. 276–285.

Vapnik, V. (1995). *The Nature of Statistical Learning Theory*, (Springer Verlag, NY).

Syed, N. A. Liu, H. and Sung, K. K. (1999). *Incremental Learning with Support Vector Machines*, In Proceedings of the Workshop on Support Vector Machines at the International Joint Conference on Artificial Intelligence (IJCAI-99), Stockholm, Sweden,
http://citeseer.ist.psu.edu/syed99incremental.html

http://www.mathworks.com

Chapter 18

Portfolio Risk Management Using Support Vector Machine

Sanjeet Singh
Indian Institute of Management Calcutta
D.H. Road, Joka, Kolkata-700104, India
e-mail: sanjeet@ iimcal.ac.in

Abstract

Portfolio diversification (i.e., possessing shares of many companies at the same time for reducing risks) is considered to be an important task in the investor's community to reduce the risk of a portfolio without not necessarily reducing the returns. This chapter will present a classification study of different categories of companies on the basis of their various financial attributes using a quadratic optimization based classifier namely Support Vector Machine (SVM). We have also used this model for sector wise classification of the company. To validate the performance, we compared the results with the ratings for companies provided by ICICI direct, a well known trading website in Indian stock market. The comparison shows that the model generated by SVM is efficient and the results obtained using this technique are quite impressive.

Key Words: Portfolio management, support vector machine, quadratic programming, machine learning

18.1 Introduction

Data Mining, the extraction of hidden predictive information from large databases, is a new powerful technology with great potential to help companies focus on the most important information in their data warehouses. Data mining is the ideal tool to model different business applications, such as investment return prediction, market fluctuation simulation, stock or mutual fund analysis, consumption categorization etc. Data mining can predict future trends and behaviors, allowing business to make proactive, knowledge driven decisions [Bonchi, Giannotti, Mainetto and Pedreschi (1999); Thearling (11); Wang and Weigend (2004)]. The finance industry has been relying on various data mining tools over two decades [Cohen,Zeneca Pharmaceuticals and Olivia Parr Rud (2005)]. But these days various machine learning algorithms are also studied to understand market severity and predict the future trends. Neural networks are extensively used for stock predictions. [Hochreter and Schmidhurber (1997)] has presented an efficient algorithm called "flat minimum search" that outperforms other widely used methods on stock market prediction tasks. [Saad, Prokhorov and Wunsch (1998)] compared three neural network models for low false alarm stock trend predictions. Neural networks have become the standard for detecting credit-card fraud. Other machine learning algorithms that are studied for market analysis are genetic programming [Markose, Tsang, Er and Salhi (2001)] and decision trees [Harries and Horn (1995)]. They predicted stock share indices and showed improved computational efficiency of data mining approaches. [Ahmed, Warsi and Ahmed (2004)] presented an unsupervised learning method to group Indian companies of same industry sector into different categories on the basis of their annual balance sheets to help investors identifying companies with maximum return. In this chapter, we used optimization based classifier namely SVM to predict the rating of an unrated company as well as to predict the sector of the company just by looking at its various financial attributes. The remainder of this chapter is organized as follows. Section 18.2 presents the fundamentals of Portfolio diversification and review the related literature. Section 18.3 gives the detail of SVM models relevant to our study. Section 18.4 gives the detail of various experiments conducted on the input data and its results Section 18.5 concludes with the findings of this study and the future research directions.

18.2 Portfolio Risk Management

We have divided the risk of holding shares into its two parts-unique risk and market risk [Brealey, Myers, and Allen (2006)]. The risk that potentially can be eliminated by diversification is called unique risk or firm specific risk. Unique risk stems from the fact that many of the perils that surround an individual company are peculiar to that company and perhaps its immediate competitors. But there is also some risk that we can't avoid, regardless of how much we diversify. This risk is generally known as market risk. Market risk stems from the fact that there are other economy wide perils that threaten all businesses. That is why stocks have a tendency to move together. And that is why investors are exposed to market uncertainties, no matter how many stocks they hold. Even a little diversification can provide a substantial reduction in variability. Suppose we calculate and compare the standard deviations of randomly chosen one-stock portfolios, two-stock portfolios, five-stock portfolios, etc. A high proportion of the investments would be in the stocks that are individually very risky. However diversification can cut the variability of returns about in half. Notice also that we can get most of this benefit with relatively few stocks. The improvement is slight when the number of securities is increased beyond, say, 20 or 30. Diversification works because prices of different stocks do not move exactly together. In modern portfolio theory, [Sharpe, Alexander and Bailey (1999)] has shown that holding a diversified portfolio of many shares can eliminate firm specific risks. Over time some firms will perform better than others in properly diversified portfolio; then differences will balance. The gain in one stock is offset by loss in other, stabilizing the overall earnings of the investment. Portfolio Diversification is an important strategy in shares business to reduce risk. Investor should diversify their investment i.e. they should include shares of more than one company in their investment baskets. There are two steps in selection of shares of good companies that generate good return on investment

- Selection of good Industrial Sectors
- Selection of companies in those good Industrial sectors.

Performance of the companies depends upon different factors, internal factors (companies management, R & D, good marketing, patents etc) and external factors (cost of raw material, government policies and other macroeconomic factors). Every sector has few external factors associated with them that affect the profitability of companies of those sectors. It is very difficult to use computational methods to compare companies of different Industrial sectors because sector related factors cannot be easily quantified. But in a given sector where most of the external factors affect the company in the similar manner, various attributes based on the company's performance can be used to compare them. Based on the various factors companies are rated as overweight, equal weight and underweight.

Ratings: Stocks have also been given a three rating scale - Overweight, Equal weight and Underweight. To arrive at these rating we have considered factors like; the past performance of the company, dividend record and return ratios, price performance, future expectations, with some thought given to the stocks market fancy.

Overweight: If an investor is looking to add a stock to his or her portfolio, stocks with this rating may be the best candidates for consideration. This stock is expected to perform better than the broader market over the next 9-12 months.

Equal weight: An investor who has stocks with an Underweight rating can consider continuing to hold the stock. But should monitor the stock's ongoing performance and compare the potential benefits of owning a stock with higher ratings. An investor looking to add a stock to his or her portfolio may also consider stock with this rating, however preference should be given to the stocks rated 'Overweight'. Price could be a factor here.

Underweight: An investor would not usually consider stocks with this rating. An investor who has stocks with an Underweight rating should consider the benefits of owning a stock with higher ratings. An Underweight rating to a share signifies an expectation that the stock will perform worse than the broader market over the next 9-12 months.

18.3 Support Vector Machine

A classification task usually involves with training and testing data which consist of some data instances. Each instance in the training set contains one "target value" (class labels) and several "attributes" (features). The goal of SVM is to produce a model which predicts target value of data instances in the testing set which are given only the attributes. Support Vector Machines [Cortes and Vapnik (1995)] are powerful methods for learning a classifier, which have been applied successfully to many NLP tasks such as base phrase chunking [Kudo and Matsumoto (2000)] and part-of-speech tagging [Nakagawa, Kudoh and Matsumoto (2001)]. SVM is primarily designed for binary classification. The SVM constructs a binary classifier that outputs $+1$ or -1 given a sample vector $x \in \Re^n$. The decision is based on the separating hyper plane as follows:

$$c(x) = \begin{cases} +1, & if\, w \cdot x + b > 0, w \in \Re^n, b \in \Re \\ -1, & otherwise \end{cases}$$

The class for an input x, $c(x)$, is determined by seeing which side of the space separated by the hyper plane, $wx + b = 0$, the input lies on. Given a set of labeled training samples

$$\{(y_1, x_1), (y_2, x_2), \ldots, (y_N, x_N)\}, x_i \in \Re^n, y_i \in \{+1, -1\}$$

the SVM training tries to find the optimal hyper plane, i.e., the hyper plane with the maximum margin. Margin is defined as the distance between the hyper plane and the training samples nearest to the hyper plane. Maximizing the margin insists that these nearest samples, called as support vectors (SVs), exist on both sides of the separating hyper plane and the hyper plane lies exactly at the midpoint of these support vectors. This margin maximization tightly relates to the fine generalization power of SVMs. Assuming that $|w \cdot x_i + b| = 1$ at the support vectors without loss of generality, SVM training can be formulated as the following optimization problem:

$$\text{Minimize } \frac{1}{2}\|w\|^2$$
$$\text{subject to } y_i(w \cdot x_i + b) \geq 1, i = 1, 2, \ldots, N.$$

The solution of this problem is known to be written as follows, using only support vectors and weights for them:

$$f(x) = wx + b = \sum_{i \in SVs} y_i \alpha_i x \cdot x_i + b$$

where $\alpha'_i s$ are the Lagrange multipliers in dual to the above optimization problem. In the SVM learning, we can use a function $k(x_i, x_j)$ called a *kernel function* instead of the inner product in the above equation. Introducing a kernel function means mapping an original input x using a transformation function

$$\Phi(x), \text{ s.t. } \Phi(x_i) \cdot \Phi(x_j) = k(x_i, x_j)$$

to another, usually a higher dimensional, feature space. We construct the optimal hyper plane in that space. By using kernel functions, we can construct a non-linear separating surface in the original feature space. Fortunately, such non-linear training does not increase the computational cost if the calculation of the kernel function is as cheap as the inner product. Some of the kernel functions popular in applications of SVM's [Kudo and Matsumoto (2001, 2000); Yamada, Kudo and Matsumoto (2000)] are given

below:
- linear: $k(x_i, x_j) = x_i^T \cdot x_j$.
- polynomial: $k(x_i, x_j) = (\gamma x_i^T \cdot x_j + r)^d, \gamma > 0$.
- radial basis function(RBF): $k(x_i, x_j) = exp(-\gamma \|x_i - x_j\|)^2, \gamma > 0$.
- sigmoid : $k(x_i, x_j) = \tanh(\gamma x_i^T \cdot x_j + r)$.

Here, γ, r and d are kernel parameters.
Depending upon the nature of the data, we will now discuss some of the support vector machine models.

18.3.1 Linearly or Non-linearly Separable SVM

SVM is one kind of learning machine based on statistical learning theory. The basic idea of applying SVM for classification can be stated briefly as follows. First, map the input vectors into one feature space (possible with a higher dimension), either linearly or non-linearly, which is relevant with the selection of the kernel function. Both the basic linear separable case and the most useful linear non-separable case for most real life problems are considered here.

18.3.2 The Linear Separable Case

In this case, there exists a separating hyper plane $w \cdot x + b = 0$, which implies $y_i(w \cdot x_i + b) \geq 1, i = 1, 2, \ldots, N$. By minimizing $\frac{1}{2}\|w\|^2$ subject to the constraint $y_i(w \cdot x_i + b) \geq 1, i = 1, 2, \ldots, N$, the SVM approach tries to find a unique separating hyper plane. Here $\|w\|^2$ is the Euclidean norm of w, which maximizes the distance between the hyper plane, i.e. Optimal Separating Hyperplane or OSH [Cortes and Vapnik (1995)], and the nearest data points of each class. The classifier is called the largest margin classifier. SVM training procedure amounts to solving a convex Quadratic Programming (QP) problem. The solution is a unique globally optimized result can be shown having the following expansion:

$$w = \sum_{i=1}^{N} y_i \alpha_i x_i$$

only if the corresponding $\alpha_i > 0$, these x_i are called Support Vectors. When SVM is trained, the decision function can be written as:

$$f(x) = sgn(\sum_{i=1}^{N} y_i \alpha_i x \cdot x_i + b)$$

where sgn () in the above formula is the given sign function.

18.3.3 The Linear Non-separable Case

SVM performs a nonlinear mapping of the input vector x from the input space $\Re^d$ into a higher dimensional Hilbert space, where the mapping is determined by the kernel function. It finds the OSH in the space H corresponding to a non-linear boundary in the input space. Two generally used kernel functions are listed below:

$$k(x_i, x_j) = (\gamma x_i^T \cdot x_j + 1)^d$$
$$k(x_i, x_j) = exp(-\gamma \|x_i - x_j\|^2)$$

and the form of the decision function is

$$f(x) = sgn\left(\sum_{i=1}^{N} y_i \alpha_i k(x_i, x_j) + b\right)$$

For a given data set, only the kernel function and the regularity parameter C must be selected to specify one SVM.

18.3.4 Multi-class SVMs

As described above, the standard SVM learning constructs a binary classifier. To make a named entity recognition system based on the BIO representation, we require a multi-class classifier. Among several methods for constructing a multi-class SVM [Hsu and Lin (2002)], here we describe the one-vs-rest method and the pair wise method. Both one-vs-rest and pair wise methods construct a multi-class classifier by combining many binary SVMs. In the following explanation, K denotes the number of the target classes.

One Vs Rest: Construct K binary SVMs, each of which determines whether the sample should be classified as class i or as the other classes. The output is the class with the maximum f (x) in the following equation:

$$f(x) = sgn\left(\sum_{i=1}^{N} y_i \alpha_i K(x_i, x_j) + b\right)$$

Pair wise: Construct K (K- 1) = 2 binary SVMs, each of which determines whether the sample should be classified as class i or as class j. Each binary SVM has one vote, and the output is the class with the maximum votes.

18.4 Experimental Results

We have used by ICICI direct [Content.icicidirect (9)], a well known financial services company in Indian capital market, to get the required input data. This input data then have been analyzed using well known SVM software SVMLight [8]. We have considered sectors namely Textile, Healthcare, IT, FMCG, Engineering, Metals and Automobiles for our analysis. There are a total of 190 rated companies in these 7 sectors. And then we look at the financial attributes which can affect the performance of the company. We have shortsighted 12 attributes which generally affect the companies rating. The number of financial attributes used to rate a company in any sector is not fixed. It all depend upon how important that particular attribute is for that sector and thus for the classification. We now define class labels which will be used for the sake of our analysis. The rating of the company can be underweight (class label 1), equal weight (class label 2) or overweight (class label 3). Then we find the type of Kernel function which will give us the most accurate results. This should be done at the first place because all the latter results will depend upon whether we have selected the kernel function appropriately or not. In the following table 1. We have shown that how different kernels perform when used on same input and test data:

Table 1. Performance comparison of different kernels.

Kernel Option	Training Data	Testing Data	Features	Classes	Accuracy
Linear	155	35	12	3	84.15
Polynomial	155	35	12	3	73.71
Sigmoid	155	35	12	3	70.85

From the above table we come to know the linear kernel function should be used as if the number of features is large, one may not need to map data to a higher dimensional space. Now we have tried to increase the amount of training data as well as change the number of features and see whether it has any effect on the efficiency or not. In the previous example, we have used all the 12 attributes as till now we haven't studied whether all the attributes should be used or not. We found that out of the 12 features which we are initially using 2 features are actually decreasing our models efficiency rather then increasing it. So we have used only 10 features for rating the unknown companies using our model. Moreover as we increase the training data the efficiency of the model increased. This is due to the reason that this data is used by the SVM module to learn and understand the classification characteristics. The more the input data more it will learn. The following table shows the accuracy of linear SVM when varying data sizes and different no of features are used

Table 2. Performance of linear SVM for varying data sizes.

Kernel Option	Training Data	Testing Data	Features	Classes	Accuracy
Linear	155	35	12	3	84.15
Linear	155	35	10	3	89.82
Linear	190	30	10	3	92.32

As already mentioned it is difficult to use computational methods to compare companies of different Industrial sectors because sector related factors cannot be easily quantified. But in a given sector where most of the external factors affect all the companies in the similar manner, various attributes based on the company's performance can be used to compare them. So next we tried to generate model using the sector wise data and see whether we are able to improve our accuracy level or not. We considered 7 sectors namely Textile, Healthcare, IT, FMCG, Engineering, Metals and Automobiles.

The number of financial attributes used to rate the companies in any sector depends upon how important that particular attribute is for classification. We first enlisted those attributes which are important for all irrespective of the sector that we want to study. After that we tried various permutations and combinations for the remaining financial attributes to come across the features which are most important to generate ratings for that particular sector firms. Like for example dividend paid is totally irrelevant in case of IT and healthcare as these sectors hardly pay any dividends and reinvest all the earning in R & D etc. It's irrespective of the financial condition of the company. Even the better firms like Infosys will pay very little in the form of dividends. So the dividend paid is not affecting the ratings at all. Similarly the best possible attributes for all the sectors are first identified and then they are used to generate the model. From the experiments we found that we get different set of classification rules for companies of different sector, which is in conformity that external factors vary for different industrial sectors. Now in the following table we will present the classification results of different companies in different sectors.

Table 3. Experimental results for classifying companies in different sectors.

Sector	Training Data	Testing Data	Features	classes	Accuracy
Textile	30	12	9	3	83.33
Healthcare	40	12	10	3	91.66
IT	20	12	9	3	75.00
FMCG	22	12	10	4	75.00
Engineering	30	12	11	3	83.33
Metals	18	12	10	3	75.00
Automobiles	15	12	11	3	66.67

After looking at the results the first reaction that comes to our mind is that the efficiency have not improved much by generating models for sector wise data. But if we analyze it we can easily see a correlation between the training data for any sector and the accuracy of that sector specific model. As the input data increases the accuracy of the SVM model also increases along with it. We can see for the healthcare sector which has more training data gives the most accurate result while results for sectors like IT and metals are not very good as the input data in these sectors is small.

From this we can generalize that it's better to first see the sector to which the unrated company belongs. If the sector to which the company belongs didn't have much input data i.e. the number of rated companies in that sector are low, then the accuracy of the sector specific SVM model might not be very high and in those cases the general SVM model developed will be more useful and expected to give more accurate result.

We also analyzed which companies are classified wrongly by our models. We found that there is negligible misclassification for overweight companies. But sometimes Overweight companies are classified as equal weight and some equal weight are rated as Under Weight. But there is no case where overweight are classified as underweight or visa versa. So we can safely assume that any company classified as overweight by our model will generally by overweight or in few cases equal weight. This can be used for taking investment decisions.

But if we do not know even the sector of the company which we want to classify. Then one thing we can do is use the general model we have developed. The other thing is we should identify the sector of the company and then use that company specific model for it. But the main problem is how to identify it. We can use this model for sector wise classification of the company. The classes which we have defined in the previous models according to the ratings of the company can now be generated using the sector of the company and then using that we can generate the class for the unknown company. In this study we have considered only some of the financial attributes which affect the companies in all the sectors as the components of the input data vector for the SVM model.

Previously we have defined classes as overweight, equal weight or underweight. Now we can redefine classes again to generate a new model. In this model we gave a same class to companies belonging to same sector. For example - each IT firm is given a class of 5 irrespective of the individual rating of the company. The assumption we have made during generating this model is as some external factors affect all the companies in certain sector similarly. So the financial ratios of companies in particular sector have certain correlation in them and thus they fall in some range. So if we enter the financial attributes of the company whose sector wise classification is required, depending upon the ratios and certain correlation among same sector firms we can predict the sector of the firm. Following table 4 shows the class labels assigned by us to different industrial sectors:

Table 4. Class labels for different sectors.

Sector Type	Class	Sector Type	Class
Automobile	1	IT	5
Metals	2	Healthcare	6
Engineering	3	Textile	7
FMCG	4		

But while using this model we have to be bit careful. The 7 sectors we have considered doesn't have equal representation in the input data i.e. for certain sectors we have as many as 40 rated companies while for others we have only 15. So if the result after using this sector classification model shows the sector of the firm as the one which have inadequate representation in the input data, we should be bit more careful as the probability of error in that case is higher. And in those cases try to use the general SVM model to rate the company rather then predicting its sector first and then using its sector specific model to classify it.

After doing all this study we generated SVM models that can predict rating of some of the unrated companies with us. We have also used this model for sector wise classification of the company.

18.5 Conclusion

While analyzing financial credibility of the companies, individual preferences and projections play an important role and can have adverse effects on return on investment. By using machine-learning algorithms we can reduce such kind of biased decision possibility. In this chapter we have shown the use and applicability of optimization based classifier namely Support Vector Machine for Portfolio diversification problem for Indian Share market. From the analysis we found that there is negligible misclassification for overweight companies. But sometimes Overweight companies are classified as equal weight and some equal weight are rated as Under Weight. But there is no such case where overweight are classified as underweight or visa versa. This analysis can help an investor to identify companies for achieving optimum returns on investment by chalking out Overweight/ Underweight companies in a given industrial sector. An investor can acquire portfolios of just not one company but a congregation of profit making Overweight rated companies and consider absconding Underweight rated companies from his basket. After studying the classification rules we observed that there is a need for more informative attributes and more input data to increase the prediction rate. We also observed that sometimes only few of the attributes played a pivoting role in learning classification rules from the data. A few of the attributes did not contribute much in classification and hence may be considered less important while taking investment decisions. This technique can be used with other conventional investment analysis tools for better financial analysis of companies.

Acknowledgements. The author is thankful to the editors and unknown referees for their critial evaluation of the chapter and suggestions for the improvement in the presentation of the chapter.

Bibliography

Ahmed, K., Warsi, S. and Ahmed, A. (2004). *Application of K-means Clustering for Portfolio Diversification in Indian Share Market*, KBCS -2004, Fifth International Conference on Knowledge Based Computer Systems, India.

Bonchi, F., Giannotti, F., Mainetto, G. and Pedreschi, D. (1999). *Using Data mining techniques in Fiscal Fraud Detection*, In Proc. DaWak'99, First Int'l Conference on Data Warehousing and Knowledge Discovery.

Brealey, R., Myers, S. and Allen, F. (2006). *Corporate Finance*, 8th Edition, (McGraw- Hill).

Cohen, J. J., Zeneca Pharmaceuticals and Olivia Parr Rud, C. (2005). *Data Mining of Market Knowledge in the Pharmaceutical Industry*, (NESUG, Philadelphia, PA).

Cortes, C. and Vapnik, V. (1995). Support vector networks, *Machine Learning* **20**, pp. 273–297.
Harries, M. and Horn, K. (1995). *Detecting Concept Drift in Financial Time Series Prediction using Symbolic Machine Learning*, In Xin Yao, editor, *English Australian Joint Conference on Artificial Intelligence*, pp. 91–98, World Scientific Publishing, Singapore.
Hochreter, S. and Schmidhurber, J. (1997). Flat Minima, *Neural Computation*, **9**, pp. 1–42.
Hsu, C-W and Lin, C.-J. (2002). A comparison of methods for multiclass Support Vector Machines, *IEEE Transactions on Neural Networks* **13**, pp. 415 - 425.
Content.icicidirect(9), http://content.icicidirect.com/research/research.asp
Svmlight(10), $http://svmlight.joachims.org/svm_m ulticlass.html$
Thearling (11), http://www.thearling.com/text/dmwhite/dmwhite.htm
Kudo, T. and Matsumoto, Y. (2001). Chunking with Support Vector Machines, *Proc. of NAACL*, pp. 192–199.
Kudo, T. and Matsumoto, Y. (2000). Use of support vector learning for chunk identification, In *Proc. of CoNLL-2000 and LLL-2000*, pp. 142-144
Markose, S., Tsang, E., Er, H. and Salhi, A. (2001). Evolutionary Arbitrage for FTSE-100 Index Options and Futures, In *Proceedings of the 2001 Congress of Evolutionary Computations*, pp. 275–282.
Nakagawa, T., Kudoh, T. and Matsumoto, Y. (2001). Unknown word guessing and part-of-speech tagging using support vector machines, In *Proc. of the 6th NLPRS*, pp. 325–331.
Saad, E. W., Prokhorov, D. V. and Wunsch, D. C. (1998). Comparative study of stock trend prediction using time delay, recurrent and probabilistic neural networks. *IEEE Transactions on Neural Networks* **9**, pp. 1456–1470.
Sharpe, W. F., Alexander, G.J. and Bailey, J.V. (1999). *Investment*, 6th Edition, (Prentice-Hall).
Wang, H. and Weigend, A. (2004). Data mining for financial decision making, Editorial, *Journal of Decision Support Systems* **37**, pp. 457–460.
Yamada, H., Kudo, T. and Matsumoto, Y. (2000). Using substrings for technical term extraction and classification. *IPSJ SIGNotes* (**NL-140**), pp. 77–84.

Chapter 19

Weak Convergence of an Iterative Scheme with a Weaker Coefficient Condition

Yasunori Kimura
Department of Mathematical and Computing Sciences
Tokyo Institute of Technology
Tokyo 152-8552, Japan
e-mail: yasunori@is.titech.ac.jp

Abstract

We deal with a generalized proximal point algorithm for a sequence of m-accretive operators in Banach spaces. We investigate the condition of coefficients more deeply, and obtain weak convergence of an iterative scheme with a weaker coefficient condition.

Key Words: Accretive operator, resolvent, m-accretive operator, proximal point algorithm, iterative scheme, weak convergnece

19.1 Introduction

The theory of monotone operators is one of the most important fields of convex analysis, which has close relationship to convex optimization theory, fixed point theory, and others. In particular, the problem of finding zeros of maximal monotone operators can be applied to many kinds of problems such as equilibrium problems, variational inequalities, convex minimization problems, and others.

Let A be a monotone operator defined on a real Hilbert space H and let us consider the problem of finding a zero of A, that is, a solution $z \in H$ of an operator inclusion $0 \in Az$. One of the most popular methods of approximating this solution is the proximal point algorithm, which was introduced by [Martinet (1970)] and generally studied by [Rockafellar (1976)].

Namely, for an arbitrary initial point $x_1 \in H$, generate a sequence $\{x_n\}$ by
$$x_{n+1} = (I + \rho_n A)^{-1} x_n + e_n$$
for $n \in \mathbb{N}$, where $\{\rho_n\}$ is a positive real sequence bounded away from 0 and $\{e_n\}$ is a sequence of H satisfying $\sum_{n=1}^{\infty} \|e_n\| < \infty$. This type of approximating methods has been studied with various types of additional conditions. For the studies in Hilbert spaces, see [Brézis and Lions (1978)], [Pazy (1979)], [Kamimura and Takahashi (2000a)], and others. For the studies in Banach spaces, see [Bruck and Reich (1977)], [Nevanlinna and Reich (1979)], [Reich (1980)], [Jung and Takahashi (1991)], [Reich and Zaslavski (2000)], [Kimura and Takahashi (to appear)], and others.

In 1992, [Eckstein and Bertsekas (1992)] considered a generalized proximal point algorithm for a maximal monotone operator A given by the following: $x_1 \in H$ and
$$x_{n+1} = \alpha_n x_n + (1 - \alpha_n)(I + \rho_n A)^{-1} x_n + e_n$$
for $n \in \mathbb{N}$. They assumed for a sequence of coefficients $\{\alpha_n\}$ that $|\alpha_n| < \alpha < 1$ for all $n \in \mathbb{N}$, and obtained weak convergence to a zero of A. [Kamimura and Takahashi (2000b)] dealt with this scheme in the setting of a uniformly convex Banach space as follows:

Theorem 19.1 ([Kamimura and Takahashi (2000b)]). *Let E be a uniformly convex Banach space which satisfies either the Opial property or the Fréchet differentiablity of the norm. Let A be an m-accretive operator on E such that $A^{-1}0$ is nonempty. For a sequence $\{\alpha_n\} \subset [0, 1[$ and $\{\rho_n\} \subset]\rho_0, \infty[$ with $\rho_0 > 0$, generate an iterative sequence by $x_1 \in E$ and*
$$x_{n+1} = \alpha_n x_n + (1 - \alpha_n)(I + \rho_n A)^{-1} x_n + e_n$$
for $n \in \mathbb{N}$, where $\{e_n\}$ is a sequence of E such that $\sum_{n=1}^{\infty} \|e_n\| < \infty$. If $\{\alpha_n\}$ satisfies that $\limsup_{n \to \infty} \alpha_n < 1$, then $\{x_n\}$ converges weakly to a point of $A^{-1}0$.

They used an m-accretive operator A defined on a Banach space, which is a generalization of a maximal monotone operator defined on a Hilbert space, and obtained weak convergence to a zero of A. The assumption for $\{\alpha_n\}$ is that $0 \leq \alpha_n < \alpha < 1$ for $n \in \mathbb{N}$.

In this chapter, we investigate the condition of coefficients $\{\alpha_n\}$ more deeply, and obtain weak convergence of an iterative scheme to a zero of an m-accretive operator with a weaker condition. We also treat a sequence of m-accretive operators for generating an iterative scheme. This idea follows that disscussed in [Kimura and Takahashi (to appear)].

19.2 Preliminaries

In what follows, a Banach space E will always be over the real scalar field. We denote its norm by $\|\cdot\|$ and its dual space by E^*. The value of $x^* \in E^*$ at $y \in E$ is denoted by $\langle y, x^* \rangle$, and the normalized duality mapping of E to E^* is denoted by J, that is, $J(x) = \{x^* \in E^* : \|x\|^2 = \|x^*\|^2 = \langle x, x^* \rangle\}$ for $x \in E$.

The modulus δ_E of convexity of a Banach space E is defined as follows: δ_E is a function of $[0, 2]$ into $[0, 1]$ such that

$$\delta_E(\epsilon) = \inf\{1 - \|x + y\|/2 : \|x\| \leq 1, \|y\| \leq 1, \|x - y\| \geq \epsilon\}$$

for each $\epsilon \in [0, 2]$. E is said to be uniformly convex if $\delta_E(\epsilon) > 0$ for each $\epsilon > 0$. From the definition of δ_E, one has that

$$\left\| \frac{x+y}{2} \right\| \leq \rho \left(1 - \delta_E \left(\frac{\epsilon}{\rho} \right) \right)$$

for all $\epsilon \geq 0$, $\rho > 0$ and $x, y \in E$ satisfying that $\|x\| \leq \rho$, $\|y\| \leq \rho$, and $\|x - y\| \geq \epsilon$. It is known that a uniformly convex Banach space is reflexive.

A Banach space E is said to satisfy the Opial property [Opial (1967)] if a sequence $\{x_n\}$ of E converges weakly to x, then $x \neq y$ implies

$$\liminf_{n \to \infty} \|x_n - x\| < \liminf_{n \to \infty} \|x_n - y\|.$$

A norm of E is said to be Fréchet differentiable if, for each $x \in E$ with $\|x\| = 1$, $(\|x + ty\| - \|x\|)/t$ converges uniformly for y with $\|y\| = 1$ as $t \to 0$.

A mapping T from a nonempty closed convex subset C of a Banach space E into E is said to be nonexpansive if $\|Tx - Ty\| \leq \|x - y\|$ for all $x, y \in C$. We denote the set of all fixed points of T by $F(T)$, that is, $F(T) = \{x \in C : x = Tx\}$.

Let E be a Banach space and A a set-valued operator on E. A is called an accretive operator if

$$\|x_1 - x_2\| \leq \|(x_1 - x_2) + \lambda(y_1 - y_2)\|$$

for every $\lambda > 0$ and $x_1, x_2, y_1, y_2 \in E$ with $y_1 \in Ax_1$ and $y_2 \in Ax_2$. If an accretive operator A satisfies that the range of $I + \rho A$ is the whole space E for any $\rho > 0$, we call A an m-accretive operator, where I is the identity mapping on E.

For an accretive operator A, it is known that the operator $I + \rho A$ has a single-valued inverse operator. We call $(I + \rho A)^{-1}$ the resolvent of A. By definition, we have that $\mathrm{dom}(I + \rho A)^{-1} = \mathrm{ran}(I + \rho A)$ and $\mathrm{ran}(I +$

$\rho A)^{-1} = \text{dom } A$, where the domain of a set-valued mapping S is denoted by $\text{dom } S = \{x \in E : Sx \neq \emptyset\}$, and the range of S by $\text{ran } S = \bigcup_{x \in \text{dom } S} Sx$. Thus it follows that, if A is m-accretive, then $(I + \rho A)^{-1}$ is defined on the whole space E. We also know that $(I + \rho A)^{-1}$ is a nonexpansive mapping with $F((I + \rho A)^{-1}) = A^{-1}0$ for all $\rho > 0$, where $A^{-1}0 = \{z \in E : 0 \in Az\}$. We note that, if the underlying space E is a Hilbert space, then A is m-accretive if and only if A is maximal monotone. For more details see, for example, [Takahashi (2000)].

The following lemmas show some fundamental properties for sequences of real numbers, which play important roles in our main results.

Lemma 19.1 ([Tan and Xu (1993)]). *Let $\{\lambda_n\}$ and $\{\mu_n\}$ be nonnegative real sequences satisfying $\lambda_{n+1} \leq \lambda_n + \mu_n$ for all $n \in \mathbb{N}$. If $\sum_{n=1}^{\infty} \mu_n < \infty$, then $\{\lambda_n\}$ has a limit in $[0, \infty[$.*

Lemma 19.2 ([Suzuki and Takahashi (1999)]). *Let $\{\lambda_n\}$ and $\{\mu_n\}$ be nonnegative real sequences satisfying $\sum_{n=1}^{\infty} \lambda_n = \infty$ and $\sum_{n=1}^{\infty} \lambda_n \mu_n < \infty$. Then, for any $\kappa > 0$, there exists a subsequence $I = \{n_i\} \subset \mathbb{N}$ such that $\sum_{j \in \mathbb{N} \setminus I} \lambda_j \leq \kappa$ and $\lim_{i \to \infty} \mu_{n_i} = 0$.*

19.3 Weak Convergence of an Iterative Scheme under the Opial Property

In this section, we deal with an iterative scheme whose underlying space has the Opial property. This is a generalized result of the theorem proved by [Kamimura and Takahashi (2000b)]. We assume that the coefficients $\{\alpha_n\}$ used in generating an iterative scheme satisfy that $\sum_{n=1}^{\infty}(1 - \alpha_n)$ is divergent instead of $\limsup_{n \to \infty} \alpha_n < 1$.

Theorem 19.2. *Let E be a uniformly convex Banach space which satisfies the Opial property. Let $\{A_n\}$ be a sequence of m-accretive operators on E such that $C_0 = \bigcap_{n=1}^{\infty} A_n^{-1}0$ is nonempty. Suppose the following:*

(WLS) For any sequences $\{u_n\}$ and $\{v_n\}$ in E satisfying that $u_n \in A_n v_n$ for every $n \in \mathbb{N}$ and that $\{u_n\}$ converges strongly to 0, every subsequential weak limit point of $\{v_n\}$ belongs to C_0.

For a sequence $\{\alpha_n\} \subset [0, 1[$, generate an iterative sequence by $x_1 \in E$ and

$$x_{n+1} = \alpha_n x_n + (1 - \alpha_n) J_n x_n + e_n$$

for $n \in \mathbb{N}$, where $J_n = (I + A_n)^{-1}$ for $n \in \mathbb{N}$ and $\{e_n\}$ is a sequence of E such that $\sum_{n=1}^{\infty} \|e_n\| < \infty$. If $\sum_{n=1}^{\infty}(1 - \alpha_n) = \infty$, then $\{x_n\}$ converges weakly to a point of C_0.

Remark 19.1. The name of the condition (WLS) stands for "weak limit supremum." Using the notation of set-convergence, we can write this condition as follows:

(WLS) w-Ls$_{n\to\infty} A_n^{-1} u_n \subset C_0$ for any sequence $\{u_n\}$ converging strongly to 0.

We note that, for a sequence $\{C_n\}$ of subsets of E, w-Ls$_{n\to\infty} C_n$ is defined as the set of all subsequential weak limit points of sequences $\{x_n\}$ satisfying $x_n \in C_n$ for all $n \in \mathbb{N}$.

It is known that the condition (WLS) has a close relationship to convergence of sequence of resolvents generated by a sequence of maximal monotone operators; see [Kimura (2006)].

To prove the theorem, we divide its proof with three lemmas.

Lemma 19.3. Let $\{A_n\}$ be a sequence of m-accretive operators on a Banach space E with a common zero in C_0. If a sequence $\{A_n\}$ of m-accretive operators satisfies the condition (WLS) in the theorem above, then every subsequence $\{A_{n_i}\}$ of $\{A_n\}$ also satisfies this condition.

Proof. Let $\{u_i\}$ and $\{v_i\}$ be sequences in E satisfying that $u_i \in A_{n_i} v_i$ for all $i \in \mathbb{N}$ and suppose that $\{u_i\}$ converges strongly to 0. Let $z \in C_0 \cap \bigcap_{n=1}^{\infty} A_n^{-1} 0$, that is, $z \in C_0$ and $0 \in A_n z$ for all $n \in \mathbb{N}$. Then, we define sequences $\{u'_n\}$ and $\{v'_n\}$ by

$$u'_n = \begin{cases} u_i, & (n = n_i \text{ for some } i \in \mathbb{N}) \\ 0 & (\text{otherwise}) \end{cases}$$

and

$$v'_n = \begin{cases} v_i, & (n = n_i \text{ for some } i \in \mathbb{N}) \\ z & (\text{otherwise}) \end{cases}$$

for every $n \in \mathbb{N}$. Then we have that $u'_n \in A_n v'_n$ for all $n \in \mathbb{N}$ and that $\{u'_n\}$ converges strongly to 0. Since $\{A_n\}$ satisfies (WLS), every subsequential weak limit point of $\{v'_n\}$ belongs to C_0. Since $\{v_i\}$ is a subsequence of $\{v'_n\}$, every subsequential weak limit point of $\{v_i\}$ is also that of $\{v'_n\}$ and hence it belongs to C_0, which is a desired result. □

Lemma 19.4. *Let E be a uniformly convex Banach space. Let $\{A_n\}$ be a sequence of m-accretive operators on E such that $C_0 = \bigcap_{n=1}^{\infty} A_n^{-1} 0$ is nonempty, and suppose that the condition (WLS) in Theorem 19.2 holds. For a sequence $\{\alpha_n\} \subset [0, 1[$, generate an iterative sequence $\{x_n\}$ as in Theorem 19.2. Then both $\{x_n\}$ and $\{J_n x_n\}$ are bounded, and $\{\|x_n - z\|\}$ has a limit for every $z \in C_0$. Further, if $\sum_{n=1}^{\infty}(1 - \alpha_n) = \infty$, then there exists a subsequence $I = \{n_i\} \subset \mathbb{N}$ such that $\sum_{j \in \mathbb{N} \setminus I}(1 - \alpha_j) < \infty$ and that every subsequential weak limit point of $\{x_{n_i}\}$ belongs to C_0.*

Proof. Let z be an arbitrary point of C_0. Since $z = J_n z$ and J_n is nonexpansive for every $n \in \mathbb{N}$, we have that

$$\|x_{n+1} - z\| = \|\alpha_n x_n + (1 - \alpha_n) J_n x_n + e_n - z\|$$
$$\leq \alpha_n \|x_n - z\| + (1 - \alpha_n) \|J_n x_n - z\| + \|e_n\|$$
$$\leq \alpha_n \|x_n - z\| + (1 - \alpha_n) \|x_n - z\| + \|e_n\|$$
$$= \|x_n - z\| + \|e_n\|$$

for $n \in \mathbb{N}$. Since $\sum_{n=1}^{\infty} \|e_n\| < \infty$, by Lemma 19.1 we have that $c = \lim_{n \to \infty} \|x_n - z\|$ exists and hence $\{x_n\}$ is bounded. Since $\|J_n x_n - z\| \leq \|x_n - z\|$ for $n \in \mathbb{N}$, $\{J_n x_n\}$ is also bounded. If $c = 0$, then $\{x_n\}$ converges strongly to $z \in C_0$ and it is the desired result. Let us assume that $c > 0$. Then, without loss of generality, we may assume that $\{\|x_n - z\|\}$ is a positive real sequence. Since $0 \in A_n z$ and $x_n - J_n x_n \in A_n J_n x_n$, we have

$$\|J_n x_n - z\| \leq \left\| J_n x_n - z + \frac{1}{2}(x_n - J_n x_n - 0) \right\|$$
$$= \frac{1}{2} \|(x_n - z) + (J_n x_n - z)\|$$
$$\leq \|x_n - z\| \left(1 - \delta_E \left(\frac{\|x_n - J_n x_n\|}{\|x_n - z\|}\right)\right),$$

and thus

$$\|x_{n+1} - z\| \leq \alpha_n \|x_n - z\| + (1 - \alpha_n) \|J_n x_n - z\| + \|e_n\|$$
$$\leq \alpha_n \|x_n - z\|$$
$$+ (1 - \alpha_n) \|x_n - z\| \left(1 - \delta_E \left(\frac{\|x_n - J_n x_n\|}{\|x_n - z\|}\right)\right) + \|e_n\|$$
$$= \|x_n - z\| - (1 - \alpha_n) \|x_n - z\| \delta_E \left(\frac{\|x_n - J_n x_n\|}{\|x_n - z\|}\right) + \|e_n\|.$$

It follows that

$$(1 - \alpha_n) \|x_n - z\| \delta_E \left(\frac{\|x_n - J_n x_n\|}{\|x_n - z\|}\right) \leq \|x_n - z\| - \|x_{n+1} - z\| + \|e_n\|.$$

Hence we have

$$\sum_{n=1}^{\infty}(1-\alpha_n)\|x_n - z\|\delta_E\left(\frac{\|x_n - J_n x_n\|}{\|x_n - z\|}\right) \leq \|x_1 - z\| - c + \sum_{n=1}^{\infty}\|e_n\| < \infty.$$

Then, by Lemma 19.2, there exists a subsequence $I = \{n_i\} \subset \mathbb{N}$ such that

$$\sum_{j\in\mathbb{N}\setminus I}(1-\alpha_j) \leq 1 < \infty \text{ and } \lim_{i\to\infty}\|x_{n_i} - z\|\delta_E\left(\frac{\|x_{n_i} - J_{n_i} x_{n_i}\|}{\|x_{n_i} - z\|}\right) = 0.$$

Since E is uniformly convex and $\lim_{i\to\infty}\|x_{n_i} - z\| = \lim_{n\to\infty}\|x_n - z\| = c > 0$, we have

$$\lim_{i\to\infty}\|x_{n_i} - J_{n_i} x_{n_i}\| = 0.$$

Let $y \in E$ be a subsequential weak limit point of $\{x_{n_i}\}$. That is, there exists a subsequence $\{x_{n_{i_j}}\}$ of $\{x_{n_i}\}$ converging weakly to y. Then, since $\{x_{n_{i_j}} - J_{n_{i_j}} x_{n_{i_j}}\}$ converges to 0, it follows that $\{J_{n_{i_j}} x_{n_{i_j}}\}$ also converges weakly to y. Again using that $x_n - J_n x_n \in A_n J_n x_n$ for $n \in \mathbb{N}$ and (WLS) with Lemma 19.3, we have that $y \in C_0$, which completes the proof. □

Lemma 19.5. *In addition to the assumptions in Lemma 19.4, suppose that E has the Opial property. Then, $\{x_{n_i}\}$ converges weakly to a point $x_0 \in C_0$.*

Proof. Let y_0 and y_0' be subsequential weak limit points of $\{x_{n_i}\}$. Then, there exist subsequences $\{y_j\}$ and $\{y_j'\}$ of $\{x_{n_i}\}$ converging weakly to y_0 and y_0', respectively. By Lemma 19.4, we have that y_0 and y_0' belong to C_0 and hence both $\{\|x_{n_i} - y_0\|\}$ and $\{\|x_{n_i} - y_0'\|\}$ have limits. Suppose for a contradiction that $y_0 \neq y_0'$. Then, by the Opial property, we have that

$$\lim_{i\to\infty}\|x_{n_i} - y_0\| = \lim_{j\to\infty}\|y_j - y_0\| < \lim_{j\to\infty}\|y_j - y_0'\|$$
$$= \lim_{i\to\infty}\|x_{n_i} - y_0'\| = \lim_{j\to\infty}\|y_j' - y_0'\|$$
$$< \lim_{j\to\infty}\|y_j' - y_0\| = \lim_{i\to\infty}\|x_{n_i} - y_0\|,$$

which is a contradiction. Hence we have that $\{x_{n_i}\}$ converges weakly to a point $x_0 \in C_0$. □

Proof. [**Theorem 19.2**] Using Lemmas 19.3, 19.4, and 19.5, we have that there exists a subsequence $I = \{n_i\} \subset \mathbb{N}$ such that $\sum_{j\in\mathbb{N}\setminus I}(1-\alpha_j) < \infty$ and $\{x_{n_i}\}$ converges weakly to some point $x_0 \in C_0$. Let $m, l \in \mathbb{N}$ satisfying

that $n_{m-1} < l \leq n_m$. Then, if $n_{m-1} < l < n_m$, then we have that
$$\begin{aligned}x_{n_m} &= \alpha_{n_m-1}x_{n_m-1} + (1-\alpha_{n_m-1})J_{n_m-1}x_{n_m-1} + e_{n_m-1}\\ &= x_{n_m-1} + (1-\alpha_{n_m-1})(J_{n_m-1}x_{n_m-1} - x_{n_m-1}) + e_{n_m-1}\\ &= x_{n_m-2}\\ &\quad + (1-\alpha_{n_m-2})(J_{n_m-2}x_{n_m-2} - x_{n_m-2}) + e_{n_m-2}\\ &\quad + (1-\alpha_{n_m-1})(J_{n_m-1}x_{n_m-1} - x_{n_m-1}) + e_{n_m-1}\\ &= \cdots\\ &= x_l + \sum_{k=l}^{n_m-1}((1-\alpha_k)(J_k x_k - x_k) + e_k).\end{aligned}$$

Since $\{x_n\}$ and $\{J_n x_n\}$ are both bounded, so is $\{\|J_n x_n - x_n\|\}$. Thus we have that

$$\begin{aligned}\|x_{n_m} - x_l\| &\leq \sum_{k=l}^{n_m-1}((1-\alpha_k)\|J_k x_k - x_k\| + \|e_k\|)\\ &\leq \sum_{k=n_{m-1}+1}^{n_m-1}((1-\alpha_k)\|J_k x_k - x_k\| + \|e_k\|)\\ &\leq \sum_{k=n_{m-1}+1}^{n_m-1}((1-\alpha_k)\kappa + \|e_k\|),\end{aligned}$$

where $\kappa = \sup_{n\in\mathbb{N}}\|J_n x_n - x_n\|$. Letting $d_m = \sum_{k=n_{m-1}+1}^{n_m-1}((1-\alpha_k)\kappa + \|e_k\|)$, we have that $\|x_{n_m} - x_l\| \leq d_m$ for $n_{m-1} < l \leq n_m$. It also follows that

$$\sum_{m=1}^{\infty} d_m = \sum_{j\in\mathbb{N}\setminus I}((1-\alpha_j)\kappa + \|e_j\|) < \infty,$$

and hence $\lim_{m\to\infty} d_m = 0$. Let $f^* \in E^*$ and $\epsilon > 0$ arbitrarily. Then, since $\{x_{n_m}\}$ converges weakly to $x_0 \in C_0$, there exists $m_0 \in \mathbb{N}$ such that

$$|\langle x_{n_m} - x_0, f^*\rangle| < \frac{\epsilon}{2} \text{ and } 0 \leq d_m \|f^*\| < \frac{\epsilon}{2},$$

for all $m > m_0$. For any $l > n_{m_0}$, there exists $m_1 \in \mathbb{N}$ such that

$$m_0 < m_1 \text{ and } n_{m_1-1} < l \leq n_{m_1},$$

and we get that

$$\begin{aligned}|\langle x_l - x_0, f^*\rangle| &\leq |\langle x_l - x_{n_{m_1}}, f^*\rangle| + |\langle x_{n_{m_1}} - x_0, f^*\rangle|\\ &\leq \|x_l - x_{n_{m_1}}\| \|f^*\| + \frac{\epsilon}{2}\\ &< d_{m_1}\|f^*\| + \frac{\epsilon}{2}\\ &< \frac{\epsilon}{2} + \frac{\epsilon}{2} = \epsilon.\end{aligned}$$

Since $\epsilon > 0$ and $f^* \in E^*$ are arbitrary, we conclude that $\{x_n\}$ converges weakly to a point $x_0 \in C_0$. □

19.4 Weak Convergence of an Iterative Scheme under the Fréchet Differentiability of the Norm

The main result in the previous section is still valid if we replace the Opial property in the assumptions with the Fréchet differentiability of the norm.

We begin this section with the following important lemma essentially proved by [Reich (1979)]. See also [Takahashi and Kim (1998)].

Lemma 19.6 ([Reich (1979)]). *Let E be a uniformly convex Banach space which has a Fréchet differentiable norm, C a nonempty closed convex subset of E. Let $\{T_n\}$ be a sequence of nonexpansive mappings on C having a common fixed point. Let $S_n = T_n T_{n-1} \cdots T_1$ for every $n \in \mathbb{N}$. Then, for $x \in C$, the set $\bigcap_{n=1}^{\infty} \operatorname{clco}\{S_m x : m \geq n\} \cap \bigcap_{n=1}^{\infty} F(T_n)$ consists of at most one point.*

Now we show the main result of this section. In the proof, we employ the idea shown in [Brézis and Lions (1978)]. See also [Kamimura and Takahashi (2000b)].

Theorem 19.3. *Let E be a uniformly convex Banach space which has a Fréchet differentiable norm. Let $\{A_n\}$ be a sequence of m-accretive operators on E such that $C_0 = \bigcap_{n=1}^{\infty} A_n^{-1} 0$ is nonempty. Suppose that $\{A_n\}$ satisfies the condition (WLS). For a sequence $\{\alpha_n\} \subset [0, 1[$, generate an iterative sequence by $x_1 \in E$ and*

$$x_{n+1} = \alpha_n x_n + (1 - \alpha_n) J_n x_n + e_n$$

for $n \in \mathbb{N}$, where $J_n = (I + A_n)^{-1}$ for $n \in \mathbb{N}$ and $\{e_n\}$ is a sequence of E such that $\sum_{n=1}^{\infty} \|e_n\| < \infty$. If $\sum_{n=1}^{\infty} (1 - \alpha_n) = \infty$, then $\{x_n\}$ converges weakly to a point of C_0.

Proof. Firstly, let us assume that $e_n = 0$ for all $n \in \mathbb{N}$. Then, letting $T_n = \alpha_n I + (1 - \alpha_n) J_n$ for each $n \in \mathbb{N}$, we have that $\{T_n\}$ is a sequence of nonexpansive mappings on E having a common zeros C_0. We also have that $x_{n+1} = T_n T_{n-1} \cdots T_1 x$ for each $n \in \mathbb{N}$. By Lemma 19.4, every subsequential weak limit point of $\{x_{n_i}\}$ belongs to C_0 and by Lemma 19.6, such a point is unique. Therefore, $\{x_{n_i}\}$ converges weakly to a point $x_0 \in C_0$. Using the method in the proof of Theorem 19.2, we obtain that $\{x_n\}$ converges weakly to a point $x_0 \in C_0$.

For a general case, using the method shown in Theorem 6 of [Kamimura and Takahashi (2000b)], we have the desired result. Namely, if a sequence $\{y_n\}$ generated by $y_1 = y \in E$ and

$$y_{n+1} = \alpha_n y_n + (1 - \alpha_n) J_n y_n$$

for $n \in \mathbb{N}$ converges weakly to a point of C_0 for all $y \in E$, then a sequence $\{x_n\}$ given by $x_1 \in E$ and

$$x_{n+1} = \alpha_n x_n + (1 - \alpha_n) J_n x_n + e_n$$

for $n \in \mathbb{N}$ also converges weakly to a point of C_0. □

19.5 Iterative Scheme for a Single m-accretive Operator

Finally, we consider an iterative scheme generated by a single m-accretive operator. The following theorem proved by Browder will be used in the proof of our result to show that a certain sequence of m-accretive operators satisfies the condition (WLS).

Theorem 19.4 ([Browder (1968)]). *Let C be a nonempty closed convex subset of a uniformly convex Banach space E and T a nonexpansive mapping on C with $F(T) \neq \emptyset$. If $\{x_n\}$ converges weakly to $z \in C$ and $\{x_n - Tx_n\}$ converges strongly to 0, then z is a fixed point of T.*

The following is a weak convergence theorem of an iterative scheme for a single m-accretive operator with the coefficient condition which is weaker than that of the theorem proved by [Kamimura and Takahashi (2000b)].

Theorem 19.5. *Let E be a uniformly convex Banach space which satisfies either the Opial property or the Fréchet differentiability of the norm. Let A be an m-accretive operator on E such that $A^{-1}0$ is nonempty. For a sequence $\{\alpha_n\} \subset [0, 1[$ and $\{\rho_n\} \subset]\rho_0, \infty[$ with $\rho_0 > 0$, generate an iterative sequence by $x_1 \in E$ and*

$$x_{n+1} = \alpha_n x_n + (1 - \alpha_n)(I + \rho_n A)^{-1} x_n + e_n$$

for $n \in \mathbb{N}$, where $\{e_n\}$ is a sequence of E such that $\sum_{n=1}^{\infty} \|e_n\| < \infty$. If $\sum_{n=1}^{\infty}(1 - \alpha_n) = \infty$, then $\{x_n\}$ converges weakly to a point of $A^{-1}0$.

Proof. It is sufficient to show that a sequence of m-accretive operators $\{\rho_n A\}$ satisfies the condition (WLS) with $C_0 = A^{-1}0$. Let $\{u_n\}$ and $\{v_n\}$ be sequences satisfying that $u_n \in \rho_n A v_n$ for every $n \in \mathbb{N}$ and $\{u_n\}$ converges strongly to 0. Then, we have that

$$v_n + \frac{1}{\rho_n} u_n \in (I + A) v_n$$

and therefore

$$v_n = (I + A)^{-1} \left(v_n + \frac{1}{\rho_n} u_n \right)$$

for every $n \in \mathbb{N}$. Since $(I + A)^{-1}$ is nonexpansive, we have

$$\left\| v_n - (I + A)^{-1} v_n \right\| = \left\| (I + A)^{-1} \left(v_n + \frac{1}{\rho_n} u_n \right) - (I + A)^{-1} v_n \right\|$$
$$\leq \left\| v_n + \frac{1}{\rho_n} u_n - v_n \right\|$$
$$= \frac{1}{\rho_n} \|u_n\| \leq \frac{1}{\rho_0} \|u_n\|$$

for $n \in \mathbb{N}$. Tending $n \to \infty$, we have that $\lim_{n \to \infty} \left\| v_n - (I + A)^{-1} v_n \right\| = 0$. Suppose that a subsequence $\{v_{n_i}\}$ of $\{v_n\}$ converges weakly to $v_0 \in E$. Using the fact that $(I + A)^{-1}$ is a nonexpansive mapping whose set of fixed points is $A^{-1}0$, by Theorem 19.4, we get $v_0 \in A^{-1}0$. Hence the condition (WLS) is satisfied, and applying Theorem 19.2 and Theorem 19.3, we obtain the result. □

Acknowledgements. The author is supported by Grant-in-Aid for Scientific Research No. 19740065 from Japan Society for the Promotion of Science.

Bibliography

Brézis, H. and Lions, P. -L. (1978). Produits infinis de résolvantes, *Israel J. Math.* **29**, pp. 329–345.

Browder, F. E. (1968). *Nonlinear operators and nonlinear equations of evolution in Banach spaces*, Nonlinear functional analysis (*Proc. Sympos. Pure Math.*, Vol. XVIII, Part 2, Chicago, Ill., 1968), *Amer. Math. Soc.*, Providence, R. I., 1976, pp. 1–308.

Bruck, R. E. and Reich, S. (1977). Nonexpansive projections and resolvents of accretive operators in Banach spaces, *Houston J. Math* **3**, pp. 459–470.

Eckstein, J. and Bertsekas, D. P. (1992). On the Douglas-Rachford splitting method and the proximal point algorithm for maximal monotone operators, *Math. Programming* **55**, pp. 293–318.

Jung, J. S. and Takahashi, W. (1991). Dual convergence theorems for the infinite products of resolvents in Banach spaces, *Kodai Math. J.* **14**, pp. 358–365.

Kamimura, S. and Takahashi, W. (2000). Approximating solutions of maximal monotone operators in Hilbert spaces, *J. Approx. Theory* **106**, pp. 226–240.

Kamimura, S. and Takahashi, W. (2000). Weak and strong convergence of solutions to accretive operator inclusions and application, *Set-Valued Anal.* **8**, pp. 361–374.

Kimura, Y. (2006). *A characterization of strong convergence for a sequence of resolvents of maximal monotone operators*, Fixed point theory and its applications, *Yokohama Publ.*, Yokohama, pp. 149–159.

Kimura, Y. and Takahashi, W. (to appear). A generalized proximal point algorithm and implicit iterative schemes for a sequence of operators on Banach spaces, *Set-Valued Anal.*

Martinet, B. (1970). Régularisation d'inéquations variationnelles par approximations successives, *Rev. Française Informat. Recherche Opérationnelle* **4**, pp. 154–158.

Nevanlinna, O. and Reich, S. (1979). Strong convergence of contraction semigroups and of iterative methods for accretive operators in Banach spaces, *Israel J. Math.* **32**, pp. 44–58.

Opial, Z. (1967). Weak convergence of the sequence of successive approximations for nonexpansive mappings, *Bull. Amer. Math. Soc.* **73**, pp. 591–597.

Pazy, A. (1979). Remarks on nonlinear ergodic theory in Hilbert space, *Nonlinear Anal.* **3**, pp. 863–871.

Reich, S. (1979). Weak convergence theorems for nonexpansive mappings in Banach spaces, *J. Math. Anal. Appl.* **67**, pp. 274–276.

Reich, S. (1980). Strong convergence theorems for resolvents of accretive operators in Banach spaces, *J. Math. Anal. Appl.* **75**, pp. 287–292.

Reich, S. and Zaslavski, A. J. (2000). Infinite products of resolvents of accretive operators, *Topol. Methods Nonlinear Anal.* (Dedicated to Juliusz Schauder, 1899–1943), **15**, pp. 153–168.

Rockafellar, R. T. (1976). Monotone operators and the proximal point algorithm, *SIAM J. Control Optim.* **14**, pp. 877–898.

Suzuki, T. and Takahashi, W. (1999). On weak convergence to fixed points of nonexpansive mappings in Banach spaces, *Nonlinear analysis and convex analysis* (Niigata, 1998), *World Sci. Publ.*, River Edge, NJ, pp. 341–347.

Takahashi, W. (2000). *Nonlinear functional analysis: fixed point theory and its applications*, (Yokohama Publishers, Yokohama).

Takahashi, W. and Kim, G.-E. (1998). Approximating fixed points of nonexpansive mappings in Banach spaces, *Math. Japon.* **48**, pp. 1–9.

Tan, K. K. and Xu, H. K. (1993). Approximating fixed points of nonexpansive mappings by the Ishikawa iteration process, *J. Math. Anal. Appl.* **178**, pp. 301–308.

Chapter 20

Complementarity Modeling and Game Theory: A Survey

S. K. Neogy
Indian Statistical Institute
7, S. J. S. Sansanwal Marg, New Delhi-110016, India
e-mail: skn@isid.ac.in

A. K. Das
Indian Statistical Institute
203, B. T. Road, Kolkata-700108, India
e-mail: akdas@isical.ac.in

Abstract

Modeling using a complementarity framework arises naturally for games, economics, engineering and management decision making problems. This chapter presents a survey on complementarity models in non-cooperative games and certain classes of structured stochastic game problems. This framework is implemented as a linear complementarity, vertical block linear complementarity and extended linear complementarity problem.

Key Words: Non-cooperative game, generalized bimatrix game, generalized polymatrix game, structured stochastic game, switching control property, ARAT property, vertical linear complementarity model, extended linear complementarity model

20.1 Introduction

Complementarity model provides a unifying framework for several optimization problems. The use of a complementarity framework is well known for problems in operations research ([Glassey (1978)], mathematical economics [Pang and Lee (1981)]), geometry and engineering ([Cryer (1971)], [Fridman and Chernina (1967)] and [Pang, Kaneko and Hallman (1979)]).

Complementarity problem is well studied in the literature and various generalizations of the linear complementarity problem have been proposed for modeling more complicated real life problems in different fields. For recent books on the linear complementarity problem and its applications, see [Cottle, Pang, and Stone (1992)], [Murthy (1988)] and [Facchinei and Pang (2003)]. The algorithm presented by [Lemke and Howson (1964)] to compute an equilibrium pair of strategies to a bimatrix game tiggered many researchers to model non-cooperative and stochastic game problems as complementarity problem.

In this survey, we consider the modeling of non-cooperative and structured stochastic game problems as a complementarity problem. In Section 20.2, we present various complementarity problems and algorithms available in the literature. Complementarity models for bimatrix games, polymatrix games and its generalizations are presented in Section 20.3. Complementarity models in stochastic games are discussed in Section 20.4. Section 20.5 contains concluding remarks and areas of further research.

In what follows we first describe linear complementarity problem and its various generalizations.

20.2 Various Complementarity Problems and its Algorithms

20.2.1 *Linear Complementarity Problem*

The *linear complementarity problem* can be stated as follows:

Given a square matrix M of order n with real entries and an n dimensional vector q, find n dimensional vectors w and z satisfying

$$w - Mz = q, \; w \geq 0, \; z \geq 0 \tag{20.1}$$

$$w^t z = 0. \tag{20.2}$$

If a pair of vectors (w, z) satisfies (20.1), then the problem $\text{LCP}(q, M)$ is said to have a feasible solution. A pair (w, z) of vectors satisfying (20.1) and (20.2) is called a solution to the $\text{LCP}(q, M)$. This problem is denoted as $\text{LCP}(q, M)$. The LCP is normally identified as a problem of mathematical programming and provides a unifying framework for several optimization problems like linear programming, linear fractional programming, convex quadratic programming and the bimatrix game problem.

20.2.2 Vertical Linear Complementarity Problem

A number of generalizations of the linear complementarity problem have been proposed to accomodate more complicated real life problems as well as to diversify the field of applications. The concept of a vertical block matrix was introduced by [Cottle and Dantzig (1970)] in connection with the generalization of the linear complementarity problem introduced by them. Consider a rectangular matrix A of order $m \times k$ with $m \geq k$. Suppose A is partitioned row-wise into k blocks in the form

$$A = \begin{bmatrix} A^1 \\ A^2 \\ \vdots \\ A^k \end{bmatrix}$$

where each $A^j = ((a^j_{rs})) \in R^{m_j \times k}$ with $\sum_{j=1}^{k} m_j = m$. Then A is called a *vertical block matrix of type* $(m_1, \ldots, m_k)$. If $m_j = 1$, $\forall\, j = 1, \ldots, k$, then A is a square matrix. Thus a vertical block matrix is a natural generalization of a square matrix. Cottle-Dantzig's generalization involves a system $w - Az = q$, $w \geq 0$, $z \geq 0$ where $A \in R^{m \times k}$, $m \geq k$ and the variable $w_1, w_2, \ldots, w_m$ are partitioned into k nonempty sets $\mathcal{S}_j$, $j = 1, 2, \ldots, k$. Let $\mathcal{T}_j = \mathcal{S}_j \cup \{z_j\}$, $j = 1, 2, \ldots, k$. The problem is to find a solution $w \in R^m$ and $z \in R^k$ of the system such that exactly one member of each set $\mathcal{T}_j$ is nonbasic. This problem reduces to the standard linear complementarity problem when $k = m$ and $\mathcal{T}_j = \{w_j, z_j\}$. This generalization was formally introduced later in a paper of [Cottle and Dantzig (1970)]. The formal statement of the problem is as follows:

Given an $m \times k$ ($m \geq k$) vertical block matrix A of type $(m_1, m_2, \ldots, m_k)$ and a vector $q \in R^m$ where $m = \sum_{j=1}^{k} m_j$, find $w \in R^m$ and $z \in R^k$ such that

$$w - Az = q, \quad w \geq 0,\ z \geq 0 \qquad (20.3)$$

$$z_j \prod_{i=1}^{m_j} w_i^j = 0, \text{ for } j = 1, 2, \ldots, k. \qquad (20.4)$$

Cottle-Dantzig's generalization was designated later by the name *vertical linear complementarity problem* [Cottle, Pang, and Stone (1992)] and this problem is denoted as VLCP(q, A).

Referring to the above generalization by [Cottle and Dantzig (1970)], [Lemke (1970)] observes that "It is reasonable to expect that there will be valuable applications of their results forthcoming." In recent years, a number of applications of the complementarity framework proposed by [Cottle and Dantzig (1970)] have been reported in the literature. [Ebiefung and Kostreva (1993)] introduce a generalized Leontief input-output linear model and formulate it as a vertical linear complementarity problem. This model can be effectively used for the problem of choosing a new technology and also for solving problems related to energy commodity demands, international trade, multinational army personnel assignment and pollution control. [Gowda and Sznajder (1996)] report an interesting extension of the bimatrix game model and the problem of computing a pair of equilibrium strategies for this extended model has a VLCP formulation. It is expected that this generalized bimatrix game will have applications in economics. This sort of applications and the potential future applications have motivated the researchers to consider the modeling of various game problems as complementarity problem.

20.2.3 Extended Linear Complementarity Problem

[Schutter and De Moor (1995)] consider an extension of the linear complementarity problem, namely, Extended Linear Complementarity Problem (ELCP) so that various extensions can be viewed in an unifying framework. The *extended linear complementarity problem* can be described as follows:

Given two matrices $C \in R^{p \times t}$, $D \in R^{r \times t}$, two vectors $c \in R^p$, $d \in R^r$ and s subsets θ_j of $\{1, 2, \ldots, p\}$, find a vector $x \in R^t$ such that

$$Cx \geq c \tag{20.5}$$

$$Dx = d \tag{20.6}$$

$$\prod_{i \in \theta_j} (Cx - c)_i = 0, \ \forall\, j \in \{1, 2, \ldots, s\} \tag{20.7}$$

or show that no such vector exists.

We denote this problem by ELCP(C, D, c, d, Θ) where $\Theta = \{\theta_1, \ldots, \theta_s\}$ is the collection of subsets θ_j of $\{1, 2, \ldots, p\}$ and $|\Theta| = s$. If $x \in R^t$ satisfies (20.5) and (20.6) then the problem ELCP(C, D, c, d, Θ) is said to have a *feasible solution*. The *complementarity condition* (20.7) says that each set $\theta_j, j \in \{1, 2, \ldots, s\}$ corresponds to a subgroup of inequalities of $Cx \geq c$ and for each θ_j at least one inequality should hold as equality. If the feasible solution $x \in R^t$ satisfies the complementarity condition (20.7) then we say that it is a solution of ELCP(C, D, c, d, Θ). [Schutter and De Moor (1995)] also studied the general solution set of an ELCP and develop an algorithm to find all its solutions.

20.2.4 Lemke's Algorithm

The complementary pivot scheme due to [Lemke (1965)] (also known as Lemke's algorithm) for solving (20.1) and (20.2) has stimulated a considerable amount of research for the classes of matrices M for which it can process LCP(q, M). The steps of the algorithm are given below.

The initial solution to (20.1) and (20.2) is taken as

$$w = q + d\, z_0$$

$$z = 0$$

where $d \in R^n$ is any given positive vector which is called *covering vector* and z_0 is an artificial variable which takes a large enough value so that $w > 0$. The ray is called *primary ray* [Lemke (1970)].

Step 1: Decrease z_0 so that one of the variables w_i, $1 \leq i \leq n$, say w_r is reduced to zero. We now have a basic feasible solution with z_0 in place of w_r and with exactly one pair of complementary variables (w_r, z_r) being nonbasic.

Step 2: At each iteration, the complement of the variable which has been removed in the previous iteration is to be increased. In the second iteration, for instance, z_r will be increased.

Step 3: If the variable corresponding to the selected column in step 2 that enters the basis can be arbitrarily increased, then the procedure terminates in a *secondary ray*. If a new basic feasible solution is obtained with $z_0 = 0$, we have solved (20.1) and (20.2). If in the new basic feasible solution $z_0 > 0$, we have obtained a new basic pair of complementary variables (w_s, z_s). We repeat step 2.

Lemke's algorithm consists of the repeated applications of steps 2 and 3. If nondegeneracy is assumed, the procedure terminates either in a secondary ray or in a solution to (20.1) and (20.2). If degenerate almost complementary solutions are generated, then cycling can be avoided using the methods discussed by [Eaves (1971)]. See [Cottle, Pang, and Stone (1992)] for a detailed discussion on Lemke's algorithm.

20.2.5 Cottle-Dantzig Algorithm for Vertical Linear Complementarity Problem

Lemke's algorithm for the LCP(q, M) has been extended with some modifications to the VLCP(q, A) in [Cottle and Dantzig (1970)] and it is presented below.

The Cottle-Dantzig algorithm for the VLCP(q, A) starts with the initial solution to (20.3) and (20.4) as

$$w = q + d\, z_0$$

$$z = 0$$

where z_0 is large enough so that $w > 0$ and $d \in R^m$ is any positive vector.

Step 1: Decrease z_0 to $\bar{z}_0 = \min\{z_0 \mid q + d\,z_0 \geq 0,\ z_0 \geq 0\}$ so that one of the variables w_i, $1 \leq i \leq m$, say w_p is reduced to zero. We now have a basic feasible solution with z_0 in place of w_p. This is the initial almost proper basic feasible solution. Now let r be the unique index, $1 \leq r \leq k$, such that $p \in J_r$. We have exactly one pair of nonbasic variables (z_r, w_p) which belong to the same set of related variables.

Step 2: At each iteration, there is exactly one pair of nonbasic variables belonging to the same set of related variables. Of these, one has been eliminated from the set of basic variables in the previous iteration; the other is now selected to be included as a basic variable in the next iteration. For example, in the second iteration z_r is selected to be included in the set of basic variables.

Step 3: If the variable selected at step 2 to be included as a basic variable can be arbitrarily increased, then the procedure terminates in an almost proper ray, to be called a *secondary proper ray*. Otherwise, one or more of the existing basic variables are reduced to zero and one such variable is replaced by the entering variable to obtain a new almost proper or proper basic feasible solution. If the new basic feasible solution obtained has $z_0 = 0$ or z_0 is nonbasic, then we have solved (20.3) and (20.4) and have a solution for the VLCP(q, A). Otherwise, we have obtained a new almost proper basic feasible solution and a new pair of nonbasic variables (x_β, y_r) belonging to the same set of related variables, say the sth set, where either $(x_\beta, y_r) = (z_s, w_t)$, with $t \in J_s$ or $(x_\beta, y_r) = (w_{t_1}, w_{t_2})$, with $t_1, t_2 \in J_s$.

We repeat step 2.

The Cottle-Dantzig algorithm consists of the repeated applications of steps 2 and 3.

20.3 Complementarity Models for Bimatrix Games, Polymatrix Games and its Generalizations

20.3.1 *Bimatrix Games*

The study of the linear complementarity problem came into prominence when [Lemke and Howson (1964)] and [Lemke (1965)] showed that the problem of computing a Nash equilibrium point of a bimatrix game can be modeled as a linear complementarity problem. A bimatrix game is described as follows.

A bimatrix game is a non-cooperative nonzero-sum two person game (player I and player II) in which each player has a finite number of actions (called pure strategies). Let player I have m pure strategies and player II, n pure strategies. In a game if player I chooses strategy i and player II chooses strategy j they incur the costs a_{ij} and b_{ij} respectively where $A = ((a_{ij})) \in R^{m \times n}$ and $B = ((b_{ij})) \in R^{m \times n}$ are given cost matrices.

A mixed strategy for player I is a probability vector $x \in R^m$ whose i^{th} component x_i represents the probability of choosing pure strategy i where $x_i \geq 0$ for $i = 1, \ldots, m$ and $\sum_{i=1}^{m} x_i = 1$. Similarly, a mixed strategy for player II is a probability vector $y \in R^n$. If player I adopts a mixed strategy x and player II adopts a mixed strategy y then their *expected costs* are given by $x^t A y$ and $x^t B y$ respectively.

A pair of mixed strategies (x^*, y^*) with $x^* \in R^m$ and $y^* \in R^n$ is said to be a *Nash equilibrium pair* if

$(x^*)^t A y^* \leq x^t A y^*$ for all mixed *strategies* $x \in R^m$ and
$(x^*)^t B y^* \leq (x^*)^t B y$ for all mixed *strategies* $y \in R^n$.

It is easy to show that the addition of a constant to all entries of A or B leaves the set of equilibrium points invariant. Henceforth we assume that all entries of the matrices A and B are positive. We consider the following LCP:

$$\begin{bmatrix} u \\ v \end{bmatrix} = \begin{bmatrix} -e_m \\ -e_n \end{bmatrix} + \begin{bmatrix} 0 & A \\ B^t & 0 \end{bmatrix} \begin{bmatrix} x \\ y \end{bmatrix}, \begin{bmatrix} u \\ v \end{bmatrix}^t \begin{bmatrix} x \\ y \end{bmatrix} = 0, \begin{bmatrix} u \\ v \end{bmatrix}, \begin{bmatrix} x \\ y \end{bmatrix} \geq 0 \quad (20.8)$$

where e_m and e_n are m vectors and n vectors whose components are all 1's.

It is easy to see that if (x^*, y^*) is a Nash equilibrium pair then $(\bar{x}, \bar{y})$ is a solution to the linear complementarity problem given by (20.8) where

$$\bar{x} = x^*/(x^*)^t B y^* \text{ and } \bar{y} = y^*/(x^*)^t A y^*. \qquad (20.9)$$

Conversely, if $(\bar{x}, \bar{y})$ is a solution of the linear complementarity problem given by (20.8) then $\bar{x} \neq 0$ and $\bar{y} \neq 0$ in (20.9) are ensured from the positivity of the cost matrices A and B. Therefore (x^*, y^*) is a Nash equilibrium pair where

$$x^* = \bar{x}/e_m^t \bar{x} \text{ and } y^* = \bar{y}/e_n^t \bar{y}.$$

[Lemke and Howson (1964)] gave an efficient and constructive procedure for obtaining an equilibrium pair by solving LCP(q, M) where $M = \begin{bmatrix} 0 & A \\ B^t & 0 \end{bmatrix}$ and $q = \begin{bmatrix} -e_m \\ -e_n \end{bmatrix}$. The algorithm presented by [Lemke and Howson (1964)] to compute an equilibrium pair of strategies to a bimatrix game, later extended by [Lemke (1965)] (known as *Lemke's algorithm* presented in earlier section) to solve a linear complementarity problem.

20.3.2 Generalized Bimatrix Game

[Gowda and Sznajder (1996)] introduced a generalization of the bimatrix game presented in previous section. This generalized version of the bimatrix game is described as follows:

Let $\mathcal{A}$ and $\mathcal{B}$ be two given finite sets of matrices, $\mathcal{A}$ containing s matrices and $\mathcal{B}$ containing r matrices, each of order $m \times n$. Player I forms his payoff matrix whose i^{th} row is chosen as the ith row of some $A \in \mathcal{A}$ and then plays his choice of a mixed strategy over $\{1, 2, \ldots, m\}$. Similarly, player II (the column player) forms his payoff matrix whose j^{th} column is chosen by him as the jth column of some $B \in \mathcal{B}$ and then plays his choice of mixed strategy over $\{1, 2, \ldots, n\}$. The rest of the description of the game is the same as that of a bimatrix game.

[Mohan, Neogy and Sridhar (1996)] considered the question of computing a generalized Nash equilibrium point for the generalized bimatrix game introduced by [Gowda and Sznajder (1996)].

Suppose, $\mathcal{A} = \{A^p \mid p = 1, 2, \ldots, s\}$ and $\mathcal{B} = \{B^p \mid p = 1, 2, \ldots, r\}$. Consider the matrices $C^j, j = 1, 2, \ldots, m$ and $D^j, j = 1, 2, \ldots, n$ defined as follows:

$$C_{i.}^j = A_{j.}^i, \ 1 \leq i \leq s$$
$$D_{i.}^j = (B^i)_{j.}^t, \ 1 \leq i \leq r.$$

Without loss of generality, we may assume that each A^p, $p = 1, 2, \ldots, s$ and each B^p, $p = 1, 2, \ldots, r$ are positive matrices. Hence each C^j, $j = 1, 2, \ldots, m$ and each D^j, $j = 1, 2, \ldots, n$ are positive matrices.

$$\text{Let } X = \begin{bmatrix} C^1 \\ C^2 \\ \vdots \\ C^m \end{bmatrix} \text{ and } Y = \begin{bmatrix} D^1 \\ D^2 \\ \vdots \\ D^n \end{bmatrix}$$

where each C^j is of order $s \times n$ and each D^j is of order $r \times m$. Note that by our assumption $X > 0$, $Y > 0$.

Consider the matrix

$$A = \begin{bmatrix} 0 & C^1 \\ 0 & C^2 \\ \vdots & \vdots \\ 0 & C^m \\ D^1 & 0 \\ D^2 & 0 \\ \vdots & \vdots \\ D^n & 0 \end{bmatrix} = \begin{bmatrix} 0 & X \\ Y & 0 \end{bmatrix}$$

where the '0' blocks are of appropriate orders. This is a vertical block matrix of type $(s, \ldots, s, r, \ldots, r)$ where the number of blocks is $(m+n)$.

[Mohan, Neogy and Sridhar (1996)] showed that a generalized Nash equilibrium point as considered by [Gowda and Sznajder (1996)] of the game described above can be computed by obtaining a solution to the VLCP$(-e, A)$ where $-e$ is the column vector of order $(ms + nr)$, each of whose coordinate is -1. Here one can apply a generalized version of the algorithm of [Lemke and Howson (1964)] to this problem and compute an equilibrium point. This result is stated in the following theorem.

Theorem 20.1. *[Mohan, Neogy and Sridhar (1996)][Theorem 5.3, p. 211] A generalized Nash equilibrium point of the generalized bimatrix game can be computed by a generalized Lemke-Howson algorithm in which the complementary pivot rule is replaced by the generalized complementary pivot rule given by Cottle and Dantzig.*

Here it is possible to assume without loss of generality that the vertical block matrix A associated with the generalized bimatrix game has the form

$$A = \begin{bmatrix} 0 & C^1 \\ 0 & C^2 \\ \vdots & \vdots \\ 0 & C^m \\ D^1 & 0 \\ D^2 & 0 \\ \vdots & \vdots \\ D^n & 0 \end{bmatrix}$$

where each $C^j > 0$ and each $D^j < 0$.

The following theorem in [Mohan, Neogy and Sridhar (1996)] states that Cottle-Dantzig algorithm can compute a generalized Nash equilibrium point of the generalized bimatrix game.

Theorem 20.2. *[Mohan, Neogy and Sridhar (1996)][Theorem 5.4, p. 212] Suppose the problem of computing a generalized Nash equilibrium point of the generalized bimatrix game is formulated as finding a solution to the $VLCP(-e, A)$ where $A = \begin{bmatrix} 0 & X \\ Y & 0 \end{bmatrix}$ with $X > 0$ and $Y < 0$. Then we can compute an equilibrium point by applying Cottle-Dantzig algorithm to such a VLCP.*

20.3.3 Polymatrix Games

In this section, we consider a generalization of the well known bimatrix game presented in subsection 20.3.1 which is known as *polymatrix game*. A polymatrix game is an n-person nonzero-sum non-cooperative game. The credit for studying such a game for the first time has been attributed to [Janovskaya (1968)] by [Howson Jr. (1972)]. A description of the polymatrix game is as follows:

There are $n (\geq 2)$ players, player i with m_i pure strategies. When player i chooses his pure strategy s_i and player j his pure strategy s_j the partial payoff to player i is $a^{ij}(s_i, s_j)$ which does not depend on the choice of strategies by other players. If $(s_1, s_2, \ldots, s_n)$ is the vector of pure strategies chosen by players $1, 2, \ldots, n$, the payoff to player i is given by $\sum_{j \neq i} a^{ij}(s_i, s_j)$. Let A_{ij} denote the matrix of the partial payoffs to player i resulting from the choice of pure strategies by him and player j. Note that the order of A_{ij} is $m_i \times m_j$. A mixed strategy for player i is a probability vector $x^i = (x_1^i, x_2^i, \ldots, x_{m_i}^i)^t$.

For a given set $\bar{X} = \{\bar{x}^1, \ldots, \bar{x}^n\}$ of probability vectors or mixed strategies, the expected payoff to player i is given by

$$E_i(\bar{X}) = (\bar{x}^i)^t \sum_{j \neq i} A_{ij} \bar{x}^j.$$

We say that a set $X^* = \{x^{1*}, x^{2*}, \ldots, x^{n*}\}$ of strategies is an equilibrium set if for all i,

$$(x^{i*})^t \sum_{j \neq i} A_{ij} x^{j*} \geq (x^i)^t \sum_{j \neq i} A_{ij} x^{j*} \qquad (20.10)$$

for any set $X = \{x^1, x^2, \ldots, x^n\}$ of mixed strategies.

Let E^{ij} denote the matrix of 1's of order $m_i \times m_j$. It is easy to note that X^* is an equilibrium for the polymatrix game with payoff matrices A_{ij}, if and only if it is an equilibrium for the polymatrix game with payoff matrices $A_{ij} - \bar{k}_i E^{ij}$ where $\bar{k}_i$ is a constant for each i and that if the payoff matrices A_{ij}'s are replaced by $-A_{ij}$'s, X^* is an equilibrium for the new game if and only if the reverse inequality holds in (20.10). Given the matrices A_{ij}'s it is convenient (see [Howson Jr. (1972)]) to replace A_{ij} by $\bar{k}_i E^{ij} - A_{ij}$ for $\bar{k}_i$'s large and consider the computation of an equilibrium for the resulting polymatrix game, which is also an equilibrium for the original game. Therefore, we shall assume without loss of generality that the A_{ij}'s are positive and that $X^* = \{x^{1*}, x^{2*}, \ldots, x^{n*}\}$ is an equilibrium set if and only if for all i,

$$(x^{i*})^t \sum_{j \neq i} A_{ij} x^{j*} \leq (x^i)^t \sum_{j \neq i} A_{ij} x^{j*} \qquad (20.11)$$

for any set $X = \{x^1, x^2, \ldots, x^n\}$ of mixed strategies.

The problem of computing an equilibrium set of strategies for a polymatrix game has been considered by [Howson Jr. (1972)] who formulates this problem as a linear complementarity problem and proposes a special computational scheme which shows constructively that a polymatrix game has an equilibrium point. See also [Lemke (1970)] in this connection.

The LCP formulation of the problem of finding an equilibrium set of strategies given by [Howson Jr. (1972)] for the polymatrix game is as follows:

Let $B = \begin{bmatrix} 0 & A_{12} & A_{13} & \cdots & A_{1n} \\ A_{21} & 0 & A_{23} & \cdots & A_{2n} \\ \vdots & \vdots & \vdots & \vdots & \vdots \\ A_{n1} & A_{n2} & A_{n3} & \cdots & 0 \end{bmatrix} \geq 0$. Then a solution to $\text{LCP}(q, X)$

where

$$q = \begin{bmatrix} 0 \\ -e \end{bmatrix} \text{ and } X = \begin{bmatrix} B & -E \\ E^t & 0 \end{bmatrix} \text{ with } E = \begin{bmatrix} e^1 & 0 & \cdots & 0 \\ 0 & e^2 & \cdots & 0 \\ \vdots & \vdots & \vdots & \vdots \\ 0 & 0 & \cdots & e^n \end{bmatrix} \quad (20.12)$$

and $e^i \in R^{m_i}$ is a vector each of whose coordinates is 1, provides an equilibrium set of strategies. [Garcia (1973)] provides a computational scheme that works on an augmented LCP to solve LCP(q,X). [Miller and Zucker (1991)] also propose a nice computational scheme for finding an equilibrium set of strategies which is stated below.

Let $\tilde{B} = B + \Lambda$, where $\Lambda > 0$ is a matrix of all 1's. Let $\tilde{X} = \begin{bmatrix} \tilde{B} & -E \\ E^t & 0 \end{bmatrix}$.
Clearly, $\tilde{X}$ is contained in the copositive-plus class and hence the LCP$(q, \tilde{X})$ can be processed by Lemke's algorithm. A solution to LCP(q, X) can be easily obtained from a solution of LCP$(q, \tilde{X})$. For details see [Miller and Zucker (1991)].

20.3.4 Generalized Polymatrix Game

[Mohan and Neogy (1996b)] introduced a generalization of the polymatrix game described in previous section which is similar to the generalization of a bimatrix game presented by [Gowda and Sznajder (1996)]. In the generalized polymatrix game considered by [Mohan and Neogy (1996b)], the players not only choose their mixed strategies over their finite sets of pure strategies but also form their partial payoff matrices as follows.

Player i can form his partial payoff matrix R_{ij} with respect to player j, by choosing the rth row of R_{ij} as the rth row of a matrix in a given set of matrices $\mathcal{A}_{ij}$, $i \neq j$. We shall assume that the matrices in the set $\mathcal{A}_{ij}$, $i \neq j$ are all positive.

In the generalized polymatrix game introduced by [Mohan and Neogy (1996b)], sets $\mathcal{A}_{ij}$ of positive matrices of order $m_i \times m_j$ are given. In addition to choosing his mixed strategy $x^i = (x_1^i, \ldots, x_{m_i}^i)^t$, being a probability vector over his pure strategies $\{s_i^1, \ldots, s_i^{m_i}\}$, player i can also choose his partial payoff matrix R_{ij}, with respect to player j, $i \neq j$, by choosing $(R_{ij})_r$. as $(A_{ij})_r$. for some $A_{ij} \in \mathcal{A}_{ij}$, for $1 \leq r \leq m_i$. R_{ij} is then called a row representative of $\mathcal{A}_{ij}$. Thus, player i chooses his mixed strategy x^i as well as his row representatives R_{ij} for $i \neq j$. We refer to this game as the generalized polymatrix game with sets $\mathcal{A}_{ij}$, $j \neq i$. We use e^j to denote a vector in R^{m_j} each of whose coordinates is 1. We need the following definition for subsequent discussion.

Definition 20.1. *Given the nonempty sets $\mathcal{A}_{ij}$, $1 \leq i \leq n$, $1 \leq j \leq n$, $i \neq j$, we say that the set $\bar{X} = \{\bar{x}^1, \bar{x}^2, \ldots, \bar{x}^n\}$ of probability vectors is an ε-equilibrium set of strategies for players $1, 2, \ldots, n$ if for every $\varepsilon > 0$, there exists a row representative $M_{ij}(\varepsilon)$ of $\mathcal{A}_{ij}$ such that*

$$(\bar{x}^i)^t \sum_{j \neq i} M_{ij}(\varepsilon) \bar{x}^j \leq (u^i)^t \sum_{j \neq i} R_{ij}\, \bar{x}^j + \varepsilon, \quad 1 \leq i \leq n, 1 \leq j \leq n \quad (20.13)$$

for all probability vectors u^i of player i and for all row representatives R_{ij} of $\mathcal{A}_{ij}$, $1 \leq i \leq n$, $1 \leq j \leq n$, $i \neq j$.

We also say that $\bar{X}$ is an equilibrium set of strategies if there exists a row representative $\bar{M}_{ij}$ of $\mathcal{A}_{ij}$ such that (20.13) holds with $\varepsilon = 0$ for all probability vectors u^i and for all row representatives R_{ij} for all $1 \leq i \leq n$.

[Mohan and Neogy (1996b)] showed that when the entries of the matrices in $\mathcal{A}_{ij}$'s are bounded, there is an ε-equilibrium set of strategies $X(\varepsilon)$ for the players in the sense defined by [Gowda and Sznajder (1996)], which is a generalization of Nash's theorem. Moreover, when $\mathcal{A}_{ij}$'s are compact for all i, j, $i \neq j$, $1 \leq i \leq n$ and $1 \leq j \leq n$, then there is an equilibrium set of strategies. The proof technique by [Mohan and Neogy (1996b)] uses degree theory for a general complementarity system when $\mathcal{A}_{ij}$'s are not necessarily compact. The main theorem on the existence of an ε-equilibrium set of strategies for a generalized polymatrix game is stated below.

Theorem 20.3. *[Mohan and Neogy (1996b)][Theorem 3.1, p. 236] Suppose that the nonempty sets $\mathcal{A}_{ij}$, $1 \leq i \leq n$, $1 \leq j \leq n$, $i \neq j$ are bounded in $R^{m_i \times m_j}$. Then there exists a set $\bar{X} = \{\bar{x}^1, \bar{x}^2, \ldots, \bar{x}^n\}$ of probability vectors, where $\bar{x}^i$ is of order $m_i \times 1$ with the following property:*

For every $\varepsilon > 0$, there exists a row representative $M_{ij}(\varepsilon)$ of $\mathcal{A}_{ij}$ such that

$$(\bar{x}^i)^t \sum_{j \neq i} M_{ij}(\varepsilon) \bar{x}^j \leq (u^i)^t \sum_{j \neq i} R_{ij}\, \bar{x}^j + \varepsilon, \quad 1 \leq i \leq n, 1 \leq j \leq n \quad (20.14)$$

for all probability vectors u^i and for all row representatives R_{ij} of $\mathcal{A}_{ij}$, $1 \leq i \leq n$, $1 \leq j \leq n$, $i \neq j$.

When all the $\mathcal{A}_{ij}$'s are compact sets, the above holds with $\varepsilon = 0$.

20.3.5 Complementarity Model Associated with a Generalized Polymatrix Game

Suppose $\mathcal{A}_{ij}$'s are the given finite sets of matrices. We shall show in this section that the problem of finding an equilibrium set of strategies $\bar{X} = \{\bar{x}^1, \bar{x}^2, \ldots, \bar{x}^n\}$, where each $\bar{x}^i$ is a probability vector, for the polymatrix game with the sets $\mathcal{A}_{ij}$, $j \neq i$ can be formulated as a vertical linear complementarity problem.

Suppose the set $\mathcal{A}_{ij}$ contains s_i matrices of order $m_i \times m_j$. Define the matrices C_{ij}^r of order $s_i \times m_j$ as follows: For $i \neq j$,

$$(C_{ij}^r)_{p\cdot} = (A_{ij}^p)_{r\cdot}, \ 1 \leq p \leq s_i$$

where $\{A_{ij}^p \mid p = 1, 2, \ldots, s_i\} = \mathcal{A}_{ij}$ and for $i = j$, take

$$(C_{ij}^r)_{p\cdot} = 0, \ 1 \leq p \leq s_i.$$

In this section, let e^{s_k} denote the vector of order $s_k \times 1$ each of whose coordinate is 1 and e^{m_r} denote the vector of order $m_r \times 1$ each of whose coordinates is 1.

Let $\rho_1 = 0$ and $\rho_k = \sum_{j=1}^{k-1} m_j$. For $1 \leq k \leq n$, let $\mathcal{N}^{\rho_k}$ denote the m_k blocks of matrices $\mathcal{N}^{\rho_k + j}$ defined as
$$\mathcal{N}^{\rho_k+j} = \begin{bmatrix} C_{k1}^j & C_{k2}^j & \cdots & C_{k(k-1)}^j & 0 & C_{k(k+1)}^j & \cdots & C_{kn}^j & 0 & 0 & -e^{s_k} & 0 & \cdots & 0 \end{bmatrix}$$

for $1 \leq j \leq m_k$.

For $1 \leq r \leq n$, let $\mathcal{N}^{m+r}$ consist of exactly one row and $m + n$ columns which is given as follows:

$$\mathcal{N}^{m+r} = \begin{bmatrix} 0 & 0 & \cdots & 0 & (e^{m_r})^t & 0 & \cdots & 0 & 0 & 0 & \cdots & 0 \end{bmatrix}.$$

Now consider the vertical block matrix $\mathcal{N}$ of order $(\sum_{i=1}^{n} s_i m_i + n) \times (m + n)$ given by

$$\mathcal{N} = \begin{bmatrix} \mathcal{N}^{\rho_1} \\ \mathcal{N}^{\rho_2} \\ \vdots \\ \mathcal{N}^{\rho_n} \\ \mathcal{N}^{m+1} \\ \vdots \\ \mathcal{N}^{m+n} \end{bmatrix}$$ where $\mathcal{N}^{\rho_k}$ is the set of blocks $\mathcal{N}^{\rho_k+j}$, $1 \leq j \leq m_k$.

Complementarity Modeling and Game Theory: A Survey

The total number of blocks in $\mathcal{N}$ is $\sum_{i=1}^{n} m_i + n = m + n$.

Let q be the vector of order $(\sum_{i=1}^{n} s_i m_i + n) \times 1$ whose first $\sum_{i=1}^{n} s_i m_i$ coordinates are 0's and the last n coordinates are -1's. Then from [Mohan and Neogy (1996b)][Lemma 3.1, p.234], it follows that an equilibrium set of probability vectors $\bar{X} = \{\bar{x}^1, \bar{x}^2, \ldots, \bar{x}^n\}$ and a corresponding choice of representative matrices from $\mathcal{A}_{ij}$ can be obtained by solving the vertical block $\text{VLCP}(q, \mathcal{N})$ where q and $\mathcal{N}$ are as defined above. A modified Howson algorithm can be applied to the $\text{VLCP}(q, \mathcal{N})$ associated with the given generalized polymatrix game. The problem of finding an equilibrium point which is formulated as a VLCP is illustrated in the following example.

Example 20.1. [Mohan and Neogy (1996b)][Example 5.1, p. 238] *Suppose the sets $\mathcal{A}_{ij}$ are as follows:*

$$\mathcal{A}_{12} = \left\{ \begin{bmatrix} 2 & 0 & 4 \\ 1 & 6 & 5 \end{bmatrix}, \begin{bmatrix} 1 & 4 & 7 \\ 2 & 3 & 0 \end{bmatrix} \right\}, \quad \mathcal{A}_{13} = \left\{ \begin{bmatrix} 2 & 3 \\ 1 & 1 \end{bmatrix}, \begin{bmatrix} 1 & 3 \\ 4 & 2 \end{bmatrix} \right\},$$

$$\mathcal{A}_{21} = \left\{ \begin{bmatrix} 3 & 4 \\ 8 & 3 \\ 4 & 6 \end{bmatrix}, \begin{bmatrix} 4 & 4 \\ 3 & 7 \\ 6 & 1 \end{bmatrix} \right\}, \quad \mathcal{A}_{23} = \left\{ \begin{bmatrix} 2 & 6 \\ 6 & 5 \\ 5 & 4 \end{bmatrix}, \begin{bmatrix} 7 & 1 \\ 5 & 5 \\ 2 & 6 \end{bmatrix} \right\},$$

$$\mathcal{A}_{31} = \left\{ \begin{bmatrix} 4 & 3 \\ 3 & 6 \end{bmatrix}, \begin{bmatrix} 8 & 6 \\ 9 & 7 \end{bmatrix}, \begin{bmatrix} 3 & 5 \\ 6 & 2 \end{bmatrix} \right\},$$

$$\mathcal{A}_{32} = \left\{ \begin{bmatrix} 4 & 1 & 3 \\ 2 & 5 & 1 \end{bmatrix}, \begin{bmatrix} 3 & 5 & 6 \\ 5 & 4 & 8 \end{bmatrix}, \begin{bmatrix} 7 & 2 & 5 \\ 1 & 6 & 4 \end{bmatrix} \right\}.$$

The related matrices C_{ij}^r are as follows:

$$C_{11}^1 = \begin{bmatrix} 0 & 0 \\ 0 & 0 \end{bmatrix}, \quad C_{12}^1 = \begin{bmatrix} 2 & 0 & 4 \\ 1 & 4 & 7 \end{bmatrix}, \quad C_{13}^1 = \begin{bmatrix} 2 & 3 \\ 1 & 3 \end{bmatrix},$$

$$C_{11}^2 = \begin{bmatrix} 0 & 0 \\ 0 & 0 \end{bmatrix}, \quad C_{12}^2 = \begin{bmatrix} 1 & 6 & 5 \\ 2 & 3 & 0 \end{bmatrix}, \quad C_{13}^2 = \begin{bmatrix} 1 & 1 \\ 4 & 2 \end{bmatrix},$$

$$C_{21}^1 = \begin{bmatrix} 3 & 4 \\ 4 & 4 \end{bmatrix}, \quad C_{22}^1 = \begin{bmatrix} 0 & 0 & 0 \\ 0 & 0 & 0 \end{bmatrix}, \quad C_{23}^1 = \begin{bmatrix} 2 & 6 \\ 7 & 1 \end{bmatrix},$$

$$C_{21}^2 = \begin{bmatrix} 8 & 3 \\ 3 & 7 \end{bmatrix}, \quad C_{22}^2 = \begin{bmatrix} 0 & 0 & 0 \\ 0 & 0 & 0 \end{bmatrix}, \quad C_{23}^2 = \begin{bmatrix} 6 & 5 \\ 5 & 5 \end{bmatrix},$$

$$C_{21}^3 = \begin{bmatrix} 4 & 6 \\ 6 & 1 \end{bmatrix}, \quad C_{22}^3 = \begin{bmatrix} 0 & 0 & 0 \\ 0 & 0 & 0 \end{bmatrix}, \quad C_{23}^3 = \begin{bmatrix} 5 & 4 \\ 2 & 6 \end{bmatrix},$$

$$C_{31}^1 = \begin{bmatrix} 4 & 3 \\ 8 & 6 \\ 3 & 5 \end{bmatrix}, \ C_{32}^1 = \begin{bmatrix} 4 & 1 & 3 \\ 3 & 5 & 6 \\ 7 & 2 & 5 \end{bmatrix}, \ C_{33}^1 = \begin{bmatrix} 0 & 0 \\ 0 & 0 \\ 0 & 0 \end{bmatrix},$$

$$C_{31}^2 = \begin{bmatrix} 3 & 6 \\ 9 & 7 \\ 6 & 2 \end{bmatrix}, \ C_{32}^2 = \begin{bmatrix} 2 & 5 & 1 \\ 5 & 4 & 8 \\ 1 & 6 & 4 \end{bmatrix}, \ C_{33}^2 = \begin{bmatrix} 0 & 0 \\ 0 & 0 \\ 0 & 0 \end{bmatrix}.$$

The vertical block matrix $\mathcal{N}$ of type $(\eta_1, \eta_2, \ldots, \eta_{10})$ is given by

$$\mathcal{N} = \begin{bmatrix}
0 & 0 & 2 & 0 & 4 & 2 & 3 & -1 & 0 & 0 \\
0 & 0 & 1 & 4 & 7 & 1 & 3 & -1 & 0 & 0 \\
0 & 0 & 1 & 6 & 5 & 1 & 1 & -1 & 0 & 0 \\
0 & 0 & 2 & 3 & 0 & 4 & 2 & -1 & 0 & 0 \\
3 & 4 & 0 & 0 & 0 & 2 & 6 & 0 & -1 & 0 \\
4 & 4 & 0 & 0 & 0 & 7 & 1 & 0 & -1 & 0 \\
8 & 3 & 0 & 0 & 0 & 6 & 5 & 0 & -1 & 0 \\
3 & 7 & 0 & 0 & 0 & 5 & 5 & 0 & -1 & 0 \\
4 & 6 & 0 & 0 & 0 & 5 & 4 & 0 & -1 & 0 \\
6 & 1 & 0 & 0 & 0 & 2 & 6 & 0 & -1 & 0 \\
4 & 3 & 4 & 1 & 3 & 0 & 0 & 0 & 0 & -1 \\
8 & 6 & 3 & 5 & 6 & 0 & 0 & 0 & 0 & -1 \\
3 & 5 & 7 & 2 & 5 & 0 & 0 & 0 & 0 & -1 \\
3 & 6 & 2 & 5 & 1 & 0 & 0 & 0 & 0 & -1 \\
9 & 7 & 5 & 4 & 8 & 0 & 0 & 0 & 0 & -1 \\
6 & 2 & 1 & 6 & 4 & 0 & 0 & 0 & 0 & -1 \\
1 & 1 & 0 & 0 & 0 & 0 & 0 & 0 & 0 & 0 \\
0 & 0 & 1 & 1 & 1 & 0 & 0 & 0 & 0 & 0 \\
0 & 0 & 0 & 0 & 0 & 1 & 1 & 0 & 0 & 0
\end{bmatrix}$$

where $\eta_1 = s_1 = 2$, $\eta_2 = s_1 = 2$, $\eta_3 = s_2 = 2$, $\eta_4 = s_2 = 2$, $\eta_5 = s_2 = 2$, $\eta_6 = s_3 = 3$, $\eta_7 = s_3 = 3$, $\eta_8 = 1$, $\eta_9 = 1$, $\eta_{10} = 1$.

[Cottle and Dantzig (1970)] algorithm can be used to solve the VLCP$(q, \mathcal{N})$ by generalizing Garcia's scheme [Garcia (1973)] or Miller-Zucker's scheme [Miller and Zucker (1991)]. For details see [Mohan and Neogy (1996b)].

20.4 Complementarity Models in Stochastic Games

Stochastic games were first formulated by [Shapley (1953)]. In this fundamental paper, [Shapley (1953)] proved the existence of a value and optimal stationary strategies for discounted case which gave a method for iterative computation of the value of a stochastic game with discounted payoff. [Gillette (1957)] studied the undiscounted case or limiting average payoff case. Since then there have been a number of papers on stochastic game dealing with the problem of finding sufficient conditions for the existence of their value and their optimal or ϵ-optimal strategies. As a generalization of Shapley's stochastic games, nonzero-sum stochastic games have been considered by many researchers. The theory of stochastic games has been applied to study many practical problems like search problems, military applications, advertising problems, the traveling inspector model, and various economic applications. For details see [Filar and Vrieze (1997)].

A two-player finite state/action space zero-sum stochastic game is defined by the following objects.

(1) A state space $S = \{1, 2, \ldots, N\}$.
(2) For each $s \in S$, finite action sets $A(s) = \{1, 2, \ldots, m_s\}$ for Player I and $B(s) = \{1, 2, \ldots, n_s\}$ for Player II.
(3) A reward law $R(s)$ for $s \in S$ where $R(s) = [r(s, i, j)]$ is an $m_s \times n_s$ matrix whose $(i, j)^{th}$ entry denotes the payoff from Player II to Player I corresponding to the choices of action $i \in A(s)$, $j \in B(s)$ by Player I and Player II respectively.
(4) A transition law $q = (q_{ij}(s, s') : (s, s') \in S \times S, i \in A(s), j \in B(s))$, where $q_{ij}(s, s')$ denotes the probability of a transition from state s to state s' given that Player I and Player II choose actions $i \in A(s), j \in B(s)$ respectively.

The game is played in stages $t = 0, 1, 2, \ldots$ At some stage t, the players find themselves in a state $s \in S$ and independently choose actions $i \in A(s), j \in B(s)$. Player II pays Player I an amount $r(s, i, j)$ and at stage $(t + 1)$, the new state is s' with probability $q_{ij}(s, s')$. Play continues at this new state.

The players guide the game via strategies and in general, strategies can depend on complete histories of the game until the current stage. We are however concerned with the simpler class of *stationary strategies* which depend only on the current state s and not on stages. So for Player I, a stationary strategy

$$f \in F_S = \{f_i(s) \mid s \in S, i \in A(s), f_i(s) \geq 0, \sum_{i \in A(s)} f_i(s) = 1\}$$

indicates that the action $i \in A(s)$ should be chosen by Player I with probability $f_i(s)$ when the game is in state s.

Similarly for Player II, a stationary strategy

$$g \in G_S = \{g_j(s) \mid s \in S, j \in B(s), g_j(s) \geq 0, \sum_{j \in B(s)} g_j(s) = 1\}$$

indicates that the action $j \in B(s)$ should be chosen with probability $g_j(s)$ when the game is in state s.

Here F_S and G_S will denote the set of all stationary strategies for Player I and Player II, respectively. Let $f(s)$ and $g(s)$ be the corresponding m_s- and n_s-dimensional vectors, respectively.

Fixed stationary strategies f and g induce a Markov chain on S with transition matrix $P(f,g)$ whose $(s,s')^{th}$ entry is given by

$$P_{ss'}(f,g) = \sum_{i \in A(s)} \sum_{j \in B(s)} q_{ij}(s,s') f_i(s) g_j(s)$$

and the expected current reward vector $r(f,g)$ has entries defined by

$$r_s(f,g) = \sum_{i \in A(s)} \sum_{j \in B(s)} r(s,i,j) f_i(s) g_j(s) = f^t(s) R(s) g(s)$$

With fixed general strategies f, g and an initial state s, the stream of expected payoff to Player I at stage t, denoted by $v_s^t(f,g)$, $t = 0, 1, 2, \ldots$ is well defined and the resulting discounted and undiscounted payoffs are

$$\phi_s^\beta(f,g) = \sum_{t=0}^{\infty} \beta^t v_s^t(f,g) \text{ for a } \beta \in (0,1)$$

and

$$\phi_s(f,g) = \liminf_{T \uparrow \infty} \frac{1}{T+1} \sum_{t=0}^{T} v_s^t(f,g).$$

A pair of strategies (f^*, g^*) is optimal for Player I and Player II in the undiscounted game if for all $s \in S$

$$\phi_s(f, g^*) \leq \phi_s(f^*, g^*) = v_s^* \leq \phi_s(f^*, g),$$

for any strategies f and g of Player I and Player II. The number v_s^* is called the *value of the game* starting in state s and $v^* = (v_1^*, v_2^*, \ldots, v_N^*)$ is called the *value vector*. The definition for discounted case is similar.

We require the following definition and the results established by [Filar and Schultz (1987)][Theorem 2.1, 2.2].

Definition 20.2. *A pair of optimal stationary strategies (f^*, g^*) for an undiscounted stochastic game is asymptotically stable if there exist a $\beta_0 \in$*

$(0,1)$ and stationary strategy pairs (f^β, g^β) optimal in the β discounted stochastic game for each $\beta \in (\beta_0, 1)$ such that

(i) $\lim_{\beta \uparrow 1} f^\beta = f^*$, $\lim_{\beta \uparrow 1} g^\beta = g^*$

(ii) for all $\beta \in (\beta_0, 1)$, $r(f^\beta, g^\beta) = r(f^*, g^*)$, $P(f, g^\beta) = P(f, g^*)$ for $f \in F_S$ and $P(f^\beta, g) = P(f^*, g)$ for $g \in G_S$ where $P(f, g)$ is the transition matrix and $r(f, g)$ is the current expected reward vector which are defined earlier.

Theorem 20.4. *([Filar and Schultz (1987)][Theorem 2.1]) An undiscounted stochastic game possesses value vector v^* and optimal stationary strategies f^* for Player I and g^* for Player II if and only if there exists a solution $(v^*, t^*, u^*, f^*, g^*)$ with t^*, $u^* \in R^{|S|}$ to the following nonlinear system SYS1a.*

SYS1a: Find (v, t, u, f, g) where $v, t, u \in R^{|S|}$, $f \in F_S$ and $g \in G_S$ such that

$$v_s - \sum_{s' \in S} v_{s'} \sum_{j=1}^{n_s} q_{ij}(s, s') g_j(s) \geq 0, \ i \in A(s), s \in S \tag{20.15}$$

$$v_s + t_s - \sum_{s' \in S} t_{s'} \sum_{j=1}^{n_s} q_{ij}(s, s') g_j(s) - [R(s)g(s)]_i \geq 0, \ i \in A(s), s \in S \tag{20.16}$$

$$-v_s + \sum_{s' \in S} v_{s'} \sum_{i=1}^{m_s} q_{ij}(s, s') f_i(s) \geq 0, \ j \in B(s), s \in S \tag{20.17}$$

$$-v_s - u_s + \sum_{s' \in S} u_{s'} \sum_{i=1}^{m_s} q_{ij}(s, s') f_i(s) + [f(s)R(s)]_j \geq 0, \ j \in B(s), s \in S \tag{20.18}$$

Theorem 20.5. *([Filar and Schultz (1987)][Theorem 2.2]) If a stochastic game possesses asymptotically stable stationary optimal strategies then feasibility of the nonlinear system (SYS1b) is both necessary and sufficient for existence of a stationary optimal solution.*

SYS1b: Find (v, t, f, g) where $v, t \in R^{|S|}$, $f \in F_S$ and $g \in G_S$ such that (20.15), (20.16), (20.17) are satisfied and

$$-v_s - t_s + \sum_{s' \in S} t_{s'} \sum_{i=1}^{m_s} q_{ij}(s, s') f_i(s) + [f(s)R(s)]_j \geq 0, \ j \in B(s), s \in S \tag{20.19}$$

A major area of research in this field is to identify those classes of zero-sum stochastic games for which there is a possibility of obtaining a finite step algorithm to compute a solution. Many of the results in this area are for zero-sum games with special structures. We will refer to these zero-sum stochastic games with special structure collectively as the class of *structured stochastic games*. The class of structured stochastic games contains single controller games, switching controller games, games with state independent transitions and separable rewards and additive reward and additive transitions (ARAT) games. For the above class of structured stochastic games, it is known that optimal stationary strategies exist and the game satisfies the orderfield property (i.e., the solution to the game lies in the same ordered field as the data of the game (e.g., rational)). Many of the researchers have attempted to formulate the problem of computing a value and optimal strategies as a complementarity model and obtain a finite step method. For more details see the survey paper by [Raghavan and Filar (1991)] and [Mohan, Neogy and Parthasarathy (2001)]. In the subsequent sections, we present complementarity models for various structured stochastic games.

20.4.1 *Discounted Switching Controller Stochastic Games*

The class of switching controller stochastic games was introduced by [Filar (1981)]. In a switching controller stochastic game the law of motion is controlled by Player I alone when the game is played in a certain subset of states and Player II alone when the game is played in other states. In other words, a switching controller game is a stochastic game in which the set of states is partitioned into sets S_1 and S_2 where the transition function is given by

$$q_{i,j}(s,s^{'}) = \begin{cases} q_i(s,s^{'}), & \text{for } s' \in S, s \in S_1, i \in A(s) \text{ and } \forall j \in B(s) \\ q_j(s,s^{'}), & \text{for } s' \in S, s \in S_2, j \in B(s) \text{ and } \forall i \in A(s) \end{cases} \quad (20.20)$$

For a switching controller game the transition structure (20.20) is a natural generalization of the single controller game (in which the transition probabilities depend on the actions of a single player) from the algorithmic point of view, this class of games appear to be more difficult. The game structure was used to develop a finite algorithm in [Vrieze (1983)] but that algorithm requires solving a large number of single controller stochastic games. [Mohan, Neogy and Parthasarathy (1997a,b)] formulated a single controller game as solving a single linear complementarity problem and proved that Lemke's algorithm can solve such an LCP.

[Mohan and Raghavan (1987)] proposed an algorithm for discounted switching controller games which is based on two linear programs. Even though this procedure converges to the value vector, it need not to terminate in finitely many steps since basis vectors vary continuously. [Schultz (1992)] formulated the discounted switching controller game as a linear complementarity problem. [Schultz (1992)] used the following result to establish Theorem 20.7.

Theorem 20.6. *A β-discounted zero-sum stochastic game possesses values v_s^β for $s \in S$ and optimal stationary strategies f and g for player I and II respectively if and only if (v^β, f, g) solves the following nonlinear system.*
Find (v^β, f, g) where $v_s^\beta \in R^{|S|}$ such that

$$v_s^\beta - \beta \sum_{s' \in S} v_{s'}^\beta \sum_{j=1}^{n_s} q_{ij}(s,s') g_j(s) - [R(s)g(s)]_i \geq 0, \ i \in A(s), s \in S \quad (20.21)$$

$$-v_s^\beta + \beta \sum_{s' \in S} v_{s'}^\beta \sum_{i=1}^{m_s} q_{ij}(s,s') f_i(s) + [f(s)R(s)]_j \geq 0, \ j \in B(s), s \in S$$
$$(20.22)$$

$$f \in F_s, g \in G_s \quad (20.23)$$

Remark 20.1. Note that if v_s^β, $f(s)$, $g(s)$ satisfy (20.21), (20.22) and (20.23) then

$$v_s^\beta = [\beta P(f,g) v^\beta]_s + r_s(f,g) \quad (20.24)$$

Note that the quadratic form $f(s)R(s)g(s)$ has notation $r_s(f,g)$.

[Schultz (1992)] showed that solving a discounted switching control game is equivalent to solving a particular linear complementarity problem as demonstrated in the following theorem.

Theorem 20.7. *A discounted switching control stochastic game has values v_s^β for $s \in S$ and optimal stationary strategies f and g if and only if*

$$f \in F_S, \ g \in G_S \quad (20.25)$$

$$v_s^\beta - \beta \sum_{s' \in S} v_{s'}^\beta q_i(s,s') - [R(s)g(s)]_i \geq 0, \ i \in A(s), s \in S_1 \quad (20.26)$$

$$v_s^\beta - \theta_s - [R(s)g(s)]_i \geq 0, \ i \in A(s), s \in S_2 \quad (20.27)$$

$$-v_s^\beta + \theta_s + [f(s)R(s)]_j \geq 0,\ j \in B(s), s \in S_1 \qquad (20.28)$$

$$-v_s^\beta + \beta \sum_{s' \in S} v_{s'}^\beta q_j(s,s') + [f(s)R(s)]_j \geq 0,\ j \in B(s), s \in S_2 \qquad (20.29)$$

$$f(s) \text{ is complementary in } (20.26) \text{ and } (20.27) \qquad (20.30)$$

$$g(s) \text{ is complementary in } (20.28) \text{ and } (20.29) \qquad (20.31)$$

[Schultz (1992)] establishes the above theorem by showing that (20.25) through (20.31) imply satisfaction of conditions stated in Theorem 20.6 and therefore (f,g) are optimal with values $v^\beta(s)$. Similarly if v_s^β, $f(s)$, $g(s)$ satisfy (20.21), (20.22) and (20.23) then defining

$$\theta_s = \begin{cases} \beta \sum_{s' \in S} \sum_{i=1}^{m_s} v_{s'}^\beta q_i(s,s') f_i(s),\ s \in S_1 \\ \beta \sum_{s' \in S} \sum_{j=1}^{n_s} v_{s'}^\beta q_j(s,s') g_j(s),\ s \in S_2 \end{cases} \qquad (20.32)$$

forces (20.25) through (20.31) to be satisfied. [Schultz (1992)] observes that Lemke's algorithm with a special d works for the randomly generated problems. The formulated LCP was solved using Lemke's algorithm over 100 randomly generated problem. The ad-hoc formulation and initialization have worked in practice but have not been proven to always find a complementarity solution.

20.4.2 Undiscounted Switching Controller Game

For an undiscounted switching controller game, [Filar and Schultz (1987)] formulated the problem of computing a value vector and an optimal pair of stationary strategies as a bilinear programming problem. We observe that an undiscounted zero-sum, switching controller game can be viewed as an ELCP which follows from the following theorem of [Neogy and Das (2005)].

Theorem 20.8. *For an undiscounted, zero-sum, switching controller game, the value vector and an optimal pair of stationary strategies can be derived from any solution to the following system of linear and nonlinear inequalities (SYS2). Conversely, for such a game, a solution of the SYS2 can be derived from any pair of asymptotically stable stationary strategies.*

SYS2: Find (v,t,θ,η,f,g) where $v,t,\theta,\eta, \in R^{|S|}$, $f \in F_S$ and $g \in G_S$ such that

$$v_s - \sum_{s' \in S} v_{s'} q_i(s,s') \geq 0,\ i \in A(s), s \in S_1 \qquad (20.33)$$

$$-v_s + \theta_s \geq 0, \ s \in S_1 \tag{20.34}$$

$$v_s + t_s - \sum_{s' \in S} t_{s'} q_i(s, s') - [R(s)g(s)]_i \geq 0, \ i \in A(s), s \in S_1 \tag{20.35}$$

$$-v_s - t_s + \eta_s + [f(s)R(s)]_j \geq 0, \ j \in B(s), s \in S_1 \tag{20.36}$$

$$-v_s + \sum_{s' \in S} v_{s'} q_j(s, s') \geq 0, \ j \in B(s), s \in S_2 \tag{20.37}$$

$$v_s - \theta_s \geq 0, \ s \in S_2 \tag{20.38}$$

$$v_s + t_s - \eta_s - [R(s)g(s)]_i \geq 0, \ i \in A(s), s \in S_2 \tag{20.39}$$

$$-v_s - t_s + \sum_{s' \in S} t_{s'} q_j(s, s') + [f(s)R(s)]_j \geq 0, \ j \in B(s), s \in S_2 \tag{20.40}$$

$$f \in F_S, \ g \in G_S \tag{20.41}$$

$$f_i(s)[v_s - \sum_{s' \in S} v_{s'} q_i(s, s')] = 0, \ i \in A(s), s \in S_1 \tag{20.42}$$

$$f_i(s)[-v_s + \theta_s] = 0, \ s \in S_1, i \in A(s) \tag{20.43}$$

$$f_i(s)[v_s + t_s - \sum_{s' \in S} t_{s'} q_i(s, s') - [R(s)g(s)]_i] = 0, \ i \in A(s), s \in S_1 \tag{20.44}$$

$$g_j(s)[-v_s - t_s + \eta_s + [f(s)R(s)]_j] = 0, \ j \in B(s), s \in S_1 \tag{20.45}$$

$$g_j(s)[v_s - \theta_s] = 0, \ s \in S_2, j \in B(s) \tag{20.46}$$

$$g_j(s)[-v_s + \sum_{s' \in S} v_{s'} q_j(s, s')] = 0, \ j \in B(s), s \in S_2 \tag{20.47}$$

$$f_i(s)[v_s + t_s - \eta_s - [R(s)g(s)]_i] = 0, \ i \in A(s), s \in S_2 \tag{20.48}$$

$$g_j(s)[-v_s - t_s + \sum_{s' \in S} t_{s'} q_j(s, s') + [f(s)R(s)]_j] = 0, \ j \in B(s), s \in S_2 \tag{20.49}$$

Remark 20.2. [Neogy and Das (2005)] proved Theorem 20.8 by showing that a feasible solution to SYS2 can be used to derive a solution of SYS1b and, by Theorem 20.5, this solution solves the switching controller game. Conversely, we show that any solution of SYS1b can be used to construct a solution of SYS2. Note that the existence of asymptotic stable stationary strategies for a switching controller game has been proved by [Filar (1981)].

Remark 20.3. The complementarity model which arises in Theorem 20.8 can be posed as an extended linear complementarity problem by defining the complementary set $\Theta = \{\theta_1, \ldots, \theta_s\}$ appropriately. For a definition of ELCP see Subsection 20.2.3. Note that LCP, VLCP all can be posed as an ELCP and ELCP can be solved by the algorithm proposed by [Schutter and De Moor (1995)]. [Neogy and Das (2005)] presented the problem of formulating zero-sum undiscounted switching controller games as a linear complementarity problem by using some artificial dummy constraints. However, it is easy to see that ELCP formulation follows naturally from SYS2 in Theorem 20.8.

20.4.3 Discounted Zero-sum Stochastic Game with Additive Reward and Additive Transition

ARAT games have been studied in the literature earlier by [Raghavan, Tijs and Vrieze (1985)]. Both the discounted and the limiting average criterion of evaluation of strategies have been considered. It is known for example, that for a β-discounted zero-sum ARAT game, the value exists and both players have stationary optimal strategies, which may also be taken as pure strategies. [Raghavan, Tijs and Vrieze (1985)], have shown that undiscounted ARAT game possesses uniformly discounted optimal stationary strategies and therefore asymptotically stable optimal stationary strategies. In [Raghavan, Tijs and Vrieze (1985)], a finite step method to compute a pair of pure stationary optimal strategies and the value of the game has been suggested. However this approach involves solving a series (finite number) of Markov decision problems. See also [Filar and Vrieze (1997)] and [Raghavan and Filar (1991)].

A stochastic game is said to be an *Additive-Reward- Additive Transition game (ARAT game)* if the reward
(i) $r(s, i, j) = r_i^1(s) + r_j^2(s)$ for $i \in A(s)$, $j \in B(s)$, $s \in S$
and the transition probabilities
(ii) $q_{i,j}(s, s^{'}) = q_i^1(s, s^{'}) + q_j^2(s, s^{'})$ for $i \in A(s), j \in B(s)$, $(s, s') \in S \times S$.

[Mohan, Neogy, Parthasarathy and Sinha (1999)] formulated the discounted version of ARAT game as a vertical linear complementarity problem. To formulate ARAT stochastic games they make use of the result that there is always an optimal stationary strategy among the pure strategies for both the players and the Shapley equations hold for this game. The formulation by [Mohan, Neogy, Parthasarathy and Sinha (1999)] is presented below.

We denote the matrix $((q_i^1(s,s'), s, s' \in S, i \in A_s))$ as $Q_1(s)$ where S is the set of states. This is a $m_1(s) \times k$ matrix where $m_1(s)$ is the cardinality of A_s and k is the cardinality of s. Similarly the matrix $Q_2(s)$ of order $m_2(s) \times k$ is defined where $m_2(s)$ denotes the cardinality of the set B_s.

The Shapley equations give us the following for state s, $s \in S$.

$$\text{Val}\,[r(s,i,j) + \beta \sum_t q_{i,j}(s,s')v_\beta(t)] = v_\beta(s)$$

This implies

$$r(s,i,j) + \beta \sum_t q_{i,j}(s,s')v_\beta(t) \leq v_\beta(s) \text{ for all } i \text{ and for any fixed } j.$$

In particular, suppose the optimal pure strategy in state s is i_0 for Player I and j_0 for Player II. Then

$$r_i^1(s) + r_{j_0}^2(s) + \beta \sum_t q_i^1(s,s')v_\beta(t) + \beta \sum_t q_{j_0}^2(s,s')v_\beta(t) \leq v_\beta(s) \,\forall\, i.$$

These inequalities yield

$$r_i^1(s) + \beta \sum_t q_i^1(s,s')v_\beta(t) \leq v_\beta(s) - \eta_\beta(s) = \xi_\beta(s) \,\forall\, i$$

where $\eta_\beta(s) = r_{j_0}^2(s) + \beta \sum_t q_{j_0}^2(s,s')v_\beta(t)$ and

$\xi_\beta(s) = r_{i_0}^1(s) + \beta \sum_t q_{i_0}^1(s,s')v_\beta(t)$ and

$\xi_\beta(s) + \eta_\beta(s) = v_\beta(s)$.

Thus the inequalities are

$$r_i^1(s) + \beta \sum q_i^1(s,s')\xi_\beta(t) - \xi_\beta(s) + \beta \sum q_i^1(s,s')\eta_\beta(t) \leq 0 \,\forall\, i \in A_s,\ s \in S \tag{20.50}$$

and similarly the inequalities for Player II are

$$r_j^2(s) + \beta \sum q_j^2(s,s')\eta_\beta(t) - \eta_\beta(s) + \beta \sum q_j^2(s,s')\xi_\beta(t) \geq 0 \,\forall\, j \in B_s,\ s \in S. \tag{20.51}$$

Also for each s, in (20.50) there is an $i(s)$ such that equality holds. Similarly, for each s in (20.51) there is a $j(s)$ such that equality holds. Let for $i \in A_s$,

$$w_1(s,i) = -r_i^1(s) - \beta \sum q_i^1(s,s')\eta_\beta(t) + \xi_\beta(s) - \beta \sum q_i^1(s,s')\xi_\beta(t) \geq 0, \tag{20.52}$$

and for $j \in B_s$,

$$w_2(s,j) = r_j^2(s) - \eta_\beta(s) + \beta \sum q_j^2(s,s')\eta_\beta(t) + \beta \sum q_j^2(s,s')\xi_\beta(t) \geq 0. \tag{20.53}$$

We may assume without loss of generality that $\eta_\beta(s), \xi_\beta(s)$ are strictly positive. Since there is at least one inequality in (20.52) for each $s \in S$ that holds as an equality and one inequality in (20.53) for each $s \in S$ that holds as an equality, the following complementarity conditions will hold.

$$\eta_\beta(s) \prod_{i \in A_s} w_1(s,i) = 0 \text{ for } 1 \leq s \leq k \text{ and} \qquad (20.54)$$

$$\xi_\beta(s) \prod_{j \in B_s} w_2(s,j) = 0 \text{ for } 1 \leq s \leq k. \qquad (20.55)$$

The inequalities (20.52) and (20.53) along with the complementarity conditions (20.54), (20.55) lead to the VLCP(q, A) where the matrix A is of the form

$$A = \begin{bmatrix} -\beta Q_1 & E - \beta Q_1 \\ -E + \beta Q_2 & \beta Q_2 \end{bmatrix} \text{ and } q = \begin{bmatrix} -r^1(\cdot,\cdot) \\ r^2(\cdot,\cdot) \end{bmatrix}.$$

In the above VLCP, $Q_1 = [q_i^1(s,s')]$, $Q_2 = [q_j^2(s,s')]$ and

$$E = \begin{bmatrix} e^1 & 0 & \cdots & 0 \\ 0 & e^2 & \cdots & 0 \\ \vdots & & & \vdots \\ 0 & \cdots & \cdots & e^k \end{bmatrix}$$

is a vertical block identity matrix where e^j, $1 \leq j \leq k$ is a column vector of all 1's of appropriate order.

[Mohan, Neogy, Parthasarathy and Sinha (1999)] proved the convergence of Cottle-Dantzig algorithm by showing that the vertical block matrix arising from a zero-sum discounted ARAT game belongs to a processable class under a mild assumption. The processability of the vertical linear complementarity problem was shown under the following assumption on the ARAT game:

Either for all s and for all $j \in B_s$, $q_j^2(s,s)$ is positive or for all s, and for each s' there exists a $i \in A_s$ such that $q_i^1(s,s') > 0$ and $Q_2(s)$ is not a null matrix. For details see [Mohan, Neogy, Parthasarathy and Sinha (1999)].

In other words, a pair of pure stationary optimal strategies and the corresponding value for a zero-sum discounted ARAT game with the above assumption, can be computed by solving a single vertical linear complementarity problem.

20.4.4 Undiscounted Zero-Sum Stochastic Game with Additive Reward and Additive Transition

[Mohan, Neogy and Parthasarathy (2001)] formulated the undiscounted ARAT game as a vertical linear complementarity problem. We observe that an undiscounted ARAT game can be viewed as an ELCP which follows from the following theorem of [Neogy and Das (2005)].

Theorem 20.9. *For an undiscounted zero-sum ARAT game, the value vector and an optimal pair of stationary strategies can be derived from any solution to the following system of linear and nonlinear inequalities (SYS3). Conversely, for such a game, a solution of SYS3 can be derived from any pair of asymptotically stable stationary strategies.*
SYS3: *Find* $(\theta, \eta, \phi, \gamma, f, g)$ *where* $\theta, \eta, \phi, \gamma \in R^{|S|}$, $f \in F_S$ *and* $g \in G_S$ *such that*

$$\phi_s - \sum_{s'=1}^{N}(\theta_{s'} + \phi_{s'})q_i^1(s,s') \geq 0, \ i \in A(s), s \in S \tag{20.56}$$

$$\gamma_s - \sum_{s'=1}^{N}(\eta_{s'} + \gamma_{s'} - \theta_{s'} - \phi_{s'})q_i^1(s,s') - r_i^1(s) \geq 0, \ i \in A(s), s \in S$$
$$\tag{20.57}$$

$$-\theta_s + \sum_{s'=1}^{N}(\theta_{s'} + \phi_{s'})q_j^2(s,s') \geq 0, \ j \in B(s), s \in S \tag{20.58}$$

$$-\eta_s + \sum_{s'=1}^{N}(\eta_{s'} + \gamma_{s'} - \theta_{s'} - \phi_{s'})q_j^2(s,s') + r_j^2(s) \geq 0, \ j \in B(s), s \in S$$
$$\tag{20.59}$$

$$f_i(s)[\phi_s - \sum_{s'=1}^{N}(\theta_{s'} + \phi_{s'})q_i^1(s,s')] = 0, \ i \in A(s), s \in S \tag{20.60}$$

$$f_i(s)[\gamma_s - \sum_{s'=1}^{N}(\eta_{s'} + \gamma_{s'} - \theta_{s'} - \phi_{s'})q_i^1(s,s') - r_i^1(s)] = 0, \ i \in A(s), s \in S$$
$$\tag{20.61}$$

$$g_j(s)[-\theta_s + \sum_{s'=1}^{N}(\theta_{s'} + \phi_{s'})q_j^2(s,s')] = 0, \ j \in B(s), s \in S \tag{20.62}$$

$$g_j(s)[-\eta_s + \sum_{s'=1}^{N}(\eta_{s'} + \gamma_{s'} - \theta_{s'} - \phi_{s'})q_j^2(s,s') + r_j^2(s)] = 0, \; j \in B(s), s \in S \tag{20.63}$$

$$f \in F_S, \; g \in G_S \tag{20.64}$$

Remark 20.4. This theorem is proved by following a similar approach as in Theorem 20.8. We show that a feasible solution of SYS3 also solves SYS1b and by Theorem 20.5, this solution solves the game problem. To prove the converse, we show that any solution to SYS1b — which always exists for these games, since they possess asymptotically stable optimal stationary strategies — can be used to derive a feasible solution for SYS3.

Remark 20.5. The complementarity model which arises in Theorem 20.9 can be posed as an extended linear complementarity problem by defining the complementary set $\Theta = \{\theta_1, \ldots, \theta_s\}$ appropriately. Note that [Neogy and Das (2005)] presented the problem of undiscounted zero-sum ARAT game as a linear complementarity problem by using some artificial dummy constraints. However, in this case also ELCP formulation follows naturally from SYS3 in Theorem 20.9.

20.4.5 *Mixture Class of Stochastic Game*

[Sinha (2000)] considers the generalization of the two classes of stochastic games in which the state space S is the union of 3 disjoint subsets S_1, S_2 and S_3 such that the law of transition is controlled by Player-I in S_1 and player -II in S_2 and all the state in S_3 of the game has ARAT state. More specifically, a zero-sum stochastic game is in SC/ARAT *mixture class* if
(i). $S = S_1 \cup S_2 \cup S_3$, $S_i \cap S_j = \emptyset \; \forall \, i \neq j$
(ii). $q_{i,j}(s,s') = q_i(s,s')$, for $s' \in S, s \in S_1, i \in A(s)$ and $\forall j \in B(s)$.
(iii). $q_{i,j}(s,s') = q_j(s,s')$, for $s' \in S, s \in S_2, j \in B(s)$ and $\forall i \in A(s)$
(iv). the reward $r(s,i,j) = r_i^1(s) + r_j^2(s)$ for $i \in A(s), j \in B(s), s \in S_3$ and the transition probabilities $q_{i,j}(s,s') = q_i^1(s,s') + q_j^2(s,s')$ for $i \in A(s), j \in B(s), (s,s') \in S_3 \times S$.

[Sinha (2000)] provides a nonconstructive proof to show that the above SC/ARAT mixture class of game has ordered field property and raises the question that whether a finite step algorithm can be developed in SC/ARAT mixtures. In this connection see also [Raghavan and Filar (1991)][p. 454]. [Neogy, Das and Gupta (2008)] formulate the problem of computing the value vector v_s^β and optimal stationary strategies $f^\beta(s)$

for Player I and $g^\beta(s)$ for Player II for the class of discounted stochastic game with SC/ARAT mixture as a linear complementarity problem and the class of undiscounted stochastic game with SC/ARAT mixture is formulated as a vertical linear complementarity problem. This complementarity formulation gives an alternative proof of the ordered field property. [Neogy, Das and Gupta (2008)] also discuss the possibility of obtaining a finite algorithm for computation of value vector and optimal stationary strategies.

20.5 Concluding Remarks and Areas of Further Research

In this chapter, we present a survey on modeling and computation for a class non-coopertaive and structured stochastic game problems as complementarity problems. Most of the games, processability by Lemke's algorithm or Cottle Dantzig's algorithm are established by showing that the resulting matrix A belongs to a class which are processable by these algorithms. For some of the games like switching control or ARAT, processability may depend upon the choice of a suitable vector d which needs to be explored. Therefore, to implement available pivoting algorithms on some of these formulations, a special initialization scheme may be necessary and use of suitable degeneracy resolving mechanism may be needed. Investigation for solution of game problems using algorithms proposed by [Neogy, Das and Das (2008)] and [Schutter and De Moor (1995)] as well as interior point algorithms proposed by [Kojima, et al. (1991)] are also areas of further research.

Bibliography

Cottle, R. W. and Dantzig, G. B. (1970). A generalization of the linear complementarity problem, *Journal of Combinatorial Theory*, **8**, pp. 79–90.

Cottle, R. W., Pang, J. S. and Stone, R. E. (1992). *The Linear Complementarity Problem*, (Academic Press, New York).

Cryer, C. W. (1971). The method of Christopherson for solving free boundary problems for infinite journal bearings by means of finite differences, *Mathematics of Computation* **25**, pp. 435–443.

Eaves, B. C. (1971). The linear complementarity problem, *Management Science* **17**, pp. 612–634.

Ebiefung, A. A. and Kostreva, M. M. (1993). The generalized Leontief input-output model and its application to the choice of new technology, *Annals of Operations Research* **44**, pp. 161–172.

Facchinei, F. and Pang, J.-S. (2003). *Finite-Dimensional Variational Inequalities and Complementarity Problems*, (Springer-Verlag, New York).

Filar, J. A. (1981). Orderfield property for stochastic games when the player who controls transitions changes from state to state, *JOTA* **34**, pp. 503–515.

Filar, J. A. and Schultz, T. A. (1987). Bilinear programming and structured stochastic games, *JOTA* **53**, pp. 85–104.

Filar, J. A. and Vrieze, O. J. (1997). *Competitive Markov Decision Processes*, (Springer, New York).

Fink, A. M. (1964). Equilibrium in a stochastic n-person game, *J. Sci., Hiroshima Univ., Ser. A.* **28**, pp. 89–93.

Fridman, V. M. and Chernina, V. S.(1967). An iteration process for the solution of the finite-dimensional contact problem, *U.S.S.R. Computational Mathematics and Mathematical Physics* **7**, pp. 210–214.

Garcia, C. B. (1973). Some classes of matrices in linear complementarity theory, *Mathematical Programming* **5**, pp. 299–310.

Gillette, D. (1957). Stochastic game with zero step probabilities, in *Theory of Games*, eds. A. W. Tucker, M. Dresher and P. Wolfe, (Princeton University Press, Princeton, New Jersey).

Glassey, C. R. (1978). A quadratic network optimization model for equilibrium single commodity trade flows, *Mathematical Programming* **14**, pp. 98–107.

Gowda, M. S. and Sznajder, R. (1996). A generalization of the Nash equilibrium theorem on bimatrix games, *International Journal of Game Theory* **25**, pp. 1–12.

Janovskaya, E. B. (1968). Equilibrium points in polymatrix games, (in Russian), *Latvian Mathematical collection*.

Howson, Jr. J. T. (1972). Equilibria of polymatrix games, *Management Science* **18**, pp. 312–318.

Kojima, M., Megiddo, N., Noma, T. and Yoshise, A. (1991). *A unified approach to interior point algorithms for linear complementarity problems*, Lecture Notes in Computer Science 538, Springer-Verlag: Berlin, Germany.

Lemke, C. E. (1965). Bimatrix equilibrium points and mathematical programming, *Management Science* **11**, pp. 681–689.

Lemke, C. E. and Howson, J. T. (1964). Equilibrium points of bimatrix games, *SIAM J. Appl. Math.* **12**, pp. 413–423.

Lemke, C. E. (1970). Recent results on complementarity problems, in: J.B. Rosen, O.L. Mangasarian and K. Ritter, ed. *Nonlinear Programming*, (Academic Press, New York), pp. 349–384.

Miller, D. A. and Zucker, S. W. (1991). Copositive-plus Lemke algorithm solves polymatrix games, *Operations Research Letters* **10**, pp. 285–290.

Mohan, S. R. and Neogy, S. K. (1996b). Generalized linear complementarity in a problem of n person games, *OR Spektrum* **18**, pp. 231–239

Mohan, S. R., Neogy, S. K. and Parthasarathy, T. (1997a). Linear complementarity and discounted polystochastic game when one player controls transitions, in *Complementarity and Variational Problems*, eds: M.C. Ferris and Jong-Shi Pang, SIAM, Philadelphia, pp. 284–294.

Mohan, S. R., Neogy, S. K. and Parthasarathy, T. (1997b). Linear complementarity and the irreducible polystochastic game with the average cost criterion when one player controls transitions, in *Game theoretical applications to Economics and Operations Research*, eds. T. Parthasarathy, B. Dutta, J. A. M. Potters, T. E. S. Raghavan, D. Ray and A. Sen, (Kluwer Academic Publishers, Dordrecht, The Netherlands), pp. 153–170.

Mohan, S. R., Neogy, S. K. and Sridhar, R. (1996). The generalized linear complementarity problem revisited, *Mathematical Programming* **74**, pp.197–218

Mohan, S. R. and Raghavan, T. E. S. (1987). An algorithm for discounted switching control games, *OR Spektrum* **9**, pp. 41–45.

Mohan, S. R., Neogy, S. K. and Parthasarathy, T. (2001). Pivoting algorithms for some classes of stochastic games: A survey, *International Game Theory Review* **3**, pp. 253–281.

Mohan, S. R., Neogy, S. K., Parthasarathy, T. and Sinha, S. (1999). Vertical linear complementarity and discounted zero-sum stochastic games with ARAT structure, *Mathematical Programming* **86**, pp. 637–648.

Murty, K. G. (1988). *Linear Complementarity, Linear and Nonlinear Programming*, (Heldermann Verlag, West Berlin).

Nash, J. F.(1951). Non-cooperative games, *Ann. of Math.* **54**, pp. 286–295.

Neogy, S. K. and Das A. K., (2005). Linear complementarity and two classes of structured stochastic games, in *Operations Research with Economic and Industrial Applications: Emerging Trends*, eds: S. R. Mohan and S. K. Neogy, Anamaya Publishers, New Delhi, India pp. 156–180.

Neogy, S. K., Das A. K. and Das, P. (2008). Complementarity Problem involving a Vertical Block Matrix and its Solution using Neural Network Model, *Mathematical Programming and Game Theory for decision making*, eds. S. K. Neogy, R. B. Bapat, A. K. Das and T. Parthasarathy, ISI Platinum Jubilee Series on Statistical Science and Interdisciplinary Research, **Vol. 1**, pp. 113–130.

Neogy, S. K., Das A. K. and Gupta, A. (2008). On a Mixture Class of Stochastic Game with Ordered Field Property, *Mathematical Programming and Game Theory for decision making*, eds. S. K. Neogy, R. B. Bapat, A. K. Das and T. Parthasarathy, ISI Platinum Jubilee Series on Statistical Science and Interdisciplinary Research, **Vol. 1**, 451–477.

von Neumann, J. and Morgenstern, O. (1944). *Theory of Games and Economic Behaviour*, (Princeton University Press, Princeton, NJ).

Nowak, A. S. and Raghavan, T. E. S. (1993). A finite step algorithm via a bimatrix game to a single controller non-zerosum stochastic game, *Math. Programming* **59**, pp. 249–259.

Pang, J. S.and Lee, S. C. (1981). A parametric linear complementarity technique for the computation of equilibrium prices in a single commodity spatial model, *Mathematical Programming* **20**, pp. 81–102.

Pang, J. S., Kaneko, I. and Hallman, W. P. (1979). On the solution of some (parametric) linear complementarity problems with application to portfolio selection, structural engineering and actuarial graduation, *Mathematical Programming* **16**, pp. 325–347.

Parthasarathy, T. and Raghavan, T. E. S. (1981). An orderfield property for stochastic games when one player controls transition probabilities, *JOTA* **33**, pp. 375–392.

Raghavan T. E. S.and Filar, J. A. (1991). Algorithms for stochastic games, a survey, *Zietch. Oper. Res.* **35**, pp. 437–472.

Raghavan, T. E. S., Tijs, S. H. and Vrieze, O. J. (1985). On stochastic games with additive reward and transition structure, *JOTA* **47**, pp. 375–392.

Schutter, B. De and Moor, B. De (1995). The extended linear complementarity problem, *Mathematical Programming* **71**, pp. 289–325.

Sinha, S. (1989). *A contribution to the theory of Stochastic Games*, Ph.D thesis, Indian Statistical Institute, Delhi Centre.

Sinha, S. (2000). *A new class of Stochastic Games having Ordered field property*, Unpublished Manuscript, Jadavpur University, Kolkata.

Schultz, T. A. (1992). Linear complementarity and discounted switching controller stochastic games, *JOTA* **73**, pp. 89–99.

Shapley, L. S. (1953). Stochastic games, *Proc. Nat. Acad. Sci. USA.* **39**, pp. 1095–1100.

Vrieze, O. J. (1981). Linear programming and undiscounted games in which one player controls transition, *OR Spektrum*, **3**, pp. 29–35.

Vrieze, O. J. (1983). A finite algorithm for the switching controller stochastic game, *OR Spektrum*, **5**, pp. 15–24.